CHINA THIS CENTURY

CHINA THIS CENTURY

RAFE DE CRESPIGNY

HONG KONG
OXFORD UNIVERSITY PRESS
OXFORD NEW YORK
1992

Oxford University Press

Oxford   New York   Toronto
Petaling Jaya Singapore Hong Kong Tokyo
Delhi Bombay Calcutta Madras Karachi
Nairobi   Dar es Salaam Cape Town
Melbourne Auckland

and associated companies in
Berlin Ibadan

First published 1992
Published in the United States
by Oxford University Press, Inc., New York

British Library Cataloguing in Publication Data

Champion De Crespigny, Richard Rafe, 1936–
China this century. - 2nd ed
I. Title
951
ISBN 0 19 5851641

Library of Congress Cataloging in Publication Data

De Crespigny, Rafe.
China this century / by Rafe de Crespigny.
p. cm.
Includes bibliographical references and index.
ISBN 0-19-585164-1: $28.80 (U.S.)
1. China — History — 20th century.
I. Title.
DS774.D4 1992
951.05 — dc20
91-43572
CIP

Printed in Hong Kong
Published by Oxford University Press, Warwick House, Hong Kong

72307

To Peter
with best wishes for the
next century

Preface

THIS book was first written in the early 1970s, when the People's Republic of China was still recovering from the effects of the Cultural Revolution, as the Communist regime sought tentatively an economic solution to the difficulties which faced it and at a time when foreign visitors, business-people, and journalists were gradually being offered better opportunities to travel, to study, and to work there.

Even at that time, twenty years ago, when information on China was dominated by propaganda for one side or another, and when discussion depended largely upon conjecture, the quantity of material written and published about the history of modern China all but defied description. Scholars, observers, reporters, and interpreters of every political inclination and every degree of sympathy composed analyses of the past and the present, and offered their forecasts, ranging from the greatest optimism to the utmost foreboding, on the future development or the Communist state.

Few of these prognostications have been proven exactly correct, and there are equally few studies of recent history which are not strongly coloured by the author's personal interpretations.

I do not expect that the present work can escape such unconscious and inevitable bias, but I hope that it may present the reader who has not yet studied the Chinese language, nor the details of Chinese politics, with a narrative of events in modern China free from particular concern for one party or another and, so far as I can manage, with balance between Chinese and foreign interests. I have concentrated upon political history, but I have sought to relate this to economic circumstances, and I refer also to social concerns and on occasion to literature and art.

What I want to do is to present a picture of China during its development through the twentieth century. I have offered occasional discussion on broader themes, notably at the beginning of each chapter, and I have attempted to relate events to their contemporary background, but for every period I attempt to explain how matters may have appeared to people at the time, and why they acted as they did. I have tried not to foretell their future for them.

On the other hand, as the government and people of China in the 1990s face a crisis of confidence reflecting the problems of development, the failure of the Communist model in eastern Europe, and the influence of liberal ideals in a society long accustomed to authoritarian government, so the patterns of the past, and the lessons which may be learned from them, become essential to our judgement of the present and the future.

For over a hundred and fifty years China has been faced with two great problems: the pressure of population and the challenge of foreign competition. By 1900, the force of these circumstances had brought ruin to the traditional structures of power and belief, and the history of China in the twentieth century has been dominated by the struggle for security and prosperity.

If I have emphasized political and economic concern it is because China this century, whether successfully or not, has sought answers to its problems though government action. The traditional style of imperial government, thoroughly authoritarian when its own interests were in question, was largely unwilling and unable to alter the *status quo*. It was one of the revolutionary changes of 1911, for which Sun Yatsen was greatly responsible, that governments were expected to reform things and improve them, rather than merely to exploit their subjects.

Sadly, however, the early regimes in this new style were too weak or incompetent for such a task — though as recent scholarship reveals the difficulties facing all governments in China though the twentieth century, it is fair to say there is more sympathy owed to the old Republic than is often paid to it.

And since that time, as rulers of the People's Republic during the last forty years have experimented with socialist planning, mass enterprise, cultural revolution, and liberalized economics, we have seen the misfortunes wrought by a powerful political will, with great strength to act, but with a corresponding capacity for making mistakes.

Indeed, the lack of civil rights in China, demonstrated most dramatically by the massacre at Tiananmen Square in 1989, is not just a tragedy for individual victims within the country and a source of sorrow for liberal observers outside. With no true recognition that a government should be accountable to its subjects, one authority after another has exerted its power against weak opposition, rejecting all forms of protest, criticism, or reasoned advice, and has blundered from one error and disaster to another.

So as we observe China at the end of the twentieth century, and as we consider the difficulties faced and the sufferings endured, the two greatest dangers which appear are the arrogance of the government and the numbers of the people. The history and development of these problems and their attempted solutions, moreover and the conflicts of policy and power which accompany them, are of enormous importance not only for China but also for the world outside, which must deal with that vast country in the future.

A Note on Pronunciation

I have used pinyin, now the official transcription of the People's Republic, for most proper names.

In this system, *a* is normally long, as in 'are', but is short between *i* and *n;* *e* is short, as in French '*de*'; and *o* is a medium '*u*', as in 'full'. The value of *i* varies: it is commonly like the vowel in 'tree', but *ing* resembles the English sound, and *i* is extremely short after *c, ch, s, sh, r, z* and *zh.*

Among the diphthongs, *ai* is like the name of the letter 'I' and *ei* like that of the letter 'A', while *ou* is a long 'o' and *uo* is a short 'or'. The vowel *u* is normally long, as in 'wood', but after initial *j, q, x* and *y*, it has umlaut value: *ü.*

Some initials have apecial values: *zhou* is pronounced as the first syllable in 'Joseph', *can* is sounded 'tsan', *qin* as 'chin', and *xin* as a highly aspirated 'sin'.

The novice may now practice on the name of the celebrated square in Beijing, Tiananmen, and the name of the first president of the Republic, Yuan Shikai.

Acknowledgements

Illustrations are taken from official sources of the Republic of China and the People's Republic of China, with the exception of that at the bottom of plate 3 which is presented through courtesy of Mrs. Hedda Morrison.

The maps were prepared by R A Swoboda and Winifred Mumford; the figures by G C and C A Young.

When I first published this work, I had the advice of Liu Ts'un-yan, Wang Gungwu, Audrey Donnithorne, Pierre Ryckmans, Wang Ling, and the late Chiang Yee. Since that time, I have gained particular assistance from Fred Hung, Lo Huimin, Yuan-li Wu, James Hayes, David Faure, Kevin Bucknall, David Goodman, Mark Elvin, and Jonathan Unger.

Working in Canberra I have been fortunate to have access to material in the Oriental Section of the National Library, long supervised by Sydney Wang, and to the superb collection of the Australian National University, organized by Y S Chan.

In the course of preparing the present edition I have special occasion to thank Bill Jenner, Ian Wilson, Eugene Kamenka, Colin Jeffcott, Gary Klintworth, and Gregory Young for their comments and criticisms. Through their care and kindness, I have avoided many errors of fact and judgement. Those which remain are my own.

RAFE DE CRESPIGNY
June 1991

Contents

Figures

Maps

Plates

Plates

1

Splinters of Empire

THE empire of China during the nineteenth century, in theory at least and in appearance on the map, was one of the greatest and most extensive in the world, rivalled only by the red splashes of the British and the sweeping territory of the tsar. Around the heartland of China, the Manchu rulers of the Qing dynasty claimed suzerainty over Manchuria, Mongolia, and eastern Turkestan, and they expected tribute and general obedience from Tibet.

The capital of this vast domain was at Beijing, on the northern edge of the North China Plain, just south of the ancient frontier marked by the Great Wall. To the north-east, through a narrow corridor between mountains and the sea, was Manchuria, homeland of the non-Chinese people who had conquered the empire in the seventeenth century. Southern Manchuria, the basin of the Liao River, was open, rolling plain, extending north across a low watershed to the lands drained by the Amur and the Ussuri. Largely covered at that time by grass and pasture, surrounded by mountains and forests, Manchuria was bordered on the north and east by Russian Siberia, on the south-east by the kingdom of Korea, and on the west by the high plateau of Mongolia.

Geographically Mongolia is divided into two regions of steppe, separated by the stony desert of the Gobi. Outer Mongolia, now politically independent, is based on rivers which drain north to Lake Baikal. Inner Mongolia, still governed from Beijing, extends in an open crescent of grassland across the dry north of China Proper. Further west again, beyond the Altai Mountains, the Barköl Tagh and Tian Shan ranges divide the steppes of Zungaria in the north from the Tarim basin to the south. Here, the two routes of the ancient Silk Road to the west encircle the desert country of the Takla Makan, with oasis cities watered by streams from the snow-melt of the surrounding mountains.

In this 'new region', Xinjiang, the Islamic, largely Turkish, population had been brought under full control only when the territory was established as a province in the 1880s, and south of Xinjiang the high cold plateau of Tibet, habitable only in the deep river valleys, was never ruled directly from China. The Dalai Lama, in his residence in Lhasa near the border with India, had been compelled to acknowledge the suzerainty of Beijing, and sometimes accepted instruction from imperial residents, but the Tibetans maintained their theocratic state in effective independence.

Indeed, traditional Chinese political theory described an empire composed of circles of dominion, with tight control near the capital extending more loosely to distant regions. There was no recognition of Western concepts of a nation-state, with clear frontiers of demarcation, and beyond its formal borders the government of China also regarded neighbouring countries like Korea, Vietnam, and Burma as clients and tributaries, confirming those relationships with occasional missions and an exchange of gifts.

The regions of the imperial frontier, however, Manchuria, Mongolia, Xinjiang, and Tibet, were either cold or dry or both, and the numbers of their people were far less than the Han Chinese peasants who comprised more than ninety per cent of the population of the empire. It was the eighteen provinces of China Proper, dominated by the two great streams of the Yellow River and the Yangzi, that formed the heart of the imperial state.

In ancient times, the dry hills of loess in Shanxi and Shaanxi, around the junction of the Wei with the Yellow River, had been the centre of the Chinese world, but power had long shifted to the east, and for five hundred years, the capital of the Manchu Qing and the Ming dynasty before them had been set at Beijing. Close to the northern frontier and dominating the North China Plain, this was a strategic position which could dominate east and central Asia.

Even within China Proper, however, the south had different corncerns. From the inland basin of Sichuan to the middle reaches of the Yangzi and its great tributaries the Han and the Xiang, downstream to the estuary and Hangzhou Bay, the fertile provinces of monsoon China produced a surplus of rice. The lands south of the Yangzi, however, are divided by ridges and ranges of mountains, and no local power had ever matched the political unity and authority of the North China Plain. So the prosperous south had been controlled for centuries by the imperial power of the north, and the economy of the southern provinces was drained by the Grand Canal to support the splendour of the captial and its military ambitions beyond the Great Wall.

In similar fashion, the coastlines of China were never a matter of major concern to traditional government. The land-based dynasties were designed to deal with the nomad herdsmen and hunters of inner

Map 1 China Physical

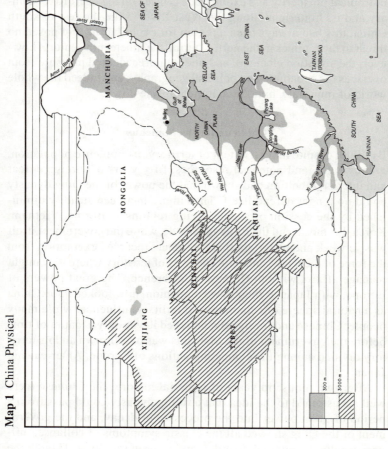

Asia, and they generally ignored the sea-borne enterprise of their Chinese subjects. There was fish and salt along the coast, but the Shandong peninsula was an obstacle to navigation from the north, and so the isolated south-east, though it raised enterprising mariners, traders, and colonists to Taiwan, the Philippines, and beyond, produced a mercantile economy too distant for the central government to control, and which it preferred indeed to discourage.

The far south, beyond the Nan Ling ranges, had long been a region of tropical prosperity and exotic trade centred on the Pearl River estuary and its hinterland along the West River. Like the rest of south China, this had always been a region for exploitation, and posed no threat to the regimes in the north. During the nineteenth century, however, the situation changed, and under unprecedented pressure from overseas the economic and political geography, and indeed the whole world of imperial China, was turned upside down.

China under the Manchus

By 1900, despite its vast extent of territory, its immense population, great wealth, and potential resources, fifty years of foreign defeat and internal rebellion had shown that the power of the Qing dynasty could no longer defend itself. The empire had been steadily diminished by the cession of such outlying regions as Burma, Vietnam, Korea, Formosa, and the Ryukyu Islands, while the government itself was so weak and uncertain that foreigners received exemption from the force of local law, and took enclaves of territory where they might manage affairs without Chinese interference. The great powers of Europe, with Japan as their new companion, now looked upon China as a natural sphere of economic and military influence, while many people, Chinese and foreigners, believed it was only a matter of time before the government of the Manchus was overthrown and its territory divided among the competing nations of Europe, America, and Japan.

After two thousand years of imperial history, the Chinese were well accustomed to the idea that a ruling dynasty could pass from a period of great power into the processes of decay, and the government of the Qing showed all the classic symptoms of collapse: military weakness, internal disorder, and peasant rebellion. Though the dynasty was originally non-Chinese, the founders had established their government with proper respect for the teachings of Chinese civilization and for the value of Chinese scholars and administrators.

Now the situation was changed: the dynasty was falling, and the foreign barbarians, who held such power within the empire, claimed their civilization was superior not only in material possessions and the weapons of war but also in the essential arts of law, politics, and philosophy.

In the past, even under dynasties of conquest, the Chinese had felt assured that they occupied the centre of the civilized world. In the nineteenth century, however, they found themselves the despised subjects of a clumsy empire, competing desperately in a world of powerful enemies. Up to this time, civilization had been determined by the teachings of traditional China; now it was claimed that success could be achieved only through the imitation of Western technology, and that the doctrines of Christianity and capitalism, preached by missionaries and merchants from the West, gave the only hope for proper development and even for survival. At the very moment their government was suffering, through internal causes, from weakness and decline, the people were faced by an aggressive alien culture which thought little of their achievements or their way of life.

The government and social structure of traditional China were based formally on the teachings of Confucianism. The family and kinship group were of central importance, both for the structure of the local community and as support against the potential oppression of government. The basic unit of imperial administration was the county, with a population of several tens of thousands, headed by a magistrate appointed from the capital. He was assisted by a large, locally recruited staff, but his administration was not greatly concerned with the day-to-day affairs of the people, and community leaders had great power within their localities. Though the ideal ruler was described as the 'father and mother' of his people, the main activity of government was to gather taxes and maintain order. In this respect, the system could tolerate a good deal of corruption and considerable low-level oppression, while it had the military machinery to crush local banditry or rebellion. What it could not do was respond effectively to major economic changes, and China in the nineteenth century had many such changes to deal with.

The Qing dynasty had been founded by Manchus, leaders of a confederation of non-Chinese tribes from the north-east, who seized power in the mid-seventeenth century when the fading Chinese dynasty of the Ming was overthrown by rebels. In the centuries which followed, the distinction between conquerors and subjects was maintained. Descendants of the original banners, regimental groups which had formed the army of conquest, held positions of privilege, and many high government posts were reserved for Manchus. Chinese men were compelled to shave their foreheads and wear the pigtail as a mark of servitude, and Manchu women were forbidden to follow the Chinese practice of foot-binding.

At the same time, however, the emperors consolidated their power by alliance with the landed gentry of the empire, traditional leaders of society and administration. The Qing dynasty continued to select officials through written examinations in the texts of Confucianism, and since leisure for this study could be secured only by men of landed wealth, the government confirmed its acceptance among the great clans of China. The system, however, emphasized literary qualities and rote learning at the expense of original thought and adaptability, and in 'Sacred Edicts' of the Qing emperors it was explained that the service owed to the state by scholars and officials was of higher priority than their duties to parents and kinsmen.

As in other matters, it was the policy of the dynasty to ensure good order and maintain the status quo. In this, it was long successful, but it achieved its ends by distorting and stultifying the living traditions of China, and by stifling the imagination and natural initiative of the people.

Most Chinese, however, had small concern with the responsibilities of government or the details of scholarship. The scholar official class numbered perhaps one per cent of the population, and the landed gentry little more. The vast majority of the people were peasants and tenant farmers. In the north of China Proper wheat, barley, and millet were grown, and in the south there was rice and a variety of dry-land crops. In both regions, much of the agriculture depended on dykes, canals, and co-operation, both to keep the lands free from the floods of the great rivers and also to bring water at suitable times to dry farmlands and to paddy-fields.

The chief revenues of the state came from land tax, tribute grain, and a monopoly on the essential salt trade,[1] while economic policy was determined largely by the price of grain and the vicissitudes of the imperial coinage. A long period of stable government had encouraged the expansion of landed estates and the transformation of small farmers into tenants. In part, this was due to the favourable bias which any man of property received from local and provincial officials, and it was common practice for land tax to be levied more harshly against small farms than against large ones. But it was also a natural economic development, as those who gained wealth from farming or trade sought security by investment in real property. Very often, peasant smallholders had insufficient resources to survive a period of poor harvests, and once they had borrowed on mortgage or as an advance against a future crop they seldom earned enough to pay off the loan. Estimates suggest that by the early nineteenth century over half the agrarian population possessed no land of their own, while three-quarters or more of the farmland was held by a few proprietors, who supplemented the limited profits of agriculture by local trade and by lending at interest to their poorer, weaker neighbours.[2]

In the middle of the nineteenth century, as internal troubles made the collection of taxation more difficult and military needs more urgent, the *lijin* duties, on internal trade, and maritime customs were added to the sources of imperial revenue. Neither the court nor the traditional bureaucracy, however, had any great interest in the development of industry and trade. It was accepted that craftsmen and artisans might support the work of farmers or supply useful trinkets for scholars and officials, but science and technology were not the concern of a gentleman. Confucian tradition gave no encouragement to new practical inventions, and had little interest in the merchant or entrepreneur. The alienation between government and finance was one reason why China never developed an industrial revolution, for no Chinese official could readily accept the concept of free, or even subsidized, enterprise, and those who sought such development had to argue against the prejudice of their colleagues. Whereas in the West, and notably in England, there was royal patronage for the new science and technology, and great merchant enterprises were supported by official banking houses and commercial law, there had been no link or understanding in China between official policy and private capital, and even as difficulties grew, when some men sought to encourage development based upon the example of the West, they had constant opposition from their colleagues.

From the viewpoint of conservative rulers and officials, the arguments of such reformers were specious and short-sighted. Since the purpose of government was to maintain order, the import of new ideas from the West could only disrupt the fabric of society without adding greatly to the strength of the empire. To a large extent they were right. On its low level of economy, China was self-sufficient and had no need for foreign trade, while under normal conditions the traditional techniques of administration were quite adequate to maintain the power of the ruling dynasty. In the nineteenth century, however, the imperial government was faced not only with the intrusion of foreigners but also with massive rebellion inside the empire; and the core of the problem was the rise in population.

At the beginning of the Qing dynasty, in the middle of the seventeenth century, the tax records of empire listed some one hundred million people. By the middle of the eighteenth century the numbers had doubled, and estimates of the population of the whole empire in about 1850 vary between 350 and 450 million, in a territory largely the same as two hundred years before.

The causes for the increase are not easy to determine. Two centuries of stable government had given favourable conditions for natural growth, but other social and economic factors must have operated to produce so many people in such comparatively short time. Among those which have been suggested are a general spread of medical knowledge and improved public health, and a decline in female infanticide. By a process of social reform the Chinese tradition which regarded daughters as useless mouths for the family to feed and a later expense for the dowry, and which had tolerated the abandoning or drowning of female infants as a form of birth control, was gradually abandoned. The balance of numbers between the sexes thus became more even, and the birth-rate inevitably rose.[3]

In many respects, the problems of population in the nineteenth century were a consequence of past prosperity. From the time of the Ming and even earlier, the Chinese people had expanded with energy and success against the internal frontiers of China Proper. In particular, south of the Yangzi, the traditional cultivation of paddy rice was joined by maize, sweet potatoes, potatoes, and peanuts, with sugar, cotton, and tobacco, and this new dry-land agriculture permitted and encouraged the destruction of native forest and its replacement by crops grown for food and cash. New areas of land were now usable, the potential for profit became greater, and the lands of the south brought fortune to China comparable to that which the discovery and exploitation of the Americas gave to Europe.[4]

Upon this basis, therefore, with a population between two and three hundred million in the seventeenth and eighteenth centuries, the empire of China had achieved an excellent balance between the numbers of people, the area of cultivable land, and the technology required to work it. Unfortunately and critically, however, though the increase of population reflected this prosperity, the expansion was based upon limited resources of available land. By the early nineteenth century the 'internal frontier' was all but completely colonized, yet the population continued to rise, through decisions made in the good times just passed. In an agrarian society the results were immediate and serious. Without any significant industrial development which might absorb excess labour or improve techniques for tilling the soil, Chinese farmers had to support more and more people from a finite area of cultivable land. Some attempts were made to take up hitherto unused ground, by terracing steep hillsides, reclaiming marshes, and by irrigating the dry regions of the north, but the results were seldom successful. High erosion and low or erratic rainfall ruined newly tilled fields, and the damage extended to existing farmland. Many areas of China still require years of restoration before they can regain their economic value.

One response to this demographic pressure was emigration overseas. There was a long-held interest in trade with South-east Asia, and some settlements had been established in that region for hundreds of years. When the Europeans arrived in the sixteenth and seventeenth centuries—the Spaniards in the Philippines, the Portuguese in Malaya and at Macau, and the Dutch in the East Indies—they learnt to respect the mercantile and on occasion the military ability of the Chinese whom they encountered. By the nineteenth century, as Western settlement expanded in Australia and North America, the Chinese 'coolie trade' provided a steady source of labour, and many immigrants established themselves prosperously in the new territories.[5] On the other hand, this southern expansion of the Chinese people was not supported by the imperial government, and there were many periods during which Chinese trade and travel overseas were restricted or explicitly forbidden.[6] The 'overseas Chinese' themselves looked back to their home country: they sought to maintain family connections and they hoped to be buried with their ancestors, but such enterprise and prosperity as they established was under the control of alien governments.

Emigration, however, was not enough to reduce the pressure of population upon arable land within China, and throughout the nineteenth century standards of living among the people declined dramatically. Some contemporary writers denounced the failures of government, but the situation was largely beyond control, and the problems which it brought would have stretched and probably broken the abilities of any regime. At the basic level, by the middle of the nineteenth century the average farmstead in north China was probably no more than three acres, and only half that area in the south. No matter how intensively they were worked, such tiny plots could barely support a family in years of good harvest, and they gave no opportunity to accumulate reserves. Both independent and tenant farmers fell further and further into debt while foreclosures and evictions consolidated the growth of great estates and added to the numbers of landless, hungry people in the villages.

The Great Rebellions

In the imperial history of China, the creed of Confucianism, with emphasis on a rational, humane attitude towards life, has held greatest influence in government and politics, but other schools of thought, far more concerned about the supernatural, have always received wide support among the common people. Many sects of Buddhism, and the whole variety of Chinese religion known as popular Taoism, have teachings of faith-healing, yoga, and sexual practices to gain longevity. These beliefs are rooted in the superstitions of China, and the priests, monasteries, and temples are only one aspect of an alternative culture which has sometimes inspired men to bloodshed and rebellion.

In much the same tradition, there have always been secret societies in China, with membership linked sometimes by religious beliefs, sometimes by political ideals, and frequently by little more than a desire for companionship and support. Under the Manchus, other motives appeared: to expel the foreign conqueror and restore the Chinese dynasty of the Ming, and the two chief organizations were the White Lotus Sect in the north and the Triads of the south. Both had wide connections among the overseas Chinese in South-east Asia, and both were loosely organized and unwieldy. In the early nineteenth century they caused major disturbances and rebellion, and even after these were defeated the secret societies continued to flourish within local communities. With activities which were often illegal and sometimes criminal, they maintained a link between village leaders and more open bandits, their contacts crossed the barriers of kinship and clan, and they could provide the cadres of insurrection.

In 1851 the combination of economic distress and government weakness culminated in the Taiping rebellion. The rebellion was centred on the God-Worshipping Society led by the preacher Hong Xiuquan, a man of Hakka background[7] who had been influenced by Protestant teaching in Guangzhou. The very phrase *taiping* 'great peace' was an ancient slogan of religious rebellion, and Hong Xiuquan interpreted his faith with Chinese rather than European concepts. His achievement, albeit short-lived, was to unify disparate groups of the discontented, bandits, pirates and smugglers, secret societies and landless men, under a single authority with a levelling creed which opposed the hierarchy of the traditional empire.

The rebellion began in Guangdong and Guangxi, spread rapidly to the middle and lower Yangzi, and there maintained a rival state until 1864. For the ordinary people of China, however, despite the weakness of the imperial government, the Taiping state offered small hope of improvement. Their conquests were marked by slaughter, for the rebels killed those captives they could not convert, and their violent hatred of educated and official classes meant there was no real likelihood of proper administration. The gentry of China, and the conservative peasants who stayed loyal to them, could play no part in the Taiping order, and the nations of the West, after vague attempts at contact, gave support to the established dynasty rather than to the rigid dogmas, the confused government, and the uncertain fortunes of the rebels.

Nevertheless, from first uprising to final destruction, the 'Heavenly Kingdom of Great Peace' extended its bounds from Nanjing north across the Huai River, west up the Yangzi into Hunan and Hubei, and east to Hangzhou, Suzhou, and almost to Shanghai. On every frontier there was constant warfare, bitterly contested ground, and a policy of attrition with a death toll in millions. Both Chinese and European travellers described regions of devastation, abandoned villages, overgrown roads and fields, and cities in ruins with no more than a dozen houses inhabited. For fifty years afterwards there was vacant land in Jiangxi, Anhui, Jiangsu, and Zhejiang to attract immigrants from surrounding territories, and even in 1953 the latter three provinces still showed a net decline in their recorded population.[8]

During the last half of the nineteenth century, similar human and natural disasters followed a comparable record of devastation and slaughter. There were major floods and in 1855 a change of course of the Yellow River from the south to the north of the Shandong peninsula, inundating thousands of square miles and drowning thousands of people with their farms, while in every region of the empire there were droughts, insect plagues, and famine. During the same period as the Taipings, though less well known to the outside world, two further rebellions also brought massive bloodshed: the Nian movement of the North China Plain and the holy war of the Muhammadans in Shaanxi and Gansu. The city of Nanjing, capital of the Taiping kingdom, was finally captured in 1864, but the Nian rebellion was not crushed until four years later and the Muslims of the north-west were destroyed only in 1873. All these campaigns were accompanied by massacre on both sides, with cities, temples, and houses looted and razed to the ground, and in nineteenth century China, when we are told that ditches were choked with bodies and that rivers ran red with blood, it is very likely the descriptions were literally true.

The great rebellions had subsided by the late 1870s, but the central and provincial governments were weak and among the villages of China there was frequent disorder and constant banditry. The dynasty was powerless either to help or to control its people, and everywhere there was unrest and discontent against the exactions of officials and landlords, or against the immigrants who came either to escape some catastrophe or to take up land which had been abandoned in the recent upheavals. More and more, however, the anger and hatred of the people was directed against the men from the West, the soldiers, the traders, and the Christian missionaries. In the Yangzi valley and on the North China Plain, the two most populous regions of China, foreigners established their presence with treaty ports and legal privileges, and they obtained permission to travel throughout the empire. Sometimes unfairly, sometimes with justice, the Chinese regarded the newcomers as responsible for their misfortunes. Secret societies were now opposed not only to the Manchus, but to all other foreigners as well, and anti-foreign riots became increasingly common as the targets became more obvious and more available.

Foreigners in China

Separated from India and the Middle East by great mountain ranges, wide deserts, and open steppe, China's contact with Europe and West Asia had never been more than marginal and sporadic. Until the nineteenth century, very few Chinese had visited Europe, and those travellers who came to China by land or by sea came primarily on embassies, trading missions, and only occasionally for small-scale settlement. Arab traders had visited Guangzhou since the eighth and ninth centuries, and in the days of the Yuan dynasty in the fourteenth century, when China was part of the Mongol world empire which stretched across central Asia, the Pope sent embassies from the Vatican to the Khan, and Marco Polo the Venetian travelled to China and served as an official of the government.

It was in the time of the Ming dynasty, successors to the Mongol Yuan, that the first major contact took place between China and the

West, and the educated Jesuits who came as missionaries to the imperial court in the sixteenth and seventeenth centuries brought a knowledge of scientific techniques from Europe which was in many ways
more important to the Chinese than the faith which they sought to
preach. For most of recorded history, the civilization of China had
been the equal and superior of any contemporary culture in the world,
but in the sixteenth and seventeenth century it happened that the
Christian nations of the West obtained the advantage in various practical fields. Explanations for this development differ widely: the
theorist may consider the influence of the Renaissance, the
Reformation, and the age of discovery in the West; he may prefer
to blame the brutal decadence of the Mongols and the Ming; or he
may regret the arrogant nationalism of Europe.

It was not until the nineteenth century, however, that the full effect
of European power was shown in China. The Jesuits, who had
impressed the Chinese with their ability as clock-makers and can-
non- founders (though gunpowder, like paper and printing, had first
been developed in China), were hamstrung in their mission by quarrels among the politicians of the Vatican, and their ultimate influence was slight. By the end of the eighteenth century, the dominant
force among the foreigners in China was the naval and trading power
of Britain, and the Catholic missionaries who had sought to convert
China by example from the top were succeeded by energetic, aggressive Protestants who emphasized personal salvation and Bible-read-
ing among the common people.

Now, for the first time, China was involved in major exchange with
the outside world. Hitherto, overseas trade had comprised the export
of luxuries, silk, tea, porcelain, and rhubarb, and the import of bullion in payment. At the end of the eighteenth century, the situation
began to change, for opium, first introduced as a medicinal drug,
became increasingly popular as a narcotic in China. It was prohibited by imperial decree in 1729, and again in 1796 and 1800, but in
1797 the British East India Company established an official monopoly
for the manufacture and sale of opium in India, annual auctions were
held in Calcutta, and the opium itself was shipped by private traders
to China. By the 1830s opium represented half the value of British
imports into China, thirty or forty thousand chests a year—and China
was exporting silver to pay for it.

Morally speaking, the trade was surely indefensible, for the British
prohibited opium within their own territories. The situation was confused, however, by the wide acceptance of opium among educated

Chinese, by the corruption of Chinese officials, and by the petty injustices of the trading system at Guangzhou which restricted the foreigners in their movement and refused them the right of permanent residence with their families. Year after year there were squabbles over customs dues and port regulations, and Chinese disapproval of foreigners was matched by the contempt which most merchants felt for the officials they had to deal with, and the sense of frustration and arbitrary exploitation which they encountered in negotiations.[9] When one upright governor, Lin Zexu, attempted to stop the opium trade, he made small attempt to negotiate with the foreign traders, and the British government was persuaded to view his policy as nothing but an insult to the flag. The greatest maritime power in the world was of no mind to accept such treatment.

In 1839, at the time of the Opium War, Chinese military equipment was not markedly inferior to that of the British. In morale and discipline, however, there was no comparison. The main imperial troops were bannermen, descendants of the hereditary garrisons established by the Manchus almost two hundred years earlier, and they proved useless. At sea the Chinese were no match, and the broadsides of British ships sank their war-junks and destroyed their shore batteries and coast defences. In the general picture of disaster a few bands of peasants gained fame for their success in petty skirmishes, but in the following year the British sailed north up the coast, occupied the Zhoushan Archipelago outside Hangzhou Bay, and advanced inland to Nanjing. By the Treaty of Nanjing in 1842 and by the supplementary Treaty of the Bogue[10] in the following year the Chinese agreed to payment of an indemnity, to cession of the island of Hong Kong, to the opening of five major 'treaty' ports to foreign trade and residence, and to the rights of British subjects to be tried by their own court and legal system. Almost incidentally, the Chinese also agreed that they would no longer attempt to prohibit the trade in opium.

This first European war, and the treaties which followed it, established a pattern of relations between China and the West for the next hundred years. Very seldom was China able to mount a convincing or successful show of force, and each defeat was followed by a dictated 'unequal' treaty, with indemnities in cash, transfers of territory and further privileges for foreigners. By the so-called 'most-favoured- nation' clause, which first appeared in the Treaty of the Bogue and became a part of all similar agreements thereafter, it was further allowed that any privileges granted to the nationals of one foreign country would also be given automatically to the subjects of all other treaty powers. By the end of the nineteenth century the concessions first acquired by the British had been extended and shared among the French, Americans, Russians, Germans, Italians, and Japanese.

In a long history of mutual dislike and disrespect few Chinese or foreign officials made any serious attempt to understand one another, and there were many instances of discourtesy and dishonesty, but it was differing concepts of justice and law that provided the most serious obstacle to any chance of agreement. The Western legal tradition provides a system for punishing those who are guilty of crimes and for solving private conflicts. In imperial China all law was criminal law, and the one great crime was to disturb the imperial peace. This approach had always guided Chinese dealings with foreign communities of merchants: within their own ghettos they could administer such rules as they liked, but if a foreigner came in official dispute with a Chinese he must submit to the terms of Chinese law.[11] To the men from the West, however, it seemed wrong that they should be compelled to live under a system so alien to their own sense of justice, administered by officials who appeared too often both incompetent and corrupt. They demanded immunity from Chinese law and the right to be tried by their own consular courts, and in later treaties the Chinese granted also that certain areas in the treaty ports, notably the International Concession at Shanghai, should be administered solely by Western officials.

Though it is a strange concept that a country's jurisdiction should not apply to all within its borders, the doctrine of 'extraterritoriality' served some useful purposes even for the Chinese. First, it relieved officials of immediate responsibility for non-Chinese and lessened the chances of political and diplomatic conflict. Secondly, as generations of Chinese merchants could testify, Western ideas of law and contract and Western administration of justice were more suitable to economic development than the traditional methods of China. The great banking houses and the beginnings of large-scale industry were all established in the neighbourhood of Shanghai, and the foreign-ruled city, with other treaty ports, became a point of contact and exchange for ideas and trade between China and the West.

None the less, the privileges held by the foreigner were a constant source of resentment. Traders and officials usually stayed in the treaty ports and had little contact with the mass of Chinese people, but missionaries, both Protestant and Catholic, travelled and settled throughout the empire and brought with them not only the doctrines of an alien creed but also the trappings of an alien power. Whether they liked it or not, all enjoyed the protection of their own government, and in time this protection and interest was extended to Chinese Christian converts. Frequently an injury offered to a missionary or to some member of his flock was used as an excuse by a foreign power to put pressure on the Chinese court. Individual missionaries could obtain great influence with local officials and their intervention was often decisive in civil disputes and even criminal cases. They naturally tended to support their own converts, but their judgement of character was not always perfect and many ordinary Chinese had good reason to resent their interference and to despise the 'rice-Christians'.

As individuals the missionaries were men and women of great courage and devotion who gave years of service to the people of China. But in the context of Chinese society their influence was often disruptive: the charity they provided broke traditional links and relationships between rich and poor, their doctrines were hostile to the patterns of Chinese belief and social organization, and even the missionaries themselves, simply and harshly though they lived by Western standards, possessed comforts and security that the Chinese they visited could never know. For all the support it received from abroad Christianity never won wide adherence in China, and it was always regarded as an alien creed.

Saddest of all in the dealings between Chinese and foreigners was the lack of recognition, on either side, that the others could be civilized human beings. A few scholars and missionaries learnt the Chinese language and admired the history, philosophy, and literature of the past, but the majority of visitors treated the people only as coolies and servants, and some wondered whether Chinese were not one step below them on the scale of evolution. For their part, many Chinese saw foreigners as barbarians and devils, and found no good reason to change their opinion. The government of the empire saw them as a nuisance and a potential danger, and policy required that they should be kept at a distance, played off one against another, and ultimately discarded when other more pressing problems had been solved and China regained its natural strength.

The Weakness of Self-strengthening

By the middle of the nineteenth century it was clear that the former system of Manchu government had failed and the survival of the dynasty depended on new forces, with new techniques to control them. The banner troops, humiliated in the Opium War, proved equally unsuccessful against the Taipings. The regular army, which had conquered all Mongolia for the Qianlong Emperor in the late eighteenth century, likewise gained no success against the rebels and in another war during 1858–60, despite a brief success at the Dagu forts, it was thoroughly defeated by British and French forces at Tianjin and then at Beijing.[12]

The destruction of the Taipings, in fact, was an achievement of Chinese arms, not of the imperial Manchu forces. Zeng Guofan, a scholar-official from Hunan, raised and trained an army of volunteers against the Taipings, and his example was followed in the eastern provinces by Li Hongzhang.[13] Zeng Guofan's troops destroyed the rebel capital at Nanjing in 1864, his protégé Zuo Zongtang combined with Li Hongzhang to crush the Nian rebellion in 1868, and in the 1870s Zuo Zongtang suppressed the Muslims of the north-west and went on to the conquest of Xinjiang in central Asia.

In practice, Zeng Guofan and his colleagues controlled the chief military forces of the empire, and though they served as officials of the Manchu dynasty they derived their basic authority from the loyalty of the officers and soldiers they commanded. Nevertheless, despite this power in Chinese hands, no single leader held such predominance that he could turn against the dynasty, and the risks were far too great for such an attempt. A civil war was almost certain, and in the aftermath of the great rebellions it was very possible the empire would fall into utter chaos. Since the new round of treaties in 1860, moreover, the foreign powers were quite content with advantages they had gained from the Manchus, and they were prepared to support them. Though the Qing dynasty was backward and ineffective, there was as yet no practical alternative.

From this time on, therefore, power was divided three ways: between the imperial court at Beijing, Li Hongzhang and his Army of the Huai, and the Hunan group of Zeng Guofan, who died in 1871, and his successor Zuo Zongtang. Taking advantage from rivalries among these Chinese leaders, the court found room to manoeuvre against them, and for the next fifty years the policies of the empire were decided on the basis of political advantage and short-term gain, while the government as a whole lurched from one crisis and one policy to another.

At the centre of the conflict was the question of westernization. When the young Tongzhi Emperor came to the throne in 1861 both the Manchu regent Yixin, Prince Gong, and the newly risen Chinese generals from the south were generally agreed on the need to adopt techniques and tools from the West, not only for use in the army and navy but also for the development of the transport system and the whole economy of the empire. For a few years, while Prince Gong controlled the government, the dynasty made some progress in negotiation with foreign powers and in the introduction of Western techniques, including the establishment of a small Foreign Office, the Zongli Yamen, with a school of interpreters, and the reorganization of the imperial maritime customs service with foreign advisers and officials. During his campaigns against the Taipings, Li Hongzhang had established arsenals to provide and maintain Western arms and equipment, and his Jiangnan Arsenal at Shanghai became a centre of heavy industry, with a language school and a unit for translating Western works on engineering, history, and international law. In similar fashion, the naval dockyard at Fuzhou founded by Zuo Zongtang, besides constructing the first modern warships for the Chinese navy, maintained an academy which used the English language for teaching navigation and French for teaching engineering. The programme was organized on Western lines, and promising students were sent overseas for further study in America and Europe.

In projects such as these the men committed to 'Self-Strengthening' sought to follow the example of the Meiji Restoration in Japan. But there were weighty arguments and strong prejudice against any such programme of reform. From the point of view of the court, the new establishments of Li Hongzhang and his fellows gave no necessary advantage to the central government, and they might contribute to a coup against the dynasty, while in the empire at large there was doubt and resentment about the new methods and techniques. Causes varied widely: many officials feared for their personal prestige and resented the men trained in Western skills; many small traders and labourers were driven from work by the new steamships and trains; and the railway lines and telegraphs were hated in the countryside for their adverse effect on the land forms used by geomancers who determined auspicious sites for ancestral graves. Behind all the reasons, however, there lay a deep conservatism, a natural suspicion of the foreigner and all his works, and a strong sense that Chinese culture was essentially superior and should be preserved against the threat of alien attitudes and new technology. Even the innovators had to defend their policy by the *tiyong* theory: that the body (*ti*) of culture should remain intact, though a few practical tools might be introduced for immediate use (*yong*).

In the long term, moreover, even without regard to the conservative opposition, projects for Western-style development proved of limited success, and many of the difficulties arose from the very nature of their organization. For large-scale work the question of capital was critical: without any effective concept of commercial law, of limited liability or of merchant banking, there was no good machinery for raising funds to support long-term investment, and a Chinese entrepreneur who sought to embark on major enterprises such as railroads, ships, or substantial factories was dependent either upon money raised through foreign markets or upon support from the imperial government. It was one thing for the Chinese government to set up arsenals and dockyards or purchase military equipment; it was another matter when it sought to encourage local investment in the infrastructure of communications and manufacturing.

The chief method used for linking private enterprise with government funding was expressed in the phrase *guandu shangban*: 'government control with merchant management'. The idea was that the government would provide capital and exercise general supervision, but the actual running of the company would be in the hands of competent businessmen. On this basis Li Hongzhang and his energetic agent Sheng Xuanhuai organised such enterprises as the China Merchants' Steam Navigation Company, founded in 1872, the Kaiping coal mine near Tianjin, established in 1876, the Shanghai Cotton Cloth Mill in 1878, and the Imperial Telegraph Administration. First results were impressive, but as time went by private shareholders became increasingly disillusioned: the official patrons regarded the enterprises as their personal concerns, the companies were more and more dominated by favourites without commercial ability, and the management was steadily less responsive to the interests of investors.[14] Later, during the 1880s and 1890s the governor Zhang Zhidong in Hubei attempted commercial development on a slogan of 'joint management'. He established cotton mills, an ironworks, and several other factories, but he was even less successful in attracting acceptable commercial management, private investors were still reluctant to entrust their money to companies dominated by official interests, and shortfalls of subscription capital meant in the long term that the projects were either maintained by government funds or sold to foreign interests.

Indeed, on their own terms, the conservatives were right, for it is not possible for one culture to absorb the technology of another without major change. Without thoroughgoing policies as were adopted at that time by Meiji Japan, the conflict between new forms of enterprise and traditional official culture meant that the 'Tongzhi Restoration' achieved no real success. Individually, Chinese who received training in Western methods and had the opportunity to read Western books were less ready to accept the dominance of traditional ways, whether in official practice or private life. In economic and social terms improved communications and the increased opportunities for development of capital by trade within the treaty ports meant that more and more Chinese money was held in foreign-controlled cities such as Shanghai, and increasing numbers of landowners had no real contact with the villagers whose farms they owned and from whom they obtained their rents.

In contrast to Japan, the fragmented response of China to the commercial and technological challenge of the West meant that the empire as a whole achieved the worst of both worlds: even the limited opening to the West created tensions in the established patterns of society and the basic peasant economy, while the government itself, disrupted by uncertainty, disagreement, jealousies, and factional infighting, was eventually rendered incapable of formulating any coherent policy. In 1875, on the death of the Tongzhi Emperor, his mother the empress-dowager placed her infant nephew on the throne and controlled the government in his name. Thereafter for most of the reign of the Guangxu Emperor, more than thirty years, the fortunes of the dynasty remained in the hands of Yehonala, the Empress-dowager Cixi, who came to be known both in China and among foreigners as 'The Old Buddha'.

The empress-dowager held her power through strength of personality, and also because her presence at court avoided difficult debate as to which man should rule as emperor or regent. A superb politician among the factions of the imperial court, she was also selfish, short-sighted, reactionary, and corrupt. She gave great authority to Li Hongzhang, but she steadily reduced the authority of Prince Gong, and in 1884 he left office altogether. Overall, the empress-dowager supported traditionalist officials and showed suspicion towards the foreigners and their ways.

Despite the hopes of Self-strengthening and the purchase of modern equipment, the policy of the imperial government towards foreign military and diplomatic pressure was one of weakness punctuated by occasional, ill-judged, firmness. There were embarrassing incidents when foreign nationals were killed in China, notably in the Tianjin Massacre of 1870, when the French consul and several missionaries were slaughtered by a mob and the imperial government was obliged to pay a large indemnity. At the same time, along the frontiers, the British from India were developing a position in the traditional tributary of Burma, and in a diplomatic coup of 1871 the Japanese forced China to abandon its claims of suzerainty over the Ryukyu islands.

In 1879, there was apparent success in negotiations against the Russians on the Ili valley in the far west of Xinjiang, but this encouraged serious miscalculation: a new faction of young, arrogant idealists, who called themselves the 'Pure Stream' (*Qingliu*), condemned the appeasement policies of Prince Gong and Li Hongzhang, and forced the prince from office. Their first opportunity to defy the foreigners came as the French extended their empire in Indochina from Saigon north to Hanoi; the French were harassed by local irregulars and in 1884 they were defeated in a major battle. When the Chinese sought to take advantage of their difficulties, however, the results were disastrous. A squadron of the French navy bombarded the Fuzhou naval yard and destroyed every ship of the Chinese southern fleet, and enemy forces invaded Taiwan and occupied the port of Jilong.[15] The imperial government sued for peace, accepted all French demands, including the loss of Vietnam, and paid another large indemnity.

Despite this humiliation, the Qing government still appeared strong enough to hold its own borders. Large sums were allocated for the rebuilding of the imperial navy, and the Chinese reacted aggressively when Japan attempted to interfere with their hegemony over the kingdom of Korea, last of the tributary states but close to the heartland of the empire and immediate neighbour to Manchuria itself. The Chinese prepared for war with confidence, and many outsiders expected them to win with ease. In 1894–5, however, the Chinese navy was smashed and sunk by smaller numbers of Japanese warships, while the army was driven in flight not only from Korea but also from southern Manchuria. In the dictated peace that followed, by the Treaty of Shimonoseki, Japan obtained not only suzerainty over Korea but also the accession of Taiwan and the right to equality of treatment with the powers of the West.

Now stories were told of how the empress-dowager had taken the naval revenues to rebuild the pavilions of the Summer Palace, and how Chinese warships had gone to battle with untrained, incompetent crews. Li Hongzhang, whose modernized Army of the Huai had been destroyed in the débâcle, was compelled to negotiate the surrender and was then dismissed from significant office. To all the world, the programme of westernization was shown as a failure, and foreigners doubted openly if the Chinese were competent to govern themselves. Inside the empire the choices were becoming more extreme, and the fall of Li Hongzhang, the compromise reformer, left little room for agreement between the conservatives and the men who demanded change.

Carving the Melon

For the nations of the West, the military collapse of China made the whole situation in the Far East uncertain and unbalanced. As the Japanese army poured across southern Manchuria, and their navy swept the Yellow Sea, their first demands required not only that the Chinese abandon all claims over the kingdom of Korea, but also that they cede their own territory in Manchuria. The occupation of Manchuria, however, was strictly opposed by Russia, France, and Germany, and pressure from these three governments compelled Japan to withdraw. For all their success, the Japanese gained no more than Korea, and resentment became all the more bitter when in 1898 Russia, as a friend of the Manchus, leased the Liaodong peninsula for herself.

In the 1950s, during the first years of Communist government in China, while the Soviet Union was a political ally and the source of loans and advice for economic development, both countries were proud to claim that Russians and Chinese had never fought a war. By and large the statement is correct. In the early days of the Manchu dynasty there had been a few skirmishes with Cossack traders and colonists along the Amur River in eastern Siberia, but neither government was prepared to place any great importance on this distant frontier of their empires, and at the Treaty of Nerchinsk in 1689 the Russians abandoned all claim to the territory, while both sides agreed to maintain peaceful trade through frontier markets and caravans. By the middle of the nineteenth century, when the British and French came in arms to Beijing to demand diplomatic and commercial rights for their people, there was already a long tradition of Russian envoys in residence, and on a limited scale the Russians approaching from the north had gained many of the privileges that other nations demanded with such small success in the south. From this time, the government of the Tsars claimed a special relationship with China and acted as 'honest broker' between the Qing empire and the West.

As they say in the proverb, however, with a friend like this, who needs an enemy? In the 1850s the energetic Russian governor of Eastern Siberia, Muraviev, settled colonies along the banks of the Amur despite Chinese claims that the original border lay on mountain ranges far to the north. In 1860, when the British and French captured Beijing and forced their treaty on the Chinese government, the Russian envoy Ignatiev claimed to use his good offices to ensure that the allies withdrew, but then extracted concessions of his own. By the Sino-Russian Treaty of Beijing, the Chinese ceded not only the territory north of the Amur but also the maritime province in the east, between the Ussuri River and the Sea of Japan. The Chinese town of Haishenwai, settled by Russian colonists, was renamed Vladivostok 'Rule the East', and was developed as a port for foreign trade and for use as a naval base.

The Russians were still not satisfied. Though Vladivostok was an excellent harbour most of the year, it was blocked by sea-ice for three or four months each winter; and though the connection from Siberia to the outside world was far closer by the sea route through Vladivostok than by land across Asia, the territory of Manchuria made an irritating salient northwards and forced Russian communications and trade into a long and wasteful detour.

Ideally, for the Russians, there were two solutions: firstly, to obtain access rights from east to west across northern Manchuria, and so shorten the route from central Siberia to Vladivostok; secondly, to control the Liaodong peninsula, which had ice-free sea all year round.[16] The situation remained in abeyance for a generation after the Treaty of Beijing, as colonists established themselves in Siberia and borders were negotiated further to the west. In the early 1890s, however, as work began on the Trans-Siberian Railway which would consolidate imperial power in central Asia and to the north of China, the Russian government looked again at Manchuria.

For a short time before 1894 the Chinese attempted to use the Russians as a counter to the aggression of Japan, and after the disaster of 1895 they were only too pleased to accept Russian intervention against the Japanese negotiators. In 1896, when Li Hongzhang was brought to St Petersburg for the coronation of Nicholas II, the Russians were rewarded for their support. By the Treaty of Moscow it was agreed that a railway across Manchuria would run north-west from Vladivostok through Harbin into Siberia, and it would be financed by the Russo-Chinese Bank. Within a year the Russian government had taken full control of this Chinese Eastern Railway, and in 1898 they secured the lease of Dalian and Port Arthur, together with a concession for the South Manchurian Railway from Harbin southwards through Mukden (Shenyang) to Dalian.[17] For the time being, at least, the Russians had obtained all they wanted.

It was widely, and correctly, rumoured that Li Hongzhang had taken bribes, but his policy had saved China from the full effect of defeat by Japan, and placed the Russians in the front line against the enemy. The overriding problem, however, was the military weakness of China and the loss of foreign confidence in its government. Each of the great powers feared its rivals might gain advantage from concessions by the Chinese, and so each required concessions for themselves in 'compensation'. In 1897, taking as its excuse the murder of two missionaries in Shandong, Germany seized Qingdao, on the south side of the peninsula, on the natural harbour of Jiaozhou Bay, and forced an agreement from the Chinese allowing the construction of two railways into the North China Plain. In response, and with claims of alliance and assistance, the Russians asked for control of Dalian and Port Arthur as counter-balance, and when that concession was granted the British demanded and obtained the lease of China's other northern naval base, Weihaiwei on the Shandong peninsula. Further south, the British also acquired a 99-year lease of the mainland 'New Territories' next to their possessions in Kowloon and Hong Kong, and the French, again as 'compensation', leased Guangzhou on the Leizhou peninsula.[18]

Still more serious was the concept of 'spheres of influence' and the jargon of 'non-alienation'. At first for reasons of trade, but increasingly for political aggrandisement, the foreigners had come to consider certain regions of the empire as their particular interest and concern. Clearly, the Russians had taken first place in Manchuria, and the French, moving north from Indochina, exacted priority in the southern provinces of Yunnan, Guangxi, and neighbouring parts of Guangdong. Japan, blocked in the north by Russia, claimed special interest in Fujian province, across the strait from their new possessions in Taiwan. All these claims the Chinese government was compelled to recognize. France and Japan received clear statements that China would cede no territorial rights to any other country in their own sphere, while Britain gained a similar assurance for the whole of central China, and also the assurance that control of the maritime customs, one of the major sources of imperial revenue, would remain in British hands.[19] Germany, though a late-comer to the competition for empire outside Europe, had entered the game with enthusiasm, and the Anglo-German Agreement of 1898, without any consultation the Chinese, acknowledged British interests in the region of the Yangzi and German concern with Shandong and the North China Plain. One way or another, lines of demarcation were drawn so that Chinese and foreigners of every persuasion looked forward to the time when those general divisions might change to political reality.

Not all the foreigners, however, were convinced that China should be divided up so neatly. In particular, the United States of America had made no claim to a specific region of interest and spoke firmly against the division of China, at least where the division might entail any limiting of trade. It was an American fleet that had forced Japan from isolation in the 1860s, and in 1898 the American government was heavily engaged in seizing the Philippines from Spain and forcing its rule upon the Filipinos, and then in a full takeover of Hawaii. It happened, however, that the United States had acquired no particular territorial concession in China, and it happened also that their trade in cotton and oil, increasing rapidly at this time, involved the region of Manchuria. In September 1899 the American Secretary of State John Hay sent his Open Door Note to Britain, Germany, and Russia, inviting them to confirm that they would not penalize the trading interests of other countries within the sphere of interest which they claimed. The policy received immediate acceptance from Britain and Germany, and later approval from France, Italy, and Japan; but the Russians replied only in the vaguest terms. Since the Open Door was concerned simply with the conditions of trade, and specifically accepted the division of China into spheres of interest, it provided small comfort to the Chinese. Yet, it was reasonably clear that the powers expected no immediate profit in a political division of the empire, and that the government might hope to survive for some time by playing one barbarian off against another.

Notes

1. In the 1890s, it was estimated that land tax provided some 28 per cent of central government revenue, tribute grain 7 per cent, and salt 15 per cent. A further 15 per cent came from *lijin* on merchandise and about 25 per cent from maritime customs. The remaining 10 per cent came from customary and other miscellaneous taxes, from the sale of degrees and various forced contributions, and also from duty on the opium trade. See, for example, *Cambridge China* 11, 63 [Feuerwerker, 'Economic trends, 1870–1911'], discussing George Jamieson, *Report on the revenue and expenditure of the Chinese empire*, London, 1897.

2. There is scholarly debate on this question, but there are no official figures: the machinery of administration was geared to collect taxes, not to investigate ownership or transfer of property..

3. For a discussion of this and other aspects of the growth of population in China, see

Ho Ping-ti, *The Population of China*, pp. 58–62, 96, 270–5. In some rural districts of nineteenth-century China men outnumbered women by as much as three to two, and the situation improved only gradually. Despite the reforms of the Republic, at the census of 1953 females comprised only 48 per cent of the population.

4. It is no coincidence that the new crops which aided the expansion of colonization in south China had been brought across the Pacific and introduced to China by Spaniards trading from their empire in America through the Philippines. The voyages of Christopher Columbus were also important to east Asia.

5. The gold rushes of California in the 1840s and Australia in the 1850s were as attractive to the Chinese as to Westerners: San Francisco at that time was known as *Jinshan* 'Gold Mountain', and Melbourne as *Xin Jinshan* 'New Gold Mountain'.

6. The one great exception to the traditional lack of interest by Chinese governments for ventures overseas was the series of voyages undertaken by Admiral Zheng He at the beginning of the fifteenth century. Under the authority of the Ming dynasty, this eunuch of the imperial court commanded fleets of great ships in seven voyages which established Chinese authority in south-east Asia, visited India and Arabia, and touched the coast of Africa. There was, however, no further development of this initiative, and the last voyage of Zheng He in 1433 marked the end of this overseas enterprise.

7. Hakka [= *kejia* 'guest families'] were people of Chinese origin who had come to the south later than the original settlers. Distinguished by language and custom (they did not, for example, enforce foot-binding upon their women), they held a marginal, inferior, position in the community.

8. Ho, *The Population of China*, pp. 242–6.

9. Guangzhou was generally known to foreigners in the nineteenth and early twentieth century as 'Canton' (a corruption of the provincial name Guangdong, of which this city was the capital). In 1757 Guangzhou was designated as the one legal port for foreign trade, and as time went on the profits of this market were transferred increasingly to the private purse of the imperial household. So the emperor gained from the trade, but his government refused to deal directly with the foreigners, and insisted that all regular contact should be maintained through private Chinese merchants, known as the Hong.

10. The Bogue is a narrows at the mouth of the Pearl River downstream from Guangzhou. The Chinese name is 'Mouth of the Tiger' *hu men*. When the Portugese arrived in the region they translated it to the Latin as *Bocca Tigris*, and it was later shortened into Bogue.

11. An immediate cause of the Opium War was the killing of a Chinese in a brawl by a British sailor. The British authorities, after what appears to have been an honest attempt to find the culprit, were unable to pin the offence down to one man and proposed to administer their own punishment to all under suspicion. The Chinese, however, paid less attention to any proof of guilt, but demanded that one British sailor should be handed over for execution as an example.

12. The conflict at this time was chiefly concerned with foreign demands for further access to trade within China, particularly along the Yangzi, and the right to maintain embassies at the imperial capital. Apart from the set-back at Dagu, the incident was best remembered for the fact that the Chinese arrested, tortured, and killed a number of foreign envoys who had come under flag and agreement of truce, and as punishment for this the allied armies destroyed the Summer Palace of the emperors outside Beijing. Beside the granting of the initial foreign demands, and the payment of an indemnity, the British also acquired the city of Kowloon, opposite Hong Kong island.

13. In the early 1860s Li Hongzhang's forces were aided by a small force of Chinese troops known as the Ever-victorious Army, which was led by foreign officers. Though useful, they played only a secondary role in the great campaigns which destroyed the Taiping rebellion. Naturally enough, however, they were a centre of attention in Western news reports, and their sometime commander, 'Chinese' Gordon, became a hero of his fellow-Englishmen.

14. One feature of Sheng Xuanhuai's administration, for example, was the manner in which he would transfer funds from one of his various enterprises to buttress the fortunes of another. The system is still used by business conglomerates of the present day,

and it is no more popular among shareholders of the company which is being stripped of its capital and assets.

15. Taiwan is the Chinese name for the province and the island, and is now generally used. In earlier Western writing, the island is often called Formosa, based on the Spanish/Portugese 'well-formed' or 'beautiful'.

16. During the period that foreign powers were dominant in the Liaodong peninsula, the Chinese town of Lüshun was generally known as Port Arthur, its Russian name. Dalian was renamed Dalny 'Distant [City]' by the Russians, but after the Japanese takeover in 1905 it was commonly known as Dairen, from the Japanese pronunciation of the Chinese characters. In modern times, Dalian and Lüshun are combined as Lüda.

As an exercise in comparative geography, it may be noted that the latitude of Port Arthur, on the edge of winter sea-ice, is $39°$ N. In Europe, this thirty-ninth parallel runs south of Naples and just north of Athens: a tribute to the warmth of the Gulf Stream, and to the opposite effect of the cooler currents and continental climate of north-east Asia.

17. The chief city of Manchuria is now generally called by its Chinese name, Shenyang. The name Mukden is Manchu, and until the late 1940s, after the defeat of Japan in the Second World War, the city was known in international usage as Mukden. I follow that convention (See note 9 above).

18. This city of Guangzhou, now the Chinese naval base of Zhanjiang, should not be confused with the Guangzhou city, formerly known as Canton, the capital of Guangdong province.

19. The Imperial Maritime Customs had been largely founded by Sir Robert Hart, an Ulsterman who first came to China in the 1850s, and who was Inspector-General of the Customs from 1863 to his death in 1911. The office was staffed partly by foreigners and partly by Chinese, and by the 1890s its revenue provided almost a quarter of the income of the central government. Hart held enormous influence, and his extraordinary career symbolized one aspect of foreign involvement in China: he was formally an official of the imperial government, but he was knighted and made a baronet by the British.

Further Reading

On the Chinese tradition:

DeBary, William Theodore, Chan, Wing-tsit, and Watson, Burton (eds.), *Sources of the Chinese Tradition,* New York, Columbia University Press, 1960.

Eastman, Lloyd E., *Family, Fields and Ancestors*, Oxford University Press 1990.

Elvin, Mark, *The Pattern of the Chinese Past*, Stanford University Press, 1973 [on the economic history of China, with emphasis on the failure to developed a Western-style industrial revolution].

Fairbank, John K., Reischauer, Edwin O., *China: Tradition and Transformation*

Note: The lists attached to each chapter do no more than provide a few suggestions of works which may be consulted as a first reference for more detailed information; they cannot be, and are not intended to becomprehensive. Each volume of the Cambridge History of China series contains a general bibliography of reasonable length. The content of a work is generally indicated by its title; I sometimes attach a brief comment in brackets: [].

[revised edition] Boston; Houghton Mifflin, 1989 [earlier works by the same authors, with Albert M. Craig, are *East Asia: The Great Tradition*, and *East Asia: The Modern Transformation*, combined as *East Asia: Tradition and Transformation*].

Loewe, M.A.N., *The Pride that was China*, London, Sidgwick and Jackson, 1990 [a survey of imperial culture].

Needham, Joseph, and others, *Science and Civilisation in China*, Cambridge University Press, 1954 — [a many-volumed work on Chinese achievements in the natural sciences: the first two volumes contain relevant discussion of history and philosophy].

Pye, Lucien. W., *China, An Introduction*, Boston, Little, Brown, 1984 [on the nature of China].

Spence, Jonathan, *The Death of Woman Wang*, Penguin, 1979 [village justice in the seventeenth century].

General works on late imperial China:

Blunden, Caroline, and Elvin, Mark, *Cultural Atlas of China*, Oxford, Phaidon, 1983.

Chao, Kang, *Man and Land in Chinese History*, Stanford University Press 1986.

Elvin, Mark, The Environmental History of China: an agenda of ideas', in *Asian Studies Review* [Journal of the Asian Studies Association of Australia] 14.2 (1990).

Fairbank, John K. and Twitchett, Denis (eds.), *The Cambridge History of China*, vols 10 and 11, *Late Ch'ing 1800–1911*, Parts 1 and 2, Cambridge University Press, 1978 and 1980 [abbreviated as *Cambridge China* 10 and 11; contains authoritative chapters by individual scholars, some of which are listed below and in later Notes on Further Reading].

Freedman, M. (ed.), *Family and Kinship in Chinese Society*, Stanford University Press, 1970.

Geelan, P.J.M., and Twitchett, D.C. (eds.), *The Times Atlas of China*, London, 1974.

Hummell, Arthur W. (ed.), *Eminent Chinese of the Ch'ing Dynasty*, 2 vols, Library of Congress, Washington 1944 [biographies of leading Chinese and Manchus].

Mackerras, Colin, *Modern China: a chronology from 1842 to the present*, London , Thames and Hudson, 1982.

Schurmann, Franz, and Schell, Orville, (eds.), *Imperial China*, Penguin, 1967.

Skinner, W.F. (ed.), *The City in Late Imperial China*, Stanford University Press, 1976.

On particular topics:

Chang Chung-li, *The Chinese Gentry*, Seattle, University of Washington Press, 1955.

Chesnaux, Jean, (ed.), *Popular Movements and Secret Societies in China, 1840–1950*, Stanford University Press , 1972.

Chesnaux, Jean, (tr. C.A. Curwen), *Peasant Revolts in China 1840–1949*, London, Thames and Hudson, 1973.

Cohen, Paul A., *China and Christianity: the missionary movement and the growth of Chinese anti-foreignism*, Harvard University Press, 1963.

Davis, Fei-Ling, *Primitive Revolutionaries of China: a study of secret societies of the late nineteenth century*, London, Routledge and Kegan Paul, 1977.

Eastman, Lloyd E., *Throne and Mandarins: China's search for a policy during the Sino-French controversy 1880 – 1885*, Harvard University Press, 1967.

Fairbank, John K., *Trade and Diplomacy on the China Coast: the opening of the treaty ports, 1842 – 54*, Harvard University Press, 1954, reprinted by Stanford University Press, 1969.

Fairbank, John K. (ed.), *The Missionary Enterprise in China and America*, Harvard University Press, 1974.

Feuerwerker, Albert, *China's Early Industrialisation: Sheng Hsüan–huai (1844 – 1916) and mandarin enterprise*, reprint of Harvard University Press, 1958, by New York, Athaneum, 1970.

Feuerwerker, Albert E., *Rebellion in Nineteenth-century China*, Ann Arbor, Mich., Center for Chinese Studies, 1975.

Feuerwerker, Albert, 'Economic trends in the late Ch'ing empire', in *Cambridge China* 11.

Fletcher, Joseph, 'Ch'ing Inner Asia *c.* 1800' and 'The heyday of the Ch'ing order in Mongolia, Sinkiang and Tibet', in *Cambridge China* 10.

Ho Ping-ti, *Studies on the Population of China, 1368–1953*, Harvard University Press, 1959.

Hsiao Kung-chuan, *Rural China: imperial control in the nineteenth century*, Seattle, University of Washington Press, 1960.

Kuhn, Philip A., *Rebellion and its Enemies in Late Imperial China: militarization and social structure, 1796–1864*, Harvard University Press, 1979.

Kuhn, Philip A., The Taiping Rebellion', in *Cambridge China* 10.

Mann, Susan, *Local Merchants and the Chinese Bureaucracy, 1750–1950*, Stanford University Press, 1987.

Michael, Franz, and Chang Chung-li (eds), *The Taiping Rebellion*, 3 volumes, Seattle, University of Washington Press, 1966–71.

Perry, Elizabeth J., *Rebels and Revolutionaries in North China 1845–1945*, Stanford University Press, 1980.

Wakeman, Frederic W., Jr, *Strangers at the Gate:social disorder in south China 1839–1861*, Berkeley, California University Press, 1966.

Wakeman, Frederic W., Jr, and Carolyn Grant, (eds.), *Conflict and Control in Late Imperial China*, Berkeley, California University Press, 1975.

Waley, Arthur, *The Opium War through Chinese eyes*, London, Allen and Unwin, 1958.

Wright, Mary C., *The Last Stand of Chinese Conservatism: The T'ung-chih Restoration 1862–1874*, Stanford, Stanford University Press, 1957.

2

Fall of a Dynasty, 1900–1911

By the end of the nineteenth century, faced by the growing disaster of excess population and the pin-pricks of foreign intervention, the Manchu court had experimented with one solution after another, and each had failed. They fought the foreigners, and lost the Opium War; they defeated the great rebellions, but only with the aid of Chinese armies; they revived aggressive morality in the Purist movement of the 1880s, and they were beaten by the French; they sought hegemony in Korea, and lost to Japan. In the last years of the century, the government tried two contradictory programmes. The first, in 1895, was rapid, iconoclastic reform, forced down from the top with imperial favour; but the tensions which developed caused the destruction of the emperor, not of his conservative opponents. The second, in 1900, came when the simplistic faith of the Boxers offered a wild hope to the capital, but the rest of the empire refused to join the war, Western invaders occupied Beijing, and the court was driven into humiliated exile.

After that fiasco, the government of the empress-dowager returned to Beijing, and embarked more seriously on a controlled programme of reform. The changes, however, were too late and too slow. Manchuria was now a bone of contention between two foreign powers, Russia and Japan, and inside China the reluctant steps towards constitutional monarchy did little more than encourage the desire for nationalist change among gentry and merchants and the officers of the new reformed army. When the empress-dowager and the Guangxu Emperor died in 1908, the problems they left were more than their successors could cope with, and the fall of the dynasty was merely a matter of time.

Railroads and Reform

After the defeat by Japan in 1895, the Qing dynasty had little freedom of action. The Treaty of Shimonoseki called for an indemnity of more than two hundred million taels,[1] but official revenues were ninety million taels a year. The powers which had intervened to persuade Japan to relinquish Liaodong now stepped in once more to raise the cash for the indemnity, but they did so at a price. France and Russia lent four hundred million francs, largely French money, with Russia guaranteeing the repayments and so holding a lien on the resources of the government. Within the next few years, Britain and Germany, separately and together, made four additional loans, and they secured their investment with similar privileges.

Government loans were now the primary tool of foreign interference in Chinese affairs. With crippling debts to Japan and other powers, the precarious balance of revenues and expenditure had been broken, and the court was compelled to borrow overseas. At the same time, the general failure of government–merchant co-operation meant that private enterprise in China was largely deprived of official support. So major projects such as railways and telegraphs were commonly paid for by foreign capital on dangerous political terms, while Chinese financial resources were too slight to compete on a substantial scale against the great trading companies and manufacturing industries of Britain, America, and Europe.

At this stage, railways became a focus of investment, modernization, and foreign incursion. When the Germans took Jiaozhou Bay in 1897, one item of the settlement provided that two railroads should be built inland to the North China Plain: finance would be provided by a Chinese–German company, and German subjects would be entitled to mining concessions in a fifteen-kilometre zone on either side of the tracks. As other countries followed the German example there was constant pressure for the award of leases and the acceptance of loans, and a maze of projected lines spread across the map, all paid for with foreign capital and mortgaged as security for the loan. In 1899, at the height of the contest, the British and Russian governments were compelled to settle their quarrel about the right to build a railroad north of Beijing by a formal Political Agreement that neither side would interfere with the other's development in the 'British' territory of the Yangzi valley and the 'Russian' lands north of the Great Wall.

Until 1895, despite conservative opposition, the imperial court had tolerated a limited westernization, and had supported the proposals of Li Hongzhang and his colleagues for the building of railroads and the opening of new mines for coal and iron. The Chinese, indeed, had been proud of their new navy and army and of their developing economic strength. But now the pride turned to bitterness, moderniza-

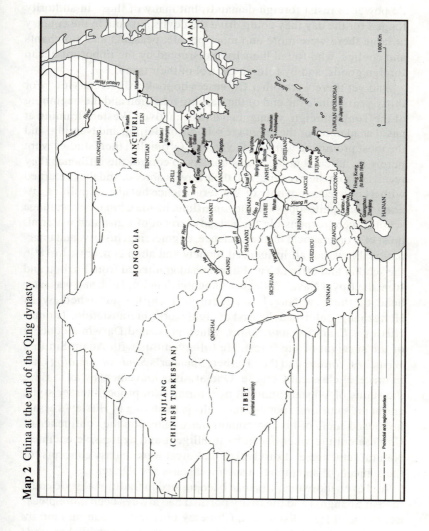

Map 2 China at the end of the Qing dynasty

tion appeared a total failure, and the arguments of reactionary councillors were supported by the clear policy of foreign powers that railways should become a means for the economic domination of China
and a spearhead of their political invasion. The government had little power to resist foreign demands, but many of those in authority
longed for the day they could drive the devils out, and in the country
at large there were more and more people, peasants and merchants
and coolies, who found their lives disrupted by the import of cheap
foreign goods, and by foreign control of their trade.

And yet there were other Chinese who took a different view. For
thirty years, since the end of the Taiping rebellion and the beginning
of the Tongzhi reign, young men had been given Western training at
naval academies or the language schools of Beijing and Nanjing, and
the most promising were sent overseas. Officially, their studies were
purely technical, but there was no way to control the influence they
received from their reading and their travels abroad, and they returned
with a knowledge not only of Western science but also of politics, literature, and philosophy. At the same time, inside China, the language
schools were responsible for a steady stream of translated material,
most of it on practical themes, science, engineering, history, and international law, but all influenced by new and alien concepts. In 1896
the scholar and official Yan Fu, who had graduated from Fuzhou and
from the Royal Naval College at Greenwich in England, and was later
head of the new Naval Academy at Tianjin established by Li
Hongzhang, published the first of eight works in translation, Thomas
Huxley's *Evolution and Ethics*, which presented Darwin's concept
of the survival of the fittest. He followed this with Adam Smith's
Wealth of Nations and then Herbert Spencer's *Study of Sociology,* all
in graceful classical Chinese. One of his achievements was to raise
the status of Western thought by treating it in proper literary style,
and at the same time, notably through his rendering of Huxley's essay,
he caused some of his contemporaries to question the confidence of
Chinese culture when faced by intelligent and aggressive enemies.
Natural selection and survival of the fittest were uneasy concepts for
the subjects of an empire in such weakness and danger.

And even those men who had been trained in traditional style came
to visit Shanghai and the treaty ports. It had always been more pleasant to travel long distance in China by boat rather than on land by
horse, foot, or sedan chair, and as foreign steamers established trading routes along the coast and up the Yangzi, officials in the south
would regularly take ship for Shanghai or Hankou or Guangzhou on

their way to their posts. These foreign cities on Chinese soil became centres of intellectual life, with newspapers and magazines in profusion and a steady stream of visitors, students, scholars, administrators, and politicians. Many preferred to visit singsong girls in a brothel than to join arguments in a teahouse, but they saw the results of Western-style government and finance, they met first-hand with Western food and dress and manners, and they could hardly escape some contact with foreign fashion and current affairs.

Thus the ideas of the West came to rival those of China, and Confucianism, closely bound to the government of the empire, faced immediate challenge and a demand for adaptation. Kang Youwei, from Guangzhou, became the leader of a reformist Confucianism. Born in 1858, by his early twenties he had read translations of Western works of science, history, and political thought. He visited Hong Kong and Shanghai, he met Chinese who had returned from overseas, and he became an ardent propagandist of westernization and reform. He was also a competent classical scholar. In his *Study of the Forged Classics,* published in 1891, he dismissed as false the greater part of the accepted canon, and in 1897, in *Confucius as Reformer*, he claimed that the books Confucius actually wrote were tracts for political reform, and that the sage should be regarded as the prophet of a new religion. Later books, notably *Utopia*, expanded these theories, and he founded a Western-Chinese school at Guangzhou, and a Society for the Study of National Self-strengthening at Beijing and Shanghai.

In 1895, after several attempts, Kang Youwei passed his *jinshi* degree, the highest in the empire, and he took the occasion of the Chinese defeat by Japan to present to the throne a long memorial, signed by six hundred fellow-candidates and widely publicized throughout the empire, urging reform and modernization. No practical results appeared, but Kang had acquired an active band of followers and he continued to send in papers. In 1897, when Germany seized Jiaozhou, he warned the emperor in specific terms that his throne and his life were in danger unless the government was changed and the foreigners held in check; the Guangxu Emperor was so impressed by this man who mentioned death and disgrace direct to the Son of Heaven that he ordered all further letters from Kang Youwei to be brought to him at once.

The emperor was now twenty-six years old. He had been on the throne since the age of three, but he had remained under the regency of the Lady Yehonala until 1889. Even then, though the empress-dowager was supposed to live in retirement at the Summer Palace, she insisted on seeing state documents, she maintained her influence among senior councillors at court, and she interfered most effectively in important appointments. After twenty years under her dominance, the emperor was terrified of her.

Despite his youth and the weakness of his position, however, the Guangxu Emperor was regarded as an honourable and conscientious ruler and his former tutor Weng Tonghe, close adviser and friend, was a respected classical scholar. After 1895, Weng Tonghe encouraged the emperor to study English and read Western history, and to consider a policy of reform. Weng had earlier opposed Kang Youwei's importunities, but he now forwarded his memorials to the throne, and under this guidance the emperor planned a programme of modernization.

On 11 June 1898 the first edict was issued to proclaim the new policy, and in the weeks that followed there was a steady stream of imperial decrees. A national board was established to develop commerce, agriculture, mining, and railways, and the traditional administration of the salt monopoly was abolished. Orders were given for the modernization of school curricula, and for the establishment of a new Western-style university at Beijing, and it was proposed that the state examination in classics should be abandoned and replaced in part by a test in economics. Further decrees provided for the adoption of Western-style military drill, additional scholarships for students to travel abroad, the abolition of sinecure offices, and permission for junior officials and common people to memorialize the throne direct.

Many of the proposals were admirable, but their presentation lacked political judgement and the reformers themselves seemed erratic and irresponsible. Having obtained the support of the emperor, they made no move to conciliate more moderate officials, and they threatened conservatives not only with the destruction of the whole traditional system but with personal humiliation and loss of rank. Kang Youwei had gained the right of direct audience with the emperor, but there was no way to ensure that the edicts so hastily issued would receive anything but passive resistance and verbal protest from members of the bureaucracy who were supposed to carry them out. And even as Kang Youwei gained the emperor's favour, the conservatives, aided by the empress-dowager, secured the dismissal of Weng Tonghe, the one senior official who really supported the idea of reform and the one man who might have provided sensible restraint to Kang and his supporters.

Kang Youwei did propose that a commission should consider the whole question of reform, and he was unhappy about the erratic, piecemeal approach. Both he and his imperial patron, however, were under immense pressure, and they were forced to accept whatever gains they could. In the process, however, the reformers and their imperial patron offended the most senior officials and scholars of the empire, they presented an apparent threat to the supremacy of the Manchu hierarchy in the state, and, worst of all, they appeared hasty and irresponsible.

By September 1898, three months after the beginning of the new policy, the conflict between the two parties had come to the point of crisis. On 5 September, when the Board of Rites refused to send on a memorial from a low-rank official, the emperor dismissed every senior member of the Board, and he appointed four junior secretaries to the Grand Council to take over their administrative duties. As opposition grew, Kang and the emperor became increasingly concerned about their future security. They sought military means to defend themselves and their programme, and General Yuan Shikai was called to audience at the palace.

Yuan Shikai was very competent and very ambitious, and he had already a remarkable record. Born in 1859 to an undistinguished branch of his clan, he gained entry to official rank only by purchase. In 1882, as China attempted to enforce a presence in Korea, Yuan Shikai was included in a military mission. In December 1884 the pro-Japanese party in Seoul staged a *coup d'état* but Yuan Shikai, commanding a local garrison, brought troops to the palace to rescue the king, and he drove the Japanese and their supporters from the city.

In 1885 Yuan was appointed Commissioner of Commerce to Korea with the status of a resident minister, and for ten years until the Sino-Japanese War he maintained China's position with authority and skill. His diplomacy was ruined by the massive military defeat, but Yuan himself emerged with nothing but credit and in 1896 he was appointed to command an army unit which was being organized under German instruction to form the first in a new programme of modernization. By 1898 his force numbered seven thousand men and included an officer training school.

To the reformers at the capital, Yuan Shikai appeared the best hope of military support. He was obviously a modernizer, he controlled a well-armed force, and he had joined Kang Youwei's Society for the Study of National Self-strengthening. In audiences with the emperor, Yuan Shikai was given ministerial rank in the central government, and was also granted independent military authority. At midnight on 17 September, Yuan Shikai was visited by Kang Youwei's close ally Tan Sitong, and it was agreed that Yuan Shikai would support the emperor against his conservative opposition, forcing the empress-dowager into full retirement, and placing real power in the emperor's hands.

However, though Yuan Shikai might be sympathetic towards the ideals of the reform party and the emperor, many reasons made him reluctant to take their part in open political warfare. Admittedly, the reformers had been hindered by the conservative party and the empress-dowager, but many of their troubles were brought on by their own mistakes and arrogance, and a sensible man like Yuan Shikai would nor wish to tie himself too closely to their cause. Still more to the point, Yuan Shikai and his troops were seriously outnumbered by other forces stationed around Beijing: against Yuan Shikai's seven thousand, the Manchu Ronglu, close friend of the empress-dowager, controlled a hundred thousand men; and Ronglu had favoured Yuan Shikai and recommended him for promotion. Further afield, in the provinces of China, the reform party had no active support among the civilian governors nor among the commanders of the army and navy. It was most unlikely that the coup would succeed, and there was no reason to believe the attempt would bring anything but a weakened central government, and perhaps final dissolution of the empire. So Yuan Shikai told Ronglu of the emperor's plans, and Ronglu sent word to the empress-dowager.

During the night of 21 September the empress-dowager returned to Beijing, and with household guards under her command she arrested the emperor and confined him to the Yingtai pavilion, on a small island in a lake of the imperial palace park. Taking the imperial seal of office, she issued decrees restoring herself to the regency and cancelling all edicts of reform. Kang Youwei, warned by the emperor who already had suspicions of Yuan Shikai, made his escape, but six of his party, including Tan Sitong, were arrested and executed. The period of reform had lasted little more than one hundred days, the empress-dowager and the conservative party now held full power, and the Guangxu Emperor never emerged again from imprisonment.

The Empress and the Boxers

For some time after the empress-dowager had forced her return to power, there were fears that the emperor might be killed in secret, but the foreign embassies and officials in the provinces sent firm warnings against any such accident. The unity of China was already precarious, and the legitimacy of the empire and the whole basis of Manchu rule depended on the sacred person of the emperor. Though people could accept the dowager as regent, few were prepared to endorse the death of the emperor, and the British in particular made it clear that they regretted the end of the programme of reform. On the other hand, the foreigners were not unduly concerned with the return of the conservative faction: as in the past, they could expect to apply suitable pressure where necessary on the regime, conscious of its weakness against their power, and the profitable game could continue as before while responsible government held any Chinese resentment under control.

Certainly, as a mature and experienced ruler, lacking any claim to the throne herself, the success of the empress-dowager was reassuring to those who had been rendered uneasy by the erratic energy of the young emperor, and she indeed possessed considerable experience in government, great skill in palace intrigue, and total ruthlessness in carrying out her policies. Many people, Chinese and foreigners, regarded her with a certain admiration, but it is hard to claim that she was in fact anything but a stupid, selfish woman, whose only talents lay in factional politics. She was interested in nothing but her own pleasure and power, she knew nothing of the empire which she insisted upon governing, and she made no attempt to produce any coherent policy for dealing with the problems of her country or for coping with the foreigners.

No government headed by a female regent would have been in a good position to embark upon any long-term programme, but the empress-dowager, now in her middle sixties, judged all proposals solely by their effect upon alliances at court, and her approval for any course of action was based primarily on the favour with which she regarded the man who sponsored it. Since the reformers had associated themselves so closely with the Guangxu Emperor, the majority of her counsellors were conservative or reactionary, but they still disagreed amongst themselves, and as one or another gained attention the policy of the government lurched this way and that.

The empress-dowager, moreover, had been humiliated by the foreigners' concern to preserve the emperor's life, and though for the time she was compelled to accept their intervention, neither she nor her chief associates held any respect or trust for their ambitions, their goodwill, or their vision of the future for China.

General Ronglu, who had been responsible for the overthrow of the emperor and who still commanded the army around Beijing, was naturally in the highest favour, and Yuan Shikai, recognized as a trusted officer, now commanded a force of twenty thousand men near the border with Shandong province. Whatever their feelings about reaction or reform, both men were concerned with the modernization of their equipment and the training of armies to operate on Western lines, and the new government at Beijing was anxious to use them. On a limited scale they achieved some success. In March 1899 the Italians demanded a sphere of interest and land for a naval base in Zhejiang, but the government refused even to discuss the question, the whole province was mobilized for war, and the Italians, humiliated, abandoned their claim. Still more impressively, soldiers from Yuan Shikai's command faced German troops in Shandong and compelled them to withdraw. Small though they were, these were the first successes since the disaster of 1895, and the government was correspondingly encouraged.

Shandong, however, remained a trouble spot. Since their first occupation of Jiaozhou in 1897 the Germans were determined to establish their presence in the whole region where they claimed a sphere of interest, and they developed Qingdao as a major commercial port. The railroad and mineral rights they held westward into the North China Plain gave opportunity for dominance of the region's overseas trade, and survey or mining parties travelled widely across the country. There were quarrels with the villagers, and German troops were very ready to take reprisals. On the north side of the peninsula, the naval base at Weihaiwei had been captured by the Japanese during the recent war and was now occupied by the British. Honouring their agreement on spheres of influence, the British were less aggressive than the Germans, but the people of Shandong had met the foreigners and many had good cause to dislike them.

The situation was made worse by a series of natural disasters and consequent famine. There are estimates that a million people died of starvation in north China between 1892 and 1894, and in Shandong in January 1898, after a poor harvest the year before, winter food ran out and the government had to send in relief supplies. In August that year the Yellow River broke its banks and flooded 2,500 square miles of farmland, with a population of more than a million people, mostly in western Shandong. In 1899 another poor crop was attacked by a plague of locusts, and the price of grain rose so high that most people could not pay it. The central government gave orders for relief with money and food, but the distribution was both inefficient and dishonest, enormous sums were stolen, and quantities of supplies never reached the people who needed them.

Under pressures such as these, the bonds of government and society began to break down. Some men took to the hill country as bandits, and even those who remained in the villages formed vigilante groups and private associations for self-protection. As early as the 1850s and 1860s, when Shandong was affected by the Nian rebellion, a variety of armed bands had sprung up, some commanded by members of the gentry, some by men of common rank, some with government authority, many more opposed to the dynasty. In one form or another these groups had maintained their existence through the troubled years since the destruction of the Nian. Often linked to secret societies, they maintained their petty authority by force among the villagers, and they fought one another, made raids against government offices, and occasionally staged riots against the foreigners and their railroads.

In 1898 and 1899 the disasters of famine brought more hungry, desperate men to join the ranks of these irregular troops. They still fought against the government, and several times murdered magistrates, but increasing numbers of them claimed to act as patriots against the foreigner. They were violently anti-Christian, and they claimed possession of magical powers which made them invulnerable to foreign weapons. Their members performed ritual exercises, moving the hands as if in shadow-boxing, and these became the mark and symbol of their association. They first called themselves 'Righteous Harmony Fists' (*Yihe quan*), from which the West adapted the name of Boxers. In 1899, however, the Manchu governor Yuxian granted them his approval, changed their name from *quan* 'Fists' to *tuan* 'militia', and urged them to adopt the slogan 'Support the Qing and destroy the foreigners'.

The techniques of 'boxing', long known in China, are practised today as a form of physical exercise and a semi-mystic ritual; and claims to invulnerability are common to many works of fiction and the teachings of popular religion. The real miracle of the Boxers came when this badly organized, superstitious, and largely illiterate rabble gained official patronage as a serious weapon against the foreigners.

After the success of confrontation against the Germans and Italians in 1899, some members of the court had become impatient with regular diplomacy, and raised hopes for a complete victory over all the foreigners in China. Ronglu and other generals knew that there was no possibility of military success against the nations of the West, but as news of the Boxers came to Beijing, some advisers accepted the tales of their magical powers. The empress-dowager, more pragmatic, argued none the less that if the Manchus could turn the movement into a national revolt against the foreigners it would give the dynasty an emotional support which it badly needed amongst the people.

Chief among the patrons of the Boxers was Zaiyi, Prince Duan, who had come to prominence for his opposition to the reforms of Guangxu and had played a leading part in support of the empress-dowager. His son Pujun had been named heir-apparent, and only the foreign refusal to accept the deposition of the Guangxu Emperor had interrupted his ambitions for his family. With this in mind, and with his supporters at court, he persuaded the empress-dowager that she should use the Boxers as allies and kill every foreigner in the country.

The idea was insane, and most officials realized it, but in the summer of 1900 the empress-dowager allowed her ministers to call the Boxers to Beijing. There was a reign of terror and disorder in the capital, the German ambassador was shot down in the street by a Manchu bannerman on his way to the Zongli Yamen, and from June to August the foreign embassies were besieged by a mixed force of Boxers, regular troops, and Muslim fanatics from the north-west, outnumbering the legation guards by ten to one and more. Ronglu, well aware of the disaster that would follow if the foreigners were killed, would not allow his men to press their attacks, and it was due to his reluctance and refusal to carry out orders that the embassies survived. On 14 August an allied army, with contingents from Britain, France, Germany, Austria-Hungary, Italy, Russia, the United States, and a strong force of Japanese, entered Beijing and raised the siege.

As most people predicted, the Boxers had been useless as a military force and proved as much of a threat to their own side as to the enemy. The brunt of the fighting against the foreigners was borne by the soldiers of the imperial army, who often fought with great courage and gained some limited success but were finally driven in total defeat away from Beijing and off to the west, while the empress-dowager, taking the emperor with her, fled west to Xi'an on a 'tour of inspection'.

The crisis for the legations at Beijing had begun with the murder of the German minister, Baron von Ketteler, and as the Kaiser farewelled the first German contingent for the army of revenge he made a celebrated speech, calling on the Germans to be as ruthless as the Huns of old, and to give the Chinese cause to fear and respect them. Each in its way, the allied contingents lived up to the Kaiser's hopes, and as their troops marched in victory through Tianjin to Beijing, they left behind them a trail of slaughter, looting, and rape. Eye-witnesses told of pillaged houses, murdered children and mutilated bodies, and every man came home with plundered souvenirs.[2]

The Boxers, however, did score one success, and this perhaps made up for all the rest: they made the foreigners afraid. In Beijing itself, though some were killed in the diplomatic missions, there was no wholesale massacre of foreigners, and in several villages, despite threats and shouting in the streets, the Christians remained unharmed. Elsewhere in the countryside, however, many missionaries lost their lives and thousands of their Chinese converts were killed by furious mobs. Regardless of what really happened, however, the foreigners never felt quite secure again in China. They still insisted on their privileges, and they insisted even more on guards within the country to protect them, but there was much less talk now of carving up the empire between the nations of the outside world.

The Last of the Old Regime

In fact, the empire was already divided. During the Boxer Crisis, as the court declared war against the foreigners, the majority of the provinces refused to accept orders. At the end of 1899, under pressure from the Western powers, Yuan Shikai was appointed Governor of Shandong in replacement of Yuxian, and despite imperial commands to support the Boxers he maintained a strong programme to put down their lawlessness and protect foreign installations. It was largely as a result of his firm measures that the Boxers fled north towards the capital. Later, as imperial armies fought the foreigners at Tianjin and Beijing, Yuan Shikai, with Li Hongzhang in Guangzhou, Zhang Zhidong at Hankou, and their colleagues in Anhui, Zhejiang, Jiangsu, and Fujian, the provinces along the Yangzi River and in the far south, sent messages to the allies that they would take no hostile action so long as their territory was not attacked .

Officially, the southern officials explained that the court had been seized by rebels and the empress-dowager was acting under duress. In accordance with this approach, they prevented Chinese forces in the south from attacking the enemy, suppressed the edicts declaring war, and continued the payment of foreign debts from their provincial treasuries.

The alliance of neutral governors saved the greater part of the empire from the miseries which the court had brought about by its misplaced faith in the Boxers, but in formal terms their conduct was surely treason: the imperial government had certainly declared war, but the provinces refused to accept their orders and entered separate arrangements with the enemy. In effect, the authority of the central government was now subject to the independent judgement and local veto of its own administrators, and in these circumstances one can hardly describe the empire as a unified state.

As the allied army neared Beijing, driving the Boxers and the court away before it, Li Hongzhang was called to head negotiations for peace. In line with the policies of his neutral colleagues in the south, he persuaded the foreign powers to accept the principle that the reason they had invaded China was not to fight the empire but to assist in time of rebellion. By this, his greatest diplomatic coup, he preserved the nominal unity of China and arranged the survival of the imperial government.

Even so, the terms of the Boxer Protocol were very harsh. At the orders of the allies, Yuxian and three members of the imperial court were required to commit suicide, and others, including Zaiyi, were banished for life to the frontiers. The Chinese government put up monuments to the foreigners who had been killed, suspended official examinations in all cities where foreigners had been mistreated, and prohibited, under pain of death, all anti-foreign societies. The Dagu forts were razed to the ground, the legation quarter at Beijing was made defensible and reserved for foreigners who would be guarded by their own troops, and the railway through Tianjin was garrisoned by foreign soldiers. China was required to pay an indemnity of 450 million taels, and since she had no means of finding this sum from her own resources a loan was arranged to cover the amount, with interest and security taken from maritime customs, internal customs, and increased taxes and duties on a wide variety of imported merchandise. The protocol was signed on 7 September 1901 and ten days later, proud of their victories and loaded with plunder, the allied armies left for home. On 7 January 1902 the empress-dowager and the emperor returned to Beijing.

The Russians, however, were not yet satisfied. Their embassy had been threatened and they sent a contingent to the allied army, but they paraded their indignation less than the others, and they supported Li Hongzhang in his negotiations for peace. They received, none the less, a larger share of the indemnity than any other power, and their troops occupied the chief cities of Manchuria and seized the railway zone. The local Chinese armies were easily driven back, and the Russians now asked, as reward for their good offices in the Boxer crisis, that the three north-eastern provinces should be placed under their control.

The Qing court was helpless. Li Hongzhang was known to be in Russian pay, but he rightly pointed out that unless the Chinese signed some agreement the Russians would continue to hold the territory. Li Hongzhang died, however, on 7 November 1901, and as the question remained in the balance the imperial government was deluged with protests from the provinces, urging that Manchuria should not be abandoned, while the British and Japanese made it clear they would regard it as a most unfriendly act should China alienate Manchuria to the tsar. On 30 January 1902 Britain entered a formal alliance with Japan and in April, though the Chinese government still refused any concessions, the Russians announced they would evacuate Manchuria. The programme for their departure, however, was very slow, the Russians saw no need for haste, and it seemed clear that an army would be needed to get them out. If the Chinese could not do it, the Japanese would.

The Anglo-Japanese alliance was a turning point of relations between the great powers. For Japan, the treaty gave immediate prestige and the prospect of great power in the future. For Britain, it marked the end of splendid isolation. By its terms Britain and Japan recognized each other's interests in China, and Britain approved Japan's position in Korea. Should either party be involved in war to defend those interests the other would remain neutral, but if another country joined the enemy the ally would then come to help. As a result, if Japan should attack Russia the two enemies would fight it out alone: though Russia had a treaty with France, and might have counted on her support, no country dared a conflict with Britain.

For China, the immediate effect of the treaty was a war fought on Chinese soil between two foreign powers to decide which should control Chinese territory. Despite the alliance with Britain, an aggressive party in Russia was confident Japan could be beaten, and even felt that a short, successful war would rally national feeling to the tsar and would quieten the growing anti-government, revolutionary activity. On 6 February 1904 the Japanese broke off diplomatic relations, and the Russians crossed the Yalu River into Korea. On the night of 8 February the Japanese made a surprise attack on the Russian Far Eastern fleet, sank several ships, and blockaded the remainder in Port Arthur. On 10 February Japan declared war on Russia.

The Russo-Japanese War was the first great conflict of the twentieth century, the land battles involved larger armies than the world had seen before, and observers from Britain, Germany, and America attended the Japanese army. Port Arthur surrendered after a five-month siege, and in March 1905, after an eight-day battle outside Mukden between armies totalling more than 600,000 men, the Russians were forced to abandon southern Manchuria. On 27 May the Russian Baltic Fleet, which had sailed for the Far East in the previous October, was caught and annihilated by the Japanese Admiral Togo in the straits of Tsushima.

Despite this succession of victories the Japanese were almost exhausted by the war, and the Russians could have continued. They were handicapped, however, by long lines of communication across the Trans-Siberian Railway, not yet in full operation, and the record of defeat had done nothing to strengthen the position of the Tsarist government at home. By a treaty signed on 5 September 1905, the Russians abandoned all interest in Korea and also transferred to the Japanese their concessions south of Mukden, including most of the South Manchurian Railway and the leases to Dairen and Port Arthur, together with the southern half of the island of Sakhalin, north of Japan. In the north, they retained control of the Chinese Eastern Railway from Vladivostok through Harbin to Lake Baikal. The settlement was recognized by the Chinese government in a special agreement with Japan. Korea was proclaimed a Japanese protectorate and in 1910 it was merged completely with the Japanese empire.

Though they still maintained formal sovereignty, the Manchus at Beijing had lost control of Manchuria, and one sign of the change was the acceptance of Chinese migration. Hitherto, Chinese-born subjects of the empire had been permitted to enter Manchuria only on restrictive terms, and few had been granted permission to settle and colonize the land. Now the barriers were removed and great numbers of Chinese settled in the north-east, either to take up and farm the open pasture-land or to serve as a work-force in the railroads and industries which were being developed with Japanese capital. In the years after 1900 the population on this last frontier increased by millions, and under Japanese hegemony the territory became the most prosperous of China. The development, however, was of small service to the rest of the country, and it was almost fifty years before Manchuria returned to the rule of a government at Beijing.

In the rest of the empire the situation was not much better. The independent stand of the southern governors at the time of the Boxer rebellion made it clear that the court could not rely upon automatic obedience from provincial authorities in all major aspects of policy, and the central government was subject to constant pressure from the rest of the empire. As foreign investment poured into the country, and railways were built throughout central China, the administrators of such provinces as Sichuan, Hubei, Anhui, Jiangsu, and Zhejiang in the Yangzi basin, and Guangdong in the far south, developed new sources of revenue independent of Beijing, while the imperial government was crippled by the need to repay the Boxer indemnity and other foreign loans.

Despite Chinese resentment of foreigners, the years after the Boxer rebellion saw modernization and foreign investment become respectable among the landed gentry as well as the merchants of China. When Zhang Zhidong was put in charge of construction for the Beijing–Hankou railroad in 1896, he attempted to sell stock to Chinese investors, but gained no success and was forced to turn instead to a foreign consortium. When the line was completed in 1906, however, several small companies, generally based on foreign capital but supported by Chinese local investors, had projects in planning or construction. Government policy would have restricted this private enterprise to favour official monopolies, but the court was not strong enough to resist the new development, and in Shanghai and the treaty ports of the middle and lower Yangzi there were a growing number of textile mills and other small industrial plants financed by joint foreign and Chinese companies.

In all this development, Japanese capital and enterprise was beginning to play a major part. As an Asian nation which was developing a modern economy and had lately achieved military victory over a great Western power, Japan held increasing influence in China both as a practical partner and as an ideal for future development. In 1904 a special committee including Zhang Zhidong presented recommendations to the throne for a national modern school system, clearly based on the Japanese model, and proposed that the Confucian examinations should be abolished.[3] In 1905 this was done, and one of the chief features of the civil service tradition came quietly to an end. The majority of young men who were able to attend school now went to those organised on Western lines, and they competed for scholarships which would take them overseas for further study. Many went to Europe and America, but great numbers travelled to Japan, where they could see at first hand the government and society which had produced such progress.

At the time of the Tongzhi period in the nineteenth century, the example of the Meiji Restoration in Japan had seemed dangerous to the authority of the Manchu dynasty in China, and that judgement remained valid forty years later. In the early 1900s, though government in Japan was still dominated by the men who seized power in 1868, a written constitution had been proclaimed in 1889 and a national parliament, the Diet, had been elected and held regular meetings since 1890. For more and more people in China it became intolerable that their nation should be controlled by an alien ruling house, with authority gained through palace intrigue. A medley of reform or revolutionary movements grew up inside China, some moderate, some radical, but all concerned with nationalism and independence, and all opposed to the existing structure of dynastic power. Many members of these opposition groups were students returned from abroad. They were often joined in China by Japanese friends and allies, and when they found themselves in trouble with the imperial authorities they fled to safety in Japan.

One provision of the Boxer settlement had required the imperial government to raise the status of the Zongli Yamen to a full Foreign Office, and the failure of the anti-foreign movement persuaded the empress-dowager to reverse her former policy. Since her return to Beijing, she had shown courtesy to the same envoys she had been trying to kill only a short time before, and many wives of foreigners were entertained with elegant tea parties in the imperial palace. Nevertheless, even if the West would accept her good graces, there was no way to disguise the political and ideological bankruptcy of her regime. Foreign policy was no more than passive acceptance of the actions of other powers, and internally, through her government introduced a number of reforms on the lines of those proposed by the ill-fated group of 1898, her chief concern was to retain her personal position and the privileges of the Manchus.

On 14 November 1908 the Guangxu Emperor died at the age of 37, and he was followed next day by the empress-dowager. It is possible that she had crowned her career by the murder of her unfortunate prisoner, but there was no firm evidence, and although she was seventy-three years old her death was not expected. As the emperor lay dying, the two-year old Puyi, son of the prince Zaifeng, a younger brother of the Guangxu emperor, and of his wife the daughter of Ronglu, was introduced as his successor, and the empress-dowager gave orders that Zaifeng should act as regent in association with her niece, the Lady Xiaoding, widow of the Guangxu Emperor. After almost half a century of power in China, it is a measure of the empress-dowager's talents that the government she established to succeed her lasted less than three years.

Neither the regent nor the new empress-dowager possessed great skill or courage, but their situation would have defied the most able politicians and statesmen. The Manchus of the imperial clan, jealous of their power and fearing for their privileges, demanded that Chinese influence in the government should be diminished. Among their targets was Yuan Shikai, who had succeeded Ronglu as commander of the Northern Army which guarded Beijing, but who was transferred to the civilian Ministry of Foreign Affairs for fear his following among the troops might make him a threat to the government. In January 1909 Yuan Shikai was attacked by the censorate, and Zaifeng ordered that he retire from office. Yuan Shikai happened to be slightly lame in one foot, and as an added humiliation it was proclaimed that the deformity made him unfit to appear in court. Other leading officials, supporters of modernization and industrial development, were likewise dismissed, and when Zhang Zhidong, Minister for Education, died in October 1909, the imperial government was left without any high officials of comparable standing and experience.

Even under the empress-dowager, though many Chinese were still basically loyal to the Manchus, there had been pressure for the government to proclaim a formal constitution. In 1909 provincial assemblies were established for the first time, and a consultative college met at Beijing with representatives from the assemblies and an equal number of nominees from the central government. The provincial assemblies had the right only to advise the local governor, and they were elected on a male franchise restricted to substantial property holders and graduates of the state examinations. Despite the lack of a popular vote the assemblies provided a forum for educated opposition to the government, and they contained many new men of considerable economic power. Under continuing pressure, the council of regency proclaimed a date, 1913, when responsible cabinet government, still under the authority of the emperor, would be introduced. In May 1911, however, as membership of the future cabinet was announced there was disappointment and disgust, for the majority of ministers would still be Manchus and almost half of them were members of the imperial house. Clearly this was no real progress, and the government itself, determined to maintain its position, expelled or exiled anyone who protested.

The final crisis came in the summer and autumn of 1911. Besides the main government railways such as the Beijing–Hankou and the Wuchang–Guangzhou lines, a number of smaller systems had been financed privately by Chinese and foreign-owned companies. The government, which had just negotiated a loan from British, German, French, and American banks to cover the costs of the main railways, now proposed to raise further money to take over all provincial lines. In the national interest it was obviously better that the chief means of transport should be firmly in central government hands, but the fact that money was being borrowed from abroad by the Manchu regime and Chinese investors were being bought out by compulsory purchase on most unfair terms caused protest throughout the empire. Sichuan became a centre of disturbance, and when riots led by the gentry were put down viciously by the local governor the unrest turned to rebellion. Politically, through its opposition to constitutional reform, the government had alienated even the moderate members of provincial assemblies. By its threat against investors in the railways, it made all men of property uncertain and suspicious.

In the city of Wuchang, on 9 October 1911, there was an accidental bomb explosion in a private house, and when police investigated they found an arsenal of weapons and ammunition, and lists of members of a revolutionary group. Some were officers of the garrison, for many junior officers had travelled overseas and shared ideals for change and reform with other, civilian, students. Faced with the threat of discovery and punishment, they determined to call the revolt at once, and in the night of 10 October they led their men to seize the arsenal and attack the local headquarters. The soldiers followed them readily, for several contingents had been sent into Sichuan to fight the rebels, and they preferred to turn against the dynasty. The rebels seized Wuchang, proclaimed a military regime and a republic for all of China, and then occupied the sister towns of Hankou and Hanyang.[4] Li Yuanhong, a colonel of the garrison who had been captured by the rebels, was forced forward as head of the government, and the provincial assembly of Hubei province was persuaded to recognise the rebellion. As the news spread through the empire, the pattern of revolt was established: junior army officers leading the rebellion, compelling their seniors to give an air of responsibility, and local assemblies providing a semblance of civilian endorsement. By the end of November every province in China Proper except Zhili (modern Hebei),[5] Henan, and Gansu had proclaimed independence of the imperial court, and 10 October, the Double Tenth, is still the Nationalist day of celebration.

The Failure of Tradition

Since early in the nineteenth century, the Manchu Qing dynasty had been faced with insoluble problems. Of these, the incursions of the foreigners, while apparently threatening to the security of the empire, were of far less significance than the crisis brought by the growth of population. There had been some respite through emigration overseas, and more recently through the opening of Manchuria, but the pressure against available resources and technology, and the social and economic tensions which accompanied it, brought endemic disturbance and massive rebellion which would have tested the competence and stability of any traditional empire, probably to breaking point.

In the struggle for solutions, the dynasty had tried various, often contradictory strategies, from foreign war to indigenous rebellion, and from ideal Confucianism to the acceptance of alien ideas. None, however, had worked, and the successful rebellions of 1911 reflected rather the moral and physical bankruptcy of the Manchu regime than the dynamism and coherence of their opponents.

For the future of China, indeed, there was a threat still greater than the decline and fall of a single alien dynasty, and that was the loss of faith in the indigenous philosophical tradition. Under the pressure of foreign example, intellectuals such as Kang Youwei and Liang Qichao turned from established concepts of political thought and sought to rebuild a new, modern justification and rationale for state and society. In doing so, their agenda was dominated by Western ideas: so that Kang Youwei presented Confucius as a reformer who anticipated concepts in European philosophy, and Liang Qichao, while praising the Chinese for 'special characteristics which are grand, noble and perfect, distinctly different from those of other races,' yet argued that their future lay in change and development on quite different lines: 'we must shatter at a blow the despotic and confused government system of some thousands of years; we must sweep away the corrupt and sycophantic learning of these thousands of years.'[6] No matter how cautiously or how strongly they presented their cases, such thinkers were rejecting the classical tradition, now identified too closely with the weakness of the old regime and the humiliations which had been suffered at the hands of the powerful foreigners.

In this, they were devastatingly effective: though the cautious conservatism of the individual Chinese remained largely untouched, and traditional concepts continued in general acceptance, the collapse of the empire and of the recognized Confucianism which had so long justified its rule left a vacuum in the heart of Chinese political philosophy.

In some respects, the discredit of a backward and complacent system of thought, as much official Confucianism had become, with an insensitive hierarchy based upon landed wealth, arbitrary scholarship, family selfishness, and the oppression of women, leaves little to regret. On the other hand, it can also be argued that even the reluctant reforms of the Manchus, with the eventual hope of a constitutional monarchy, had offered the possibility of progress on realistic lines under competent administration, while the success of the revolutionaries left open the question of what moral, legal, and political structures they would set in its place. Unlike the West, for example, where concepts of the rule of law and the rights of man have developed, however tenuously, from a liberal, Judeo-Christian tradition, the failure of Confucianism left the thinkers of China, and the political leaders, with no clear principles on which to build a future state, no reason to adopt one set or attitudes or another, nor any indigenous base upon which to make a judgement.

There was a time when Sun Yatsen, despairing of his cause among the people of China, likened them to a blank sheet. Mao Zedong answered him later: one can draw splendid characters on a blank surface. Who shall say, however, what is splendid and what is meaningless scribble?

Notes

1. The treasury tael was one Chinese ounce of silver. Its value about this time was some two-thirds of an American dollar.

2. The formal command of the allied contingent was granted to the German Count von Waldersee, though he did not arrive in China until after the fighting was over. There was also a naval contingent from the Australian colonies, which likewise saw no action; but in the national War Memorial at Canberra there are banners, arms, and personal relics such as ladies fans, brought home from Beijing.

3. The intellectual nationalist Zhang Zhidong had been a leader of the Confucianist 'Pure Stream' party in the 1880s, which embarked on the mistaken war against the French. Later, however, during a distinguished career in provincial government on the middle Yangzi, he became a notable protagonist of Westernization.

4. There are three Chinese cities at the junction of the Han River with the Yangzi. Wuchang, the provincial capital of Hubei, is on the south of the Yangzi; Hanyang is on the strip of land between the Han and the Yangzi; Hankou is north of the two rivers. In recent times, it has become customary to describe the three cities by the common name Wuhan. This practice will be followed hereafter.

5. Zhili was the name of the capital province surrounding Beijing under the empire, and the term was maintained until 1928 when it was changed to Hebei.

6. See, for example, his articles in 'A People Made New', published in Yokohama between 1902 and 1905, translated by de Bary, Chan, and Watson, *Sources of Chinese Tradition*, pp. 755–9, and discussed in *Cambridge China* 11, p. 476 [Gasster, 'Republican revolutionary movement].

Further Reading

On foreign relations:

The Cambridge History of Japan, Volume 5, *The Nineteenth Century*, edited by Marius B. Jansen, and Volume 6, *The Twentieth Century*, edited by Peter Duus, both published by Cambridge University Press, 1990 [present the perspective of China's rival].

Langer, William L., *The Diplomacy of Imperialism 1890–1902* [second edition], New York, Alfred A. Knopf, 1951.

Mason, R.H.P., and Caiger, J.G., *A History of Japan*, Melbourne, Cassell, 1972.

On the Western powers in China:

Elvin, Mark, and Skinner, G. William (eds.), *The Chinese City Between Two Worlds*, Stanford University Press, 1974.
Pearl, Cyril, *Morrison of Peking*, Sydney, Angus and Robertson, 1967.
Trevor-Roper, Hugh, *Hermit of Peking: the hidden life of Sir Edmund Backhouse*, Penguin, 1976.

On the intellectual debate:

Chang, Hao, 'Intellectual change and the reform movement 1890–1898', in *Cambridge China* 11.
Cohen, Paul A., and Schrecker, John E. (eds.), *Reform in Nineteenth-Century China*, Harvard University Press, 1976.
Gray, Jack (ed.), *Modern China's Search for a Political Form*, Oxford University Press, 1969.
Hsiao Kung-chuan, *A Modern China and a New World: K'ang Yu-wei, reformer and utopian*, 1858-1927, Seattle, University of Washington Press, 1975.
Ichiko, Chuzo, 'Political and institutional reform 1900–11', in *Cambridge China* 11.
Levenson, Joseph R., *Liang Ch'i-ch'ao and the Mind of Modern China*, Harvard University Press, 1953.
Levenson, Joseph R., *Confucian China and its Modern Fate*, 2 vols, Berkeley California University Press, 1968.
Schiffrin, Harold Z., *Sun Yat-sen and the Origins of the Chinese Revolution*, Berkeley, California University Press, 1968.
Schwartz, Benjamin, *In Search of Wealth and Power: Yen Fu and the West*, Harvard University Press, 1964.
Spence, Jonathan, *The Search For Modern China*, London, Hutchinson, 1990.

On the Boxers:

Fleming, Peter, *The Siege at Peking*, London, Hart-Davis, 1959.
Tan, Chester. C., *The Boxer Catastrophe*, New York, Columbia University Press, 1955.

On the uprising of 1911:

Esherick, Joseph W., *Reform and Revolution in China: the 1911 revolution in Hunan and Hubei*, Berkeley, California University Press, 1976.

3

Revolution Betrayed, 1911–1925

T H E destruction of Manchu power in 1911 opened new possibilities for the people of China, but it also removed the formal structure of government and opened the troubling question of philosophical and political legitimation: what was the nature and the justification for a Chinese state? Confucianism and the imperial system were both discredited, and leaders of thought and politics sought to fill the gap with ideas from the West: the nationalism, democracy, and socialism of Sun Yatsen's Three Principles for the People, based largely upon the liberal West; Communist ideals from the successful Soviet Union; and later, in the 1930s, the apparent energy of German and Italian Fascism combined with an attempt to re-establish the pride of Confucianism.

In practical terms, the government of this vast and populous country was bedevilled by the absence of any firm political infrastructure. Traditionally, authority depended upon one person, whether the emperor or his regent, and although there were conventions on how political decisions should be made, these could be overthrown at need or will. The critical question had always been that of who controlled the chief military force. In the same tradition, throughout the entire history of modern China, no significant political group has demonstrated any real understanding of the nature of a constitution or of the concept that laws might be greater than the wishes of an individual ruler.

As a result, a brief attempt at parliamentary democracy was crushed by the military authority of Yuan Shikai, and when he in turn was brought down by Japanese pressure China was divided between a factious military regime in Beijing, which maintained the formalities and profits of central government, and a multitude of warlords, who ruled the provinces in their own short-term interest, and struggled with one another to exploit the people they controlled.

Below this collection of superficial military governments, however, the people of China, and particularly the intellectuals and students influenced by concepts from abroad, sought to establish a new Chinese culture and a Chinese nationalism which might withstand the economic dominance and the military imperialism of Japan and the West. And in the far south, through the early 1920s, Sun Yatsen gradually developed a Nationalist government to replace the turmoil elsewhere. To deal with its rivals, the new regime trained an army of its own, but military power was here inspired by a political programme for revolution and reform.

Sun Yatsen: Theorist in Exile

The revolution of 1911, like the republic which followed it, was accidental, piecemeal, and erratic. Beginning with a mistaken bomb explosion, supported by a mutiny, maintained by reluctant military governors, and confirmed by alliance with the property-owning provincial assemblies, its recognized leader was a man who had not set foot on Chinese soil for the previous fifteen years. Sun Yatsen, an exile since 1896, had been the prophet and organizer of revolution among Chinese settlers and students overseas, and it was he and his fellow-conspirators who had maintained pressure against the Manchu government and held pride of place among the leaders of rebellion. Most important of all, it was Sun Yatsen's ideal of reform, republic, and democracy which gave a meaning to revolutionary movements and a rallying cause against the Qing dynasty.

Sun Yatsen was born at Xiangshan in Guangdong on 12 November 1866,[1] to a peasant family which had sufficient money and interest to send him to elementary school from the age of six until he was twelve. When Sun Yatsen was small his elder brother left for Honolulu, and in 1879 he followed there with his mother and completed his education in an Anglican missionary school and a local college. In 1883 he returned to Hong Kong, became a convert to Christianity, and married his first wife at the age of nineteen. In 1885 he enrolled in a medical school at Guangzhou, transferred to the College of Medicine for Chinese at Hong Kong in 1887 and graduated in 1892.

Up to this time, there was nothing very unusual either about Sun Yatsen's travels or his choice of a profession. Many who sailed in the 'coolie trade' ended their lives in poverty and misery under foreign rule, but others established themselves in new communities and prospered under European or American government. Sun Yatsen's brother was a successful cattleman in Hawaii, and though the islands were still officially independent their business world and society were dominated by American interests. The experience of Honolulu showed Sun Yatsen what was possible under energetic administration, and the contrast between Hong Kong and Guangzhou increased his sense of frustration with conditions in his homeland. Like other young men of his generation and later, he saw Western education and scientific knowledge as the means to achieve immediate improvement and also as a first step to understanding new ideas and realizing proposals for reform. In Sun's words, 'I set before myself the object of the over- throw of the Qing Dynasty and the establishment of a Chinese Republic in its ruins. At the very beginning I selected for my propaganda the college [in Guangzhou] at which I was studying, regarding medical science as the kindly aunt who would bring me out onto the highroad of politics.' [2]

Sun Yatsen's transfer to Hong Kong was encouraged both by the wider curriculum offered in the British school and also by the free- dom of thought which the colonial government tolerated among their Chinese subjects. He was obviously a man of great intellectual abil- ity, for besides his success in the regular examinations he read through the classic histories of China, he discussed politics constantly with friends, and he travelled widely, to Guangzhou, Macao, Xiamen (Amoy), and Shanghai, seeking support and plotting rebellion. In 1895 his group, the Revive-China Society (*Xingzhong Hui*), planned its first major coup, an uprising in Guangzhou which should be aided both by allies among the old secret societies and also by discontent- ed demobilized soldiers after the war against Japan. But their plans were betrayed, one of Sun's close friends was killed, many more were arrested, and Sun Yatsen was forced to flee not only from China but also from Hong Kong.

On the advice of Dr James Cantlie, who had been head of the med- ical school in Hong Kong and had since become a friend and advis- er, Sun Yatsen went first to Japan to seek support among the exiles and the students. His reception there confirmed his position as a lead- er against the Qing dynasty, for newspapers magnified his small and unsuccessful skirmish with the authorities in Guangzhou to a full attempt at revolution, and both Chinese and Japanese enthusiasts came to join his cause. For years to come, Japan remained his base of oper- ations, and the plans of the revolutionaries received wide support among the Japanese, particularly among young people, who were inspired very often by a sense of racial community, an alliance of the yellow men against the foreigners and the puppets, like the Manchus, who oppressed them.

With a long exile ahead of him, Sun Yatsen adopted full Western costume and cut off his pigtail, traditional sign of Chinese submission to the Manchus. He travelled from Japan to the Philippines, Hawaii, America, and then to Europe, seeking first to gain recognition among the Chinese communities and then to experience Western conditions. At first there was small response from his compatriots, but in 1896 he gained an unexpected triumph. In London on 11 October he was kidnapped by the imperial Chinese legation, on orders from Beijing, to be returned to China for execution as a rebel and traitor. With intervention from the British Foreign Office and with headlines in the London newspapers, his friend Dr Cantlie forced his release on 23 October, but the publicity had made Sun Yatsen world famous, and the back-handed tribute from his enemies gave him pride of place among the opponents of the dynasty.

The situation for a revolutionary party was by no means straight-forward, however. Among the scattered communities of Chinese exiles, students, and travellers abroad, few were confirmed opponents of the dynasty. Those, like Sun Yatsen's brother, who had migrated to take up permanent residence overseas were no longer subjects of the Chinese empire, and they were not greatly concerned with affairs outside the immediate problem of survival in a hostile, white racist society. The Chinese were never popular in many places, and in countries like Australia, America, and Canada, no matter how carefully they trod they were in constant danger from murder and riot by lynch mobs and vigilantes. Even the secret societies, whose first reason for existence was their fight against the Manchus, were now concerned primarily with the welfare of their own members in a new environment and had little thought for the cause they left behind. Sun Yatsen encouraged them to support the revolution, but his success was not great and he often complained of the apathy and 'sleepy atmosphere' among those with whom he talked.

When Sun Yatsen returned from Europe to Japan, he had the quiet support of many government ministers and officers in the army, but he also faced rivalry from alternative parties of rebellion. Most important, and potentially most dangerous, was the Emperor Protection Society (*Baohuang Hui*) established by Kang Youwei after the destruction of the Hundred Days Reform and his flight into exile in Japan. Though opposed to the rule of the empress-dowager, Kang Youwei hoped for a restoration of the Guangxu Emperor similar to that of Meiji in Japan, and this moderate policy attracted considerable support among the overseas Chinese whom Sun Yatsen was also attempting to canvass and recruit. For some years, the revolutionaries were pressed on to the defensive, but as the conservative faction within the Manchu government gathered strength, and hopes for liberal reform declined, Kang Youwei's programme became increasingly irrelevant. If change was to come in China the Manchus and their dynasty must go, and Sun Yatsen's party was the one group committed to rebellion.

In September 1900, at the time of the Boxer uprising, the Emperor Protection Society attempted a coup at Wuhan, but they failed ignominiously and Kang Youwei received much of the blame for the weakness of his administration and for his failure to provide adequate support. Sun Yatsen had planned a similar rising in Guangzhou, but the Japanese government intervened against his party and the attempt was no more successful. The failure of Kang Youwei, however, and the humiliation of the imperial court in the face of allied intervention, strengthened the position of the revolutionaries, and in the years that followed they gained increasing support both inside China and abroad. From 1902 onwards, there was a growing number of anti-Manchu groups in the cities of China, and a wide circulation of journals preaching rebellion against the dynasty. In 1904 Sun Yatsen made another tour, enrolling students and other supporters, and exhorting the secret Hong Society in America, which had lost much of its anti-Manchu fervour, to change policy again and contribute to the struggle against the empire. In 1905, after his return from Europe to Japan, Sun Yatsen became leader of a new revolutionary party, the Chinese Revolutionary United League (*Zhongguo Geming Tongmeng Hui*, known more simply as the United League, *Tongmeng Hui*), combining the great majority of petty revolutionary groups which had sprung up in recent years. Aided by conspirators and secret societies inside China, the United League sought to harass the imperial government with local uprisings, attempts at assassination, some of them successful, and a stream of propaganda in China and overseas.

Despite the prestige he had gained in the early years of exile, Sun Yatsen could well have been brushed aside by more energetic and competent rivals among the revolutionaries. His leadership of the movement, which culminated in his chairmanship of the United League, owed less to his skill as a trouble-maker than to the arguments with which he justified rebellion against the dynasty and to the programme he prepared for his cause. For those who resented the government of the Qing, Sun Yatsen not only confirmed their discontent, but also set before them a vision of the gains which revolution should bring to China and the ideal society which should be built from the ruins of empire. As early as 1897, during his first visit to Europe, he had formulated the concept of 'Three Principles for the People' (*Sanmin zhuyi*), and by 1905, in newspapers, journals, and other writings, he preached a doctrine of social and economic reform which should accompany political change in China.[3]

The three principles which Sun established to sum up his philosophy of revolution were Nationalism (*Minzu*), Authority for the People (*Minquan*) and Livelihood for the People (*Minsheng*). The concept of nationalism, naturally enough, was based on consciousness of national identity among the Chinese people, with freedom from oppression by the alien Manchus and from economic and political domination by foreign powers. In discussions, Sun Yatsen remarked on the ancient history and culture of his people, but complained of their lack of identity and purpose. To the extent that it interfered with Chinese self-confidence and independence, he opposed the fashionable acceptance of foreign ways and manners, for 'cosmopolitanism' was a luxury which an oppressed people could ill afford. In similar fashion, avoiding the faults of the imperialists, a revived Chinese nation would help the weak, oppose the aggression of great powers, and unify the world with traditional morality and love of peace.

For Sun Yatsen, people's authority meant republican democracy, with a constitution based on American models but including the old Chinese system of censorship and the right for the people not only to elect officials but to 'recall', that is, dismiss them, by a negative vote at any time. In the first years after successful revolution authority would naturally remain in the hands of military leaders and the transition to civilian rule would be gradual, but Sun did not go into details about the real difficulty of persuading soldiers to hand over government. As time would show, regardless of the excellence of his constitutional ideals, they had small chance of support in China or of acceptance among governments overseas. Though Britain, the United States, and France had democratic constitutions, they did not necessarily regard democracy as an item for export, and a country such as Britain looked askance on these experiments so close to her empire in India. Inside China it was clear to any realist that Sun Yatsen's plans must cope with the selfishness of men of power, whether imperialist or revolutionary, and also with apathy among the people, too little aware and too slightly educated to be either concerned with politics or take an effective part.

It was the principle of People's Livelihood that established the programme for economic reform, and Sun Yatsen's proposals called for a partial nationalization of land, with compensation, and a levy on all gains in the capital value of private land. His ideas are clearly based on the theories of Henry George, a reformist philosopher who had a considerable following in America, Europe, and Japan at the turn of the century, and who preached that increase in the basic value of land and rent was a function of community activity, so that profits from land should go to society as a whole rather than to any individual. On the other hand, Sun Yatsen specifically excluded Marx and his theories as irrelevant to the problems of China: the country was not suffering from inequality of wealth but from poverty and underdevelopment, and Marx's theories applied only to advanced industrialized nations. To develop China's mineral and industrial resources, he called for the build-up of state capital and industries, and for government control of private capital to prevent the formation of a great wealthy class. Most important of all, since China's economy is based upon agriculture, he called for agrarian reform on the principle of land to the tiller, and insisted that the object of food production was not to seek profit but to provide sustenance for the people:

If we can apply the *Min Sheng* Principle in this way and make the support of the people rather than profit the aim of production, then there will be hope for an abundant food supply in China. The fundamental difference, then, between the Principle of Livelihood and capitalism is this: capitalism makes profit its sole aim, while the Principle of Livelihood makes the nurture of the people its aim. With such a noble principle we can destroy the old, evil capitalistic system.[4]

In present-day terms, these three principles — national independence, popular democracy, and economic reform — present a moderate liberal programme. Sun Yatsen also preached social change: that there should be free and modern education, and that women should be emancipated from their political inequality and from the servitude of foot-binding. For a country such as China, accustomed to autocratic rule and the predominance of tradition, his proposals appeared radical and idealistic, but they presented the revolutionaries with a picture of society well worth fighting for, and conservative reformers such as Kang Youwei were faced with principles of democracy and freedom which they could hardly hope to receive even from the most enlightened dynastic government. In literary debate with Liang Qichao, Sun established a rhetorical supremacy, as his vision of the future presented revolution not just as a political coup, but as the first step towards social and economic revival. As the Qing administration lurched to its point of collapse, Sun Yatsen proclaimed the ideal state which should succeed it and, though the period after rebellion in 1911 proved a bitter disappointment, it was Sun Yatsen's writings and his dream that gave meaning to the republic and presented a challenge to every politician and every general who held power in the years which followed.

Yuan Shi kai

In October 1911, as three centuries of empire began to crumble, the regent council of the dynasty was faced with open civil war. One after another, provincial assemblies declared independence from Beijing, local military authorities abandoned their allegiance and joined the rebellion and, though in some places remnants of the old Manchu garrison attempted resistance, they were swiftly defeated and destroyed. In provincial towns and garrison cities many Manchus were killed or driven away to become refugees in the north, although some were sufficiently integrated into the local Chinese community to escape persecution. Within days of the outbreak of rebellion, the power of the Manchus was gone from the provinces of China as if they had never been.

In Beijing the imperial government still held the Northern Army, which was larger, better-trained, and far more powerful than any combination the rebels could put into the field. But the army was commanded by a Manchu, and the officers and men made it clear to the government that they resented the way their former general Yuan Shikai had been humiliated and sent into retirement two years before, and they would take orders from nobody else. In less than a week it was obvious that the Northern Army was useless to the government and was dangerously close to mutiny, and the regency implored Yuan Shikai to return and take command. With a nice touch of irony, Yuan explained that his lameness still prevented him from attending to the affairs of court. As the situation became more critical, the price offered for his support rose higher. On 14 October he became Governor-General of Hubei and Hunan, and by 1 November he was commander-in-chief of all imperial armies in North China, and also Premier in the civil administration, with the right to appoint his own cabinet.

The Manchus, in their desperation, had given full command of their government to Yuan Shikai, but it is doubtful if Yuan ever held any intention of honouring the trust. He ordered his army near Wuhan to fight one battle and defeat the rebels, but after this demonstration of military authority he pressed no further with the campaign. Inside Beijing, he gave appointments to his own supporters, replaced the imperial bodyguard with troops of his following, and took control of the treasury. On 6 December the regent Zaifeng retired from office, leaving the empress-dowager and the child emperor entirely dependent on Yuan Shikai, and from this position of power Yuan began negotiations with the enemies of the dynasty.

On 2 December, Nanjing, southern capital of the empire, had fallen to the rebels, and the exiled Sun Yatsen returned to China. On 29 December, at a conference in Nanjing among all the governments of the rebel provinces Sun Yatsen was elected the first, temporary president of the new republic, and he took office on 1 January 1912. Negotiations, however, continued with Yuan Shikai, and on 14 February Sun stepped aside so that the national assembly could elect Yuan president of the provisional government of the republic. Sun Yatsen attempted to ensure that Yuan should come to Nanjing, but Yuan evaded the requirement, and on 10 March 1912, at the age of fifty-two, he took the oath of office in Beijing. By the end of April the national assembly had also moved to the north, and Yuan Shikai's cabinet was established with all key positions held by friends and allies.

Yuan Shikai, however, had achieved his power only by a second betrayal of the dynasty he claimed to serve. The first in 1898 had condemned the Guangxu Emperor to a life of imprisonment under Empress-dowager Cixi. Now Yuan agreed with the rebels to remove the dynasty from all share in the government, compelled the Manchus to abandon their power and persuaded the empress-dowager to publish the edict of abdication. Since Yuan refused to help her or to oppose her enemies, she had no alternative. On 12 February the last edict of the Qing dynasty yielded authority to the country as a whole, and a special clause inserted by Yuan gave him full powers to organize a republican government.

Despite this record of trickery and treason, Sun Yatsen and his colleagues had likewise no choice but to accept the terms they were offered. On the one hand, the rebels at Wuhan and in other provincial centres had shown limited enthusiasm for the leadership of revolutionaries returning from exile, while Yuan Shikai clearly controlled the strongest army in the empire. Without his alliance, China would certainly be divided, with small hope of unity for years to come. By accepting presidency of the republic, Yuan also accepted, at least in theory, the provisional government of which he was nominal head and the constitution which was in preparation. In accordance with that constitution, in August 1912 the Republican Council at Nanjing made arrangements for a two-chamber parliament, with the first elections in February 1913. From his position in Beijing, Yuan Shikai saw no reason to change the current arrangements, but the threat of parliamentary control made it essential for him to form a party on which he could rely and, with wide support from many men who preferred strong and stable government to the uncertainties of democratic enthusiasms, his followers campaigned as the Republican Party.

Despite their ideals of rights for the people, the rebel leaders of the south had no experience of the techniques required for fighting an election, and their plans for social and economic reform had not yet gained wide support. As the campaign opened, a Nationalist Party, the *Guomindang,* was formed by Song Jiaoren, a young rebel leader then aged thirty. Sun Yatsen and other revolutionaries, each with his small group of conspirators and each with his own dream of the future, were proving too inflexible and too uncooperative to form a party with widespread appeal, and it was Song Jiaoren, the first man in China to show true parliamentary skill and imagination, who established an organization that reconciled the quarrelling factions and presented a coherent programme. At the elections the following February, the Nationalists defeated the Republicans in both houses of the parliament and Song Jiaoren became premier-elect.

There seemed little to choose, however, between the policies of Song Jiaoren's Nationalist Party and Yuan Shikai's Republicans. Both were conservative and ostensibly responsible, and neither made any mention of social or economic reform. The plans of the revolutionaries, for land reform, for the emancipation of women and for widespread nationalization, were passed over or disregarded in the Nationalists' platform. Their one real ambition was to control Yuan Shikai's power by establishing a counterbalance in parliament, and ideals of reform could not compete with the need to win the election in a conservative constituency dominated by property-owners.

But Song Jiaoren and his political followers were mistaken. In the wake of revolution, power in China was held by the man with the gun, and Yuan Shikai had the best and most effective weapons. On 20 March 1913 Song Jiaoren was shot at the railway station in Shanghai, and there was small doubt that Yuan Shikai had arranged the assassination. Though parliament assembled in Beijing, Yuan continued his government as before, and a loan from the five powers Britain, France, Germany, Russia, and Japan, negotiated and agreed without reference to the new assembly, ensured that he would have money for his needs. In June 1913 he forced a conflict with Sun Yatsen and the revolutionaries by transferring and replacing provincial commanders and other army officers in the south who were known to be sympathetic to their cause, and by moving units of the Northern Army to enforce his orders. In July there was a wave of rebellion and resentment and, as in 1911, some provinces declared their independence. This time, however, Yuan was fighting for himself, not for an alien dynasty, and the 'Second Revolution' sputtered and died under military threat and bribery from Beijing. Sun Yatsen fled once more from China to Japan, while the city of Nanjing, first capital of the Republic, was made an example and was looted, raped, and plundered by the reactionary General Zhang Xun.

In a sense, the failure of the 'Second Revolution' all but cancelled the effect of the first. Yuan Shikai's armies had won control of the provinces from the revolutionaries, his administration in Beijing was strong and reasonably competent, and his success in winning a loan from foreign powers showed their support for his regime. The Nationalist politicians and other supporters of parliamentary democracy protested bitterly over his rejection of their authority, but the country at large saw no reason for concern. If the dispute lay only in whether a politician such as Song Jiaoren or an experienced minister like Yuan Shikai should control the government, the weight of opinion supported Yuan. Leaving aside all questions of reform and all ideals of revolution, Song Jiaoren and his Nationalists had fought a campaign for simple political power. On those terms they had lost, and few people mourned their passing.

With military success and public acquiescence, Yuan Shikai moved towards his goal. On 6 October 1913, surrounding the parliament with men-at-arms and a mass of demonstrators and placards, he compelled the assembly to formalize his appointment as president of the republic, and arranged that the constitution which would define and limit his powers should never be approved. In November the Nationalist Party was banned and its representatives expelled from parliament, leaving it without a quorum. Early in 1914 parliament was disbanded and the local provincial assemblies were abolished by decree, and on 1 May 1914 a new constitution was proclaimed, granting the president all authority of peace and war, taxation, and official appointment which had been claimed by the last rulers of the Qing. In December a further amendment extended the term of the president from five years to ten, and gave him the right either to nominate his successor or to continue in office. In the same month, at the winter solstice, Yuan Shikai performed the traditional imperial ceremony of sacrifice at the Temple of Heaven.

Few could now have doubts of Yuan Shikai's final ambition. Since the first revolution of 1911 and his return to power, the one purpose which underlay his actions had been to destroy the Manchu dynasty and replace it with his own. For such a policy the presidency was just one step on the way, and he had no more loyalty to the republic than he had shown to the emperor and the dynasty he served before. In the first months of 1915 he moved to gain popular acceptance for a restoration of the monarchy, and an organization was established in Beijing to arrange that provincial governments and other citizen groups should submit petitions inviting him to take the throne. By the end of the year citizens' conventions in every province had voted unanimously for a return to the empire, and Yuan Shikai announced 1 January 1916 as the first day of a new reign period in a new dynasty.

There remained, however, a larger question: why did Yuan Shikai want power, and what would he do with it when he got it? Three times a turncoat, did he have any programme to strengthen China against her outside enemies and restore prosperity within? Had he any plans for reform? Or was all his strength and undoubted administrative ability devoted to nothing more than an empire for himself and a dynasty for his family? There could be some excuse for the violence with which he fought for supremacy if only people could be assured that the man would govern strongly and well once the struggle was ended. But there were few clear signs of his plans for the future: the proclamations he issued as he moved towards the throne gave no real promise, and his reliance on outside loans and foreign support left suspicion that at heart Yuan Shikai was a simple traditionalist, with no ambition beside power and glory. As if to confirm this view on a purely personal basis, Yuan's eldest son and his brothers and sisters behaved with such extravagance and arrogance that they created a constant scandal, enough to make anyone doubt the advantages of hereditary government.

Overseas, however, few of these questions were considered. From his first appointment in 1911, through all his political manoeuvres, Yuan Shikai had received consistent support from the governments of Europe and America, and particularly from Britain. If the experiment in democracy had blundered into failure, this was just what foreigners had expected, and it was generally believed that China could only be governed by a strong dictator, whatever title he might choose. In this they were surely correct, but the error the foreigners made was to assume that Yuan Shikai, strong man of the moment, could hold power merely for its own sake, without any programme to offer his people. Just as the Empress-dowager Cixi had been admired simply because she held the reins of a ramshackle government, so now there were advisers and observers quite prepared to offer Yuan their support and approval. George Ernest Morrison, the Australian who had acquired fame and prestige as correspondent in China to *The Times* of London and who was widely respected for his knowledge of the people and the country, took service with Yuan as political adviser. Other foreigners, scholars, and men of affairs, worked for him willingly and published books which described him as the man to solve the problems of the nation. In the history of the twentieth century there have been many 'strong men' in Asia whom well-informed Westerners have acclaimed as the saviours of their country; the results of such fortune-telling have often been disastrous, and Yuan Shikai is the first of an unhappy tradition.

In January 1915 the Japanese government presented Yuan Shikai with their 'Twenty-one Demands', and the bubble of his authority was broken. Japan had entered the war against Germany and Austria in August 1914 and found small difficulty seizing German territory and concessions in the Far East, including German interests in Shandong. With the European powers, including Russia, fully involved in the war, the Japanese had no effective rival in the East, and they took the opportunity to present China with a drastic set of proposals for interference in the country's administration. Besides the requirement that all Germany's treaty rights and privileges should be transferred to Japan, they also expected concessions in southern Manchuria, Fujian, and the Yangzi valley. They demanded that China should buy the bulk of her military supplies from Japan, that Japan should share in all China's iron and steel industries, that Japanese police should take part in the administration of major Chinese cities, and that the Chinese government should accept Japanese political, financial, and military advisers.

In the long run, the Twenty-one Demands remain a monument of diplomatic error. At first, they seemed to achieve some success. The United States, the only Western power not yet involved in the European war, made firm protests, but these proved of no avail, and when America did enter the war in 1917 the Lansing–Ishii pact of November that year recognized Japan's 'special relations' with China. Similar agreements in February 1917, first with Russia and with Britain, followed later by France and Italy, confirmed Japan's *démarche*. On the other hand, the Demands displayed to all the world the extent of Japanese ambition. The Western powers were compelled by circumstance to acquiesce in their ally's aggression, but they never trusted the Japanese again. By this one show of brutality, the Japanese threw away any real basis for sympathy in the West for their policy towards

China, and for twenty years to come they headed into quiet but increasing enmity.

To Yuan Shikai's government the effect of the Twenty-one Demands was devastating, and it is possible that this was the prime intention of the Japanese. Like other observers, they probably believed that Yuan Shikai was on the point of success, and they feared they could be faced with a new rival for supremacy in the Far East. Indeed, despite China's weakness and her lack of allies, the government could have faced the Japanese and refused to enter negotiations. Yuan Shikai, however, was not the man for a national war: still the politician, he hoped to keep the whole affair secret; he temporized, and on many points he gave in. Inevitably, the news spread abroad, and Yuan's failure to make a stand rendered his government pointless and empty. The revolution had not been won for this, and no one believed that a man like Sun Yatsen could have been threatened so effectively or would have responded with such weakness.[5]

In the months that followed, as Yuan was arranging the meaningless demonstrations and popular petitions which were designed to raise him to the imperial throne, he might have hoped that the Japanese would be satisfied with their first success and would allow him time to save face and gain real support among the people. But as resistance grew against him, Japanese agents, money, and arms were channelled to his opponents. On 25 December 1915 the local authorities in the far southwestern province of Yunnan demanded that his plans for the empire should be abandoned, and two days later they formally declared independence. In January 1916 the neighbouring province of Guizhou followed suit, and an Army of National Protection was sent towards the north. In February 1916 it was announced that the empire had been postponed, but the revolt continued to spread, and on 22 March Yuan Shikai abandoned his imperial claim and resumed the title of president.

Even this was not enough. The rebellions continued: the provinces of the Yangzi turned against him and Yuan Shikai's armies would do nothing to fight for him. On 22 May, when the wealthy province of Sichuan declared its independence, under the command of Chen Huan who had formerly been one of Yuan's closest friends, all hope was clearly gone, and his officers and allies pressed Yuan to resign from power. Throughout the country his imperial pretensions had become a joke among the people, and in humiliation and disappointment he became desperately ill. He died on 6 June 1916, at the age of fifty-six. The dream of empire ended with his death, and so, for years to come, did the hopes of a united republic.

Nationalism and the May Fourth Movement

The failure of Yuan Shikai removed the last pretence of unity in China, and also marked the end of the hegemony of Beijing. Yuan's authority had been based on the Northern Army, but after his death none of his generals was strong enough to succeed him, and their subordinate officers competed for profit and power. And as factional infighting in the capital made the Northern Army ineffective against the rest of the empire, other generals seized control in the provinces and fought one another, as much with bribery and treachery as with arms, for short-term gain in petty empires to oppress the people and maintain their troops.

On 7 June 1916, the day after Yuan Shikai's death, Li Yuanhong took the title of president. Li Yuanhong had been forced to act as figurehead for his mutinous troops when rebellion first broke out in Wuhan in 1911. Never a convinced revolutionary, he had acquired no power-base in Beijing from which he might deal with other factions, and within a few days, under pressure from revolutionaries in the south and the generals in Beijing, he was compelled to acknowledge the parliamentary constitution of 1912. Duan Qirui, who had served as Yuan Shikai's officer for more than fifteen years, was appointed premier.

Despite the evident weakness of the Beijing government, foreign countries still found it convenient to recognize its authority as spokesman for the nation of China and, unable to find any better alternative, they continued to do so for years to come. In 1917, under steady pressure from America, which was now prepared to join the war raging in Europe, China was forced from her position of neutrality, and on 14 May 1917 Duan Qirui declared war against Germany. He did so without parliamentary authority and without the approval of the president, and the immediate result of the announcement was that Li Yuanhong dismissed him and called in the 'Pigtail General' Zhang Xun to defend Beijing. But Zhang Xun was a complete conservative, and for two weeks in early July he restored the child emperor Puyi to the dragon throne of the Manchu Qing.[6] Zhang Xun received his nickname because he wore the pigtail himself and insisted that his soldiers wear it too. The citizens of Beijing hastened to barbers' shops to find artificial locks which would proclaim their new loyalty and perhaps save them from the brutality of a particularly ill-disciplined and violent gang of uniformed thugs. When Duan Qirui and his fellows fought their way back into Beijing on 12 July, there was a harvest of hair in the streets, thrown away by civilians or cut away by those of Zhang Xun's soldiers who changed sides. Whether worn by conservatives or discarded by revolutionaries, the pigtail must surely be one of the most inconvenient and ultimately farcical insignia ever devised.

Zhang Xun retired into private life, and Duan Qirui called a new parliament, dominated by his supporters and allies, which again declared war, this time with a semblance of legality, on 14 August. He held power with funds from a Japanese loan, and military alliance with the warlord Zhang Zuolin of Manchuria, a former bandit who ruled with Japanese support but who now extended his influence into Inner Mongolia and the area north of Beijing.[7]

In 1918 Chinese envoys attended the peace conference at Paris. Though nominally one of the victorious allies, China's contribution had been small,[8] but the Four Points of the American President Woodrow Wilson, which included the principle of national self-determination, brought many to believe that China would be treated generously by the West and that the settlement of peace would also bring an end to the humiliation of unequal treaties and foreign concessions.

From this point of view, however, the conference was a failure. Whatever theories of nationalism were used to justify the dismemberment of the Austro-Hungarian empire in eastern Europe, none of the powers had any intention of changing the situation in east Asia, and President Wilson, who might have given some help to Chinese aspirations, was outmanoeuvred by the European leaders and faced with serious opposition at home. Still more serious, it was announced by the Japanese that secret agreements with the Beijing regime had confirmed their claims in Manchuria and Shandong and some earlier concessions from the Twenty-one Demands of 1915. In April 1919, as the conference decided against China in favour of Japanese claims in Shandong, one of the delegates at Versailles sent a bitter message back to the newspapers in Beijing, criticising the selfish treachery of the government and calling for public demonstrations which might show the men of Europe the true extent of resentment against Japan.

Feeling in Beijing, already excited by false hopes of gains from the conference in Europe, now turned to fury and indignation, both among ordinary people and labour groups and, most notably, among the students of the universities. On 3 May, amidst great excitement and a mass of meetings, a demonstration was planned to protest against the government and against foreign aggression, and to demand that China refuse to ratify the Versailles treaty. On 4 May some three thousand demonstrators marched first against the Legation Quarter, where they were turned away by guards, and then against the house of the Minister of Communications, Cao Rulin, who had received much of the blame for the secret agreements with Japan. As the mob grew in numbers, rioting broke out, the house was sacked and burnt, and one of Cao's guests, the Chinese Minister to Japan, suspected of complicity in the agreements, was badly beaten. By the time police arrived, most of the demonstrators had left and only ten were arrested.

The demonstration itself was not particularly large, but the movement which began on 4 May was maintained in the next weeks with remarkable firmness and fervour. There was a general strike of students, and the Chancellor of Beijing University, Cai Yuanpei, resigned in sympathy. On 7 May the government was pressed into releasing those who had been arrested, but the disorder increased and further arrests took place. In Beijing and other major cities there was a boycott of Japanese goods by shopkeepers and their customers, dock workers refused to unload Japanese ships, and the shops and houses of Japanese companies and private citizens were destroyed or looted by rioters. Telegrams were sent by the thousand to the delegation at Versailles, telling them to reject the agreement, while the government, refusing to make any stand, left the whole affair in the hands of its envoys. Chinese students overseas blockaded the ministers' lodging places in Paris, and when the Versailles treaty with Germany was signed on 28 June 1919 no Chinese representative took part in the ceremony. The Chinese did sign the treaty with Austria, and so gained membership of the League of Nations but, as speakers and demonstrators had urged and as the refusal to sign with Germany had established, it seemed more dignified to face the threat of force than to accept one more unequal treaty.

As it turned out, however, 4 May was more than a passing enthusiasm: the May Fourth Movement, as it came to be known, involved intellectuals, students, and their teachers, in politics, philosophy, and a new style of literature. The changes had been coming since the last years of the Manchus, and they had been speeded by the excitement of the revolution, but 4 May 1919 gave the new developments wide currency, and established the pattern of debate which would dominate Chinese thought and public opinion for the next generation and which sought to provide answers for the distinctive Chinese situation in the twentieth century.

Naturally enough, the stimulus for new ideas in China came from students returned from overseas, and the first leaders among these were Chen Duxiu, Hu Shi, and Cai Yuanpei. Cai Yuanpei, who had resigned the chancellorship of Beijing University in the aftermath of 4 May, had been appointed there on his return from Germany and France in 1916, and it was his deliberate policy to encourage scholarship, argument, and thought among the students and the staff, so that the university became a centre of interest with a wide spectrum of political belief and ideology. Chen Duxiu, who had also studied in France, became Dean of the Faculty of Letters in 1917, and Hu Shi, from Cornell and Columbia in America, was Professor of Literature in the same year. At the time of 4 May a young man named Mao Zedong, who had not been overseas, was an assistant in the library.

It was Cai Yuanpei's greatest achievement that he made scholarship at the university concerned and involved with China's immediate problems. In the days of the empire students had been trained as future officials, but they were not expected to have private opinions on public affairs. Scholars were respected as the élite of society, but their academic work had to follow traditional lines, and even those who travelled abroad for further study had received a firm grounding in the traditional classics. From this time on, however, Beijing University and its imitators provided them with a place to talk and teach and with a ready audience of enthusiasts.

In 1915 Chen Duxiu founded in Shanghai a journal called 'Youth Magazine', later known as 'New Youth' (*Xin Qingnian*), or 'La Jeunesse'. From its first issues, Chen used the magazine to attack conservative attitudes, and he encouraged his audience to choose the aggressive, independent, and scientific aspects of Western culture, and to be practical rather than emptily idealist in their approach to politics. He attacked Confucianism, and blamed the failings of China on a passive acceptance of tradition. To Chen and his group the ancient culture of China, which so many people, both foreigners and Chinese, praised and sought to preserve, was better regarded as a millstone around the nation's neck, holding China back from progress in the modern world.

Quite as important as the propaganda which it published, 'New Youth' printed the majority of its articles in the vernacular *baihua* or common language, and not in classical Chinese. Both forms of the written language make use of characters, but classical Chinese was the recognized vehicle for philosophy, history, and learned respectable writing. The *baihua* style was generally regarded as clumsy and inadequate, but was widely used for popular writing, including the great novels of the Qing dynasty.

Hu Shi, already one of the most distinguished classical scholars of his time, used *baihua* language in his writings and developed a clear and effective style for literary and philosophic debate. His own ideas were liberal and pragmatic, and he sought to bring changes in society by careful study of its problems, by constant experiment and by gradual amendment. A 'Westernizer', he believed that democracy and science (which he described in a slogan as Mr De and Mr Sai) provided the best solutions to the problems of China. Most of all he was opposed to absolute ideology, whether it be conservative Confucianism or the new Marxist Communism which was now gaining popularity in the aftermath of the Russian revolution. In 1919, however, this latter debate had not yet been fully joined. In the intellectual world at least, the spokesmen for conservatism had largely been overthrown, and a wide array of new magazines and new writers were using the *baihua* style to preach modern ideas and liberalism. The demonstrations of 4 May established these scholars and thinkers, and their followers among the students, in a new position in Chinese society: still retaining the prestige of the past, but opposed to the arbitrary power of government based on military force, intensely nationalistic, and claiming to act as the conscience of the people.

One feature of the new movement was its effect upon poetry. Classical Chinese verse is constructed within a pattern of lines of regular length or set to an ancient rhythm of music, and with stress and rhyme controlled by forms of sound and tone as much as a thousand years old. The young writers of the new school, however, abandoned these restrictions, and adopted with delight the freedom of the contemporary West.

Wen Yiduo, for example, later assassinated by the Nationalists as a liberal in 1946, expressed the bitterness of China's patience and humiliation before the foreigners:

> I can wash handkerchiefs wet with salt tears;
> I can wash shirts soiled in sinful crimes.
> The grease of greed, the dirt of desire . . .
> And all the filthy things at your house,
> Give them to me — I'll wash them, given them to me![9]

Meanwhile Guo Moruo, who became the celebrated scholar of Communist China, sang with youthful energy, and in a style imitated from Walt Whitman:

> I sit alone on a rock ledge near the sea,
> I am sending off the early summer sun to the west.

Rosy clouds rise on the horizon where the sea touches the sky,
With a column of black haze in their midst, like a battle scene.
Ah, Sun, you are a burning grenade!
I want to watch you explode into blossoms, red as blood.
Your lustrous eye, unblinking, stares at me,
And I also wish to go with you, Sun.[10]

Their enthusiasm led often to excess, and to a confusion of images and allusion, but they had a freshness and romance which transformed the thoughts and expression of a whole generation.

The Years of the Warlords

Despite the storm of 4 May 1919 and the demonstrations which followed it, the rulers of Beijing faced no real danger from university students or any other civilian group. The size of the boycott and the strikes which accompanied it had threatened the country with economic collapse, and the government was certainly forced to yield on the question of China's ratification of the German peace treaty. This, however, was largely a side-issue, and the only matter of interest to Duan Qirui and his associates was the military support they could gather and maintain near the capital, and the profits they could make from its occupation. Regardless of its illegitimacy under the constitution, and its restricted control over the nation, the government in the north still obtained international recognition, negotiated for loans, and received the customs revenue excise, after the foreign powers had deducted the interest from their debts.

The chief concern of the foreigners was that their traders, missionaries, and other visitors should be permitted to travel and settle in peace, without interference from officials, from warlord armies, or from simple bandits and pirates. With this end in view, they were prepared to recognize any clique in Beijing, and in the provinces they dealt with whatever military regime happened to hold power. Exhausted by the war in Europe, and unconcerned with the chaotic politics and civil conflict of China, the Western powers accepted the situation as it was, they sought only to ensure that their own interests should not suffer, and they gained profit and influence from the trade in arms and other supplies to the rival commanders.

Duan Qirui's immediate threat came from military rivals. In July 1920 the opponents whom he had defeated two years before gained new accessions of strength from the rise of the warlord Wu Peifu, with his base at Wuhan on the middle Yangzi, and from the changing allegiance of Zhang Zuolin in Manchuria. Though Duan Qirui had served the Japanese well, Zhang Zuolin was entirely dependent on their support and they were prepared to aid him in a new invasion of China, while Wu Peifu had support from the British, Americans, and other foreigners with interests in his territory of central China. Attacked on both sides, Duan Qirui was driven from power. In August the new coalition, known as the Zhili Clique, entered Beijing and Li Yuanhong was restored as president.

In the eighteen months that followed, China gained some international success. At the end of 1921 the Washington Conference confirmed the American policy of her territorial integrity and an 'Open Door' for her trade. In February 1922 Japan agreed to return those areas in Shandong which she had seized from Germany, and Britain agreed to release her base at Weihaiwei on the north coast of the peninsula, which she had held since the wave of concessions in 1898: the transfer, however, did not take place until 1930. On the other hand, there was no real change in the status of the treaty ports, nor of Japan's position in Manchuria, and it was clear that Britain and America were prepared to accept Japanese hegemony in the Far East. The major agreement of the Conference settled the ratio of the three navies in proportion of three to Japan and five to each of the other two, but since Britain and America had fleets throughout the world and Japan concentrated her forces in the Pacific, there was little comfort for China in that balance of power.

In April 1922, however, when the Japanese client Zhang Zuolin staged a rebellion against Wu Peifu, he was sharply and surprisingly defeated and his army was bundled back across the border into Manchuria. For the time being, Wu Peifu held unchallenged control in Beijing. In a new organization of the constitution, he arranged that his former patron Cao Kun became president, but the elections were blatantly corrupt, and people were now totally disillusioned. In September 1924, Zhang Zuolin moved south again, and Wu Peifu marched north from Beijing to oppose him. With neat and decisive treachery, Wu's subordinate general Feng Yuxiang staged a mutiny behind the lines in Beijing and asked Duan Qirui and Zhang Zuolin to join him in office. Wu Peifu's position in the north collapsed, but he ferried the majority of his troops successfully south by sea and re-established himself, for the time being undisturbed, in his old base at Wuhan.

The lurches of fortune in the struggle for Beijing, with whole provinces at war, armies changing sides, and victory and defeat decided more by private betrayal than by open combat, is typical of the pattern throughout China, dominated by personal alliances, factions, and betrayal. In his last campaign north of Beijing, Wu Peifu was said to command a hundred and seventy thousand men, but the change of allegiance which brought him down could be duplicated in almost every petty battle throughout the country. Ten years after the revolution, whatever discipline there may once have been in the armies of imperial China had been lost and forgotten by the officers, and a new generation of soldiers, frequently armed with modern weapons but traditionally despised by ordinary people and generally badly trained, could find no real basis for loyalty beyond their immediate commanders. With every ally a potential enemy and every opponent a possible friend, there was no sense in pushing a battle too hard or in pursuing a defeated foe, and there were few better ways of eliminating a rival in one's own party than by forcing him to take the brunt of the fighting. An officer's only resource was the lives and the weapons of his men, and he must always use them sparingly to ensure he had adequate reserves for any future threat on flank or rear. So soldiers of every rank would engage battle carefully, would readily take time off when darkness fell or if it happened to be raining, and once it seemed clear which army had the upper hand, they were quick to retreat and very ready to change sides. Like the campaigns of the *condottieri* in renaissance Italy, war had become a matter of manoeuvres and a show of strength, and there was no room or purpose for a fight to the death.

The men who commanded these troops came from a wide variety of backgrounds, and they were not necessarily possessed of great military competence, but they did maintain sufficient personal authority to make men obey them. The more senior, such as Duan Qirui and his rival Cao Kun in Beijing, had been officers of the imperial Northern Army under Yuan Shikai. Wu Peifu was a Confucian scholar who had taken the old examinations but later served under Cao Kun's division of Yuan's forces; Yan Xishan of Shanxi province was a member of Sun Yatsen's United League before 1911 and came to power in the first revolution; and Chen Jiongming in Guangdong had edited a newspaper and sought to introduce political and educational reforms. In contrast, Zhang Zuolin had been a petty bandit in Manchuria until he joined the Japanese armies against Russia in 1904, and the 'dog's-meat' General Zhang Zongchang in Shandong, also a former bandit, was notorious for his cruelty and extortion.

And there was also the Christian General Feng Yuxiang, just come to power in the capital. Feng was a man from the north-west, and his father had served in Turkestan under Zuo Zongtang. Despite his erratic politics, and the fact that he had been bribed by the Japanese to turn against Wu Peifu, he was regarded as exceptional among the warlords for his private morality, and he expressed concern for the state of the nation and for the fate of China's lost territories in Korea and Vietnam. With the enthusiasm of a Protestant convert, he baptised whole regiments of his army with fire-hoses, but his National People's Army presented a programme of reform which was widely supported in the north.

Feng Yuxiang, however, was attacked and driven from Beijing by Zhang Zuolin, now allied with Wu Peifu, and the game appeared to continue as before. On the other hand, the years of turmoil had demonstrated that warlord structures of power were ultimately ephemeral and irrelevant. Despite their control of arms and land, none of the military chieftains had been able to establish an effective civil administration which might extend authority beyond the reach of their guns and bayonets. As an ancient sage observed, you can conquer the empire on horseback, but you cannot rule it.

Moreover, despite the weakness of the central government and the regular independence of the provinces from any national authority, and although Chen Jiongming attempted to establish a separatist state in the south, while Yan Xishan kept Shanxi for himself, there was no common acceptance that China could be divided. There are always provincial and local loyalties in China, and there have in the past been viable regional states in territories such as Sichuan, the middle and lower Yangzi, and the far south, but a thousand years of empire have created a sense of community that no provincial loyalty can rival. For better or worse, the future of the Chinese people lies in the unity of the Chinese heartland.

From both these points of view, regardless of the warlords' power, and even the patriotic goodwill of Feng Yuxiang, they now faced a more dangerous rival, Sun Yatsen and his Nationalists at Guangzhou. The dreamer of revolution was still there with his dream, and he had created a party and an authority that could no longer be ignored.

Sun Yatsen and the Republic in the South

The seizure of power by Yuan Shikai, and the struggle for succession among his former officers, had made it clear that the ideals which Sun Yatsen had proclaimed would remain still-born unless Sun and his supporters could establish some real position of strength in China. In

August 1917, Sun Yatsen proclaimed a new regime in Guangzhou, but he received no international support, he had no effective military force under his command, and in May 1918 he was driven away to take refuge in the French Concession of Shanghai.

Sun Yatsen remained in Shanghai for more than two years, publishing his memoirs, writing an *Outline of National Reconstruction*, and reorganizing his political following, now known, after the ill-fated parliamentary party of 1912, as the Chinese Nationalist Party (*Zhongguo Guomindang*).[11] In 1915, putting aside his first wife, he had married Soong Qingling, daughter of a wealthy Shanghai merchant family and a graduate of the University of California and Wellesley College in America.[12] The Soong family had given substantial support to the revolutionary cause, and though the marriage was bitterly resented, for Sun was thirty years older than his bride, and the family objected to the misalliance, the connection with the wealth of Shanghai, both legitimate and criminal, was important to the Nationalist movement. One of Sun Yatsen's chief lieutenants in the new organisation was a young man called Chiang Kaishek, a protégé of Du Yuesheng, chief of the powerful Green Gang.[13]

In April 1921, the warlord Chen Jiongming invited Sun Yatsen to return to Guangzhou as president of a separatist state in opposition to Beijing. The alliance lasted only until the following year, when Sun attempted a campaign against the north: Chen Jiongming changed his mind and seized Guangzhou, while Sun Yatsen had a narrow escape under heavy fire and retired once more to Shanghai.

In early 1923, after talks with Adolf Joffe, diplomatic representative of the Soviet Union, it was agreed that the Nationalist Party would accept alliance with the new but active Chinese Communist Party, and that they would receive Russian advice and aid on the clear understanding, as stated in the joint communiqué, that Chinese conditions did not require a Soviet-style solution. In February Sun Yatsen returned to Guangzhou for the third time to form another republican government, but on this occasion he was determined to establish his own base of loyalist, competent power. On the advice of the Comintern agent Borodin, the Nationalists were reorganized as an authoritarian party, with small cells of conspirators and workers, and with a manifesto proclaiming war against the militarists in China and the imperialists abroad.[14] The Russians also sent military advisers under Vassilii Blyukher, generally known as General Galen, to establish an officer training school, and at the Whampoa Military Academy, downstream from Guangzhou, young men were trained in the techniques of mod-

ern warfare. The cadets of Whampoa became the core of power for
the new revolution: Chiang Kaishek was commandant of the academy, and one of the men responsible for their political education was
a returned student from France, now a member of the Communist
Party, Zhou Enlai.

By 1924 Sun Yatsen's new government had secured its ground in
Guangzhou, at least against anything but a major attack, and it was
from a position of some real anthority that Sun Yatsen made his last
trip to Beijing. Chen Jiongming had been driven from the city, and
other local warlords preferred to leave the place to its own devices.
When a group of businessmen formed a Volunteer Defence Corps of
mercenaries and threatened to compete with the Nationalists for authority, the Whampoa cadets overpowered their rivals and disbanded them.
This small pin-prick had come with known British support from Hong
Kong and, whatever Sun Yatsen's original preferences might have
been, the last bitter years had shown him who his friends were. He
did not trust the Communists, and he hoped to keep them under control in his larger party, but for the time being, to a nationalist reformer,
the Soviets had proven their worth.

On 12 March 1925 Sun Yatsen died in Beijing, where he had been
invited to attend a constitutional conference. He was already crippled
by cancer when he arrived, the conference was predictably pointless,
and his plans for parliamentary reform received no support from the
rulers in the north. Yet Sun Yatsen received a hero's welcome from
the people, and later a hero's funeral. In a final political testament,
he confirmed the Nationalists as the party of revolution, with a programme of military conquest and authoritarian government before
freedom could be reached, and with a policy of friendship with Russia
and hostility to Western imperialism. Even after death he was respected as father of the revolution and founder of the republic, and both
the Nationalists and their Communist rivals claim to maintain his
work and honour him as their former leader.

It is easy to find fault with Sun Yatsen's achievements and with
his personal, erratic character. Forced out of office by Yuan Shikai,
despised by the men of power in Beijing, frequently dependent upon
foreign support and protection, and twice bundled in humiliating fashion even from his base at Guangzhou, his career was often inglorious
and he did not live to guide his party into triumph. On the other hand,
no matter what defeats he suffered, he never abandoned his hopes of
what China should become. When the ideals of the first revolution,
for which he had worked twenty years, turned to ashes at the hands

of Yuan Shikai and the warlords, he was still prepared, with perseverance and courage, to rebuild a new programme from the ruins of the old. Though the immediate policies and structure of the new Nationalist Party were more radical and authoritarian than they had been in the first days of revolution, he had no intention of accepting the Communist line of class warfare and dictatorship. In his belief that change in the future should bring not conflict and oppression but freedom, prosperity, and independence to the people, and in the consistent courage with which he maintained his ideals against odds, Sun Yatsen was a worthy man for any nation to honour.

Cities and Countryside

Despite the unrest of the years since revolution, the secondary industries of China and the balance of her overseas trade showed remarkable progress. This was due chiefly to the weakness of Europe, critically affected by World War I, but the withdrawal of the foreigners allowed new firms to develop in China and permitted great expansion in the modern section of the economy. Between 1913 and 1918 German imports fell to zero, British imports were halved and French imports were reduced to a quarter. The trade deficit fell from Hk$ 166 million in 1913 to Hk$ 16 million in 1919.[15] Freed from the competition they had suffered in the past, and largely unmolested by the government, China's private capitalists were able to establish enterprises of their own: textile companies doubled; the number of cotton spindles in factories went up almost three times; the value of silk exports rose 60 per cent; flour mills, banking companies, and merchant ships all showed increase in numbers and in capital value. Between 1914 and 1919 coal production increased from 13 million to 20 million tons a year, and iron from one to 1.8 million tons.

The success of this new development, uncontrolled by any planning or supervision, brought major social changes and considerable stress on the economy as a whole. Some families became millionaires almost overnight, while the demands of new industries and the hope of work and wages caused massive immigration from the land-hungry countryside to the increasingly overcrowded cities. In 1914 it had been estimated that the total urban population of China was in the vicinity of ten million people, but by 1919 the figure was three times as large. In a total population of some 450–500 million, this was still a small proportion, but in absolute terms it represents a staggering increase, and the governments of the day were unable to control the resultant social problems. More important still, the machinery of administration was not equipped to collect a realistic tax assessment from the new wealth of industry, and bribery or inertia was sufficient to ensure that government kept largely to its traditional sources of revenue: land tax, customs dues, and overseas loans.

The commercial respite, moreover, was short-lived. As the world war ended, Western interests returned to resume their competition, and Japan still aimed at domination of the China market. The balance of trade against China steadily increased again, industrial unemployment and the numbers of jobless in the cities began to grow, while the rest of the country, ravaged by the warlords, was held back from economic development and hard put even to feed itself.

Besides the trade deficit, the massive loans which foreign interests granted the government were a major source of capital outflow. Some loans had never been planned for the country's advantage: to repay the Boxer Indemnity China had to borrow money from the very countries which had imposed it.[16] Others, like the loans for railway development in the 1900s and the grants made to various regimes after the revolution, were formally intended to assist political unity and economic development, but the value was often lost through corruption, inefficiency, and civil war, while the securities and rates of repayment were so great that they soon came to exceed the value of new loans being negotiated. In the years from 1902 to 1936, imperialist loans and foreign debts produced a net outflow of cash from China to the West; and while money coming in was received by warlords and government ministers, the money going out either added to the burden of customs dues or was paid by direct tax on small merchants and peasants.

If the new Chinese industry of the cities now faced stiff competition from overseas, the situation in the provinces was very much more serious. For the warlords who controlled local government, the one vital interest was to maintain their supplies and the numbers of the troops who followed them. Estimates suggest that the number of men under arms in China increased from less than half a million in 1914 to more than one and a half million in 1925. No warlord could afford to lose the loyalty of his soldiers or weaken his position against his neighbours, so taxes on land, property, and trade were doubled, trebled, and multiplied by ten, and were then collected for years in advance; forced loans and government bonds without interest were pressed on the people; and money was printed regardless of backing or value. Based as it was on bribery, trickery, and a show of force, a warlord regime by its very nature was short-lived. The debts of one commander were never acknowledged by the man who drove him out, and the coming of a new regime meant sacking, looting, and occasional massacre for the citizens of the countryside through which they passed.

While the small farmer and the local Chinese businessman could be taxed into poverty or plundered into ruin, the problem was far less for the foreigners who competed against them with cheap goods from overseas. Few warlords were anxious to offend a Western power, and the countries trading or sending missionaries to China had gunboats patrolling the major rivers and the coast-line to ensure their people got protection. There was piracy along the China coast, but pirates seldom boarded a foreign ship; there was banditry among the hills, but few foreigners were attacked and it was seldom that they were killed. The disorders of China were a Chinese problem, to which foreigners were generally immune, and their trade could only benefit from such quarrelling. By the terms of the unequal treaties, once a five per cent customs due was paid, foreign imported goods were exempt from all inter-nal tolls, including the *lijin* duty levied on the transport of domes-tic goods. As a result, even after payment of transport from over-seas, a foreign importer could undercut and outsell a Chinese firm dealing in equivalent locally-manufactured goods.

Within the cities, however, despite the fluctuations of politics and economics, there was still a fair prosperity and some illusion of stability and progress. While the deposed emperor Puyi main-tained a court and palace in Beijing, the Supreme Court of the republic, staffed and attended by well-trained lawyers, tried and decided cases with which they hoped to establish a new tradition of honest modern law. While selfish quarrels rocked the warlord government, administrators at home and diplomats overseas attempted, with some success, to maintain a tradition of compe-tence and fair-dealing, and men such as Wellington Koo (later a judge of the International Court at the Hague) were respected and often listened to in foreign capitals despite the feeble govern-ments which they served. At the same time, more students were travelling to Europe, America, and Japan to gain training and qualification which would equip them for posts of responsibili-ty at home.

The coming of the republic had brought immediate reform to the education system of China. Though some classical studies were still maintained, all emphasis was now on Western languages and science, and the introduction of the *baihua* form of written Chinese in 1920 confirmed the revolution in schools. Money, however, was short, classes were large, and competition was desperately keen. From kindergarten onwards, children were pressed by parents to fight their way through examinations and gain entry to the university. Education was one clear way out of poverty, and one obvious chance to establish a place in the world. Whether driven by personal ambition or by their wish to serve the country, young people struggled for scholarships and for the limited number of grants which would enable them to take further studies overseas.

The prestige of the returned student was excessively high: where the man who stayed at home might work for years in a clerical job or assistant's post at a university, the man who came back, no matter what qualifications he had earned, could expect to receive a managerial post or a professorship. Inevitably, while some universities such as Beijing and Qinghua were respected for the quality of their staff and their students, there were other universities and colleges which produced no worthwhile work. Most serious of all, there was cause to question whether much of the training that students received, at home or overseas, was relevant and useful to the needs of the country. What was the purpose of a degree in advanced economics for a country so oppressed by foreign trade and so hamstrung by government corruption? What was the use of scientific agriculture for a peasant who could barely accumulate sufficient money for next year's seed-grain, and who received oppression, not support, from the government?

In a bitter little essay a few years later,[17] the philosopher and historian Feng Youlan compared the culture of the city with the simplicity and poverty of the countryside: men of the city saw a car as a matter of course, but men in a Chinese village would run in fear or stare in amazement. Perhaps the city showed more progress than the countryside, and the city dweller might look down on the peasant for his credulity and his poverty, just as the men of the industrialized states in the West looked down on the masses of China. But were the cities more than parasites upon the country, and was the West any better than a parasite upon the East? For the time being, none the less, political and economic power lay in the cities, where the intellectuals and students of China, desperately and genuinely nationalistic, were the spokesmen and demonstrators for modernization and reform.

Notes

1. Sun Yatsen's given name was Wen, and the name Yatsen, being the local pronunciation of characters pronounced *Yixian* in standard Mandarin, was his adopted personal name. In both Taiwan and mainland China he is commonly remembered and referred to by the special style *Zhongshan* 'Central Mountain'. The characters pronounced by the Chinese as *Zhongshan* are used to write the Japanese surname Nakayama, and Sun Yatsen used this name as a pseudonym for many of his works during his long exile in Japan.

2. Sun Yatsen, *Memoirs of a Chinese Revolutionary*, quoted in Schurmann and Schell, *Republican China*, p. 10.

3. The history of the 'Three Principles for the People' is not entirely straightforward. The first appearance of the concept in writing was in the introductory essay of *Min-pao*, official magazine of the United League, late in 1905. Thereafter, Sun Yatsen referred frequently to the Three Principles, while '*Sanmin zhuyi*/Our aim shall be' is the first couplet of the national anthem which he composed for the republic and which is still used in Taiwan. The principles were not fully elaborated until shortly before Sun's death in 1925. According to some records, he had a manuscript almost ready for publication in 1920, but it was destroyed in 1922 when he fled from Guangzhou. The actual work now known as *Sanmin zhuyi* is the transcription of a series of lectures which Sun Yatsen gave at Guangzhou in 1924, and the full text was not published until the end of that year, after he had left for Beijing. Since then, with an added introduction by Sun, these texts have been published repeatedly in China and translated into many foreign languages.

For a brief but authoritative survey of Sun Yatsen's writings and the development of his ideas on these themes, see H. Boorman, *Biographical Dictionary of Republican China 3*, pp. 186 ff.

4. From *San Min Chu I: The Three Principles of the People,* Taipei, reprinted in extract in Immanuel C.Y. Hsu, *Readings in Modern Chinese History*, pp. 410–26.

5. In fact, Sun Yatsen and the revolutionaries could not claim a completely clear conscience in their dealings with Japan and other foreigners. In 1912, when Sun Yatsen was president, he had personally approved special privileges for Japanese interests in the Hanyebing mines and ironworks in Hunan and Hubei in exchange for a loan to his new government. This was opposed by many members of his following and it remained a cause of some embarrassment to the Nationalists. On the other hand, Sun Yatsen believed that he must obtain all the support he could, from whatever source, to ensure the survival and the progress of the revolution, and had done so throughout his chequered career.

6. In a last attempt at his ideal of a constitutional monarchy, Kang Youwei gave his support to this strange and short-lived enterprise.

7. Duan Qirui was a man from Anhui province, and his loose coalition was commonly known as the Anhui Clique. Liang Qichao served briefly at this time as his minister for finance.

8. Most notable were the tens of thousands of coolies sent to the Western Front in Europe to aid in transport of supplies and maintenance behind the lines. In one of the small tragedies of World War I, they were neglected and ill-treated, abandoned by their government and died in great numbers. Very few ever found their way home again.

9. 'Laundry Song', *c.* 1925, in Hsü K'ai-yü, *Twentieth-Century Chinese Poetry*, pp. 51–2.

10. 'New Auld Lang Syne', 1920, in Hsü, *op. cit.,* pp. 31–32.

11. The name of the party appears in some transcriptions as Kuo Min Tang, and it is often referred to as the KMT.

12. The surname of the family was generally known in the Westernized form Soong:

forms of the names by which they were known in the West, I shall refer later to Wellington Koo [Pinyin *Gu*] and I shall do the same for the financier H.H. Kung [Pinyin *Kong*], connected to the Soongs by marriage.

13. Chiang Kaishek's official personal name was Zhongzheng. Kaishek was his common style. The characters of Kaishek would be rendered by the standard Pinyin transcription of Mandarin as *Jieshi*. As in the personal name of Sun Yatsen, the common English rendering reflects Cantonese pronunciation, and the name is so well known in that form I do not render it into Pinyin.

14. Borodin was the alias of Mikhail Grusenberg, of Russian Jewish birth, a close associate of Lenin, who had been a Communist agitator and agent in Chicago, Mexico and Glasgow before his assignment to China.

15. See, for example, Yu-kwei Cheng, *Foreign Trade and Industrial Development of China* (Washington, 1956) Appendix 1, based on Chinese Customs reports; and also the annual *China Year Book*. Due to the number of different coinages and currencies in circulation during the late Qing and early republican period, figures for international trade were expressed in terms of Haikwan 'Customs' Taels (*Haiguan* in Pinyin). In 1913, Hk$1 was equivalent to some 3 shillings sterling, but in 1918 Hk$1 was about 5 shillings 3 pence sterling. So the change in China's balance of trade was not so large as the raw figures would indicate.

16. Several Western powers, however, used the money they received from the Boxer Indemnity as a source of funds for the benefit of China. Various railroad and other industrial developments were financed by the British share of the Indemnity, and Britain, France, Belgium, and the United States all awarded scholarships for Chinese students and western visitors to China. Qinghua University at Peking, one of the leading educational institutions in China, was established with American money from the Boxer Indemnity.

17. 'On the distinction between the City and the Country' (*Bian Chengxiang*) in *Xin Shi Lun,* Chongqing, 1940.

Further Reading

General works dealing with the republican period:

Bianco, Lucien, *Origins of the Chinese Revolution 1915–49*, Stanford University Press, 1971.

Fairbank, John K. and Feuerwerker, Albert (eds.), *The Cambridge History of China*, vols 12 and 13, *Republican China* 1912–1949, Parts 1 and 2, Cambridge, Cambridge University Press, 1983 and 1986 [with a comprehensive bibliography and authoritative chapters by individual scholars, some of which are listed below].

Hsü, Immanuel C.Y., *Readings in Modern Chinese History*, Oxford, Oxford University Press, 1971.

Hsü, Immanuel C.Y., *The Rise of Modern China* [fourth edition], Oxford, Oxford University Press, 1990.

Moise, E.E., *Modern China — A History*, London, Longman, 1986.

Spence, Jonathan D., *The Gate of Heavenly Peace: the Chinese and their Revolution 1895–1980*, New York, Viking, 1981.

Spence, Jonathan D., *The Search for Modern China*, London, Hutchinson, 1990. Schurmann, Franz, and Schell, Orville (eds,), *Republican China*, London, Penguin,1968.

Chen, Jerome, *Yuan Shih-k'ai* [second edition], Stanford, Stanford University Press, 1972.

Isaacs, Harold R., *The Tragedy of the Chinese Revolution* [second edition], Stanford, Stanford University Press, 1951

Liew, K.S., *Struggle for Democracy: Sung Chiao-jen and the 1911 Chinese Revolution,* Canberra, Australian National University Press, 1971.

Nathan, Andrew J., 'A constitutional republic: the Peking government, 1916–28', in *Cambridge China* 12.

Wilbur, C. Martin, *The Nationalist Revolution in China, 1923–1928,* New York, 1985.

Young, Ernest P., *The Presidency of Yuan Shih-k'ai: Liberalism and Democracy in Early Republican China,* Ann Arbor, Michigan University Press, 1977.

Gillin, Donald G., Warlord: *Yen Hsi-shan in Shansi province 1911–1949,* Princeton, Princeton University Press, 1967.

McCormack, Gavan, *Chang Tso-lin in Northeast China, 1911–28: China, Japan and the Manchuria idea,* Stanford, Stanford University Press, 1978.

Pye, Lucien W., *Warlord Politics: conflict and coalition in the modernization of Republican China,* New York, Praeger, 1971.

Sheridan, James E., *Chinese Warlord: the career of Feng Yuxiang,* Stanford, Stanford University Press, 1966.

Sheridan, James E., 'The warlord era: politics and militarism under the Peking government', in *Cambridge China* 12.

Wou, Oderic Y.K., *Militarism in Modern China: the career of Wu P'ei-fu,* 1916–39, Canberra, Australian National University Press, 1978.

Chow Tse-tsung, *The May Fourth Movement: Intellectual Revolution in Modern China,* Oxford, Oxford University Press, 1960.

Goldman, Merle (ed.), *Modern Chinese Literature in the May Fourth Era,* Cambridge Mass., Harvard University Press, 1977.

Grieder, Jerome B., *Hu Shih and the Chinese Renaissance: Liberalism in the Chinese Revolution, 1917–1937,* Cambridge Mass., Harvard University Press, 1970.

Lu Hsün, "The True Story of Ah Q", in *Selected Stories,* Beijing, Foreign Languages Press, 1960.

Tan, Chester C., *Chinese Political Thought in the Twentieth Century,* New York, Doubleday, 1971.

4

Nationalist China, 1925–1936

THE death of Sun Yatsen in 1925 brought a period of uncertainty to the regime in the south, but by the middle of the following year Chiang Kaishek was confirmed as chief military commander and the Nationalist armies had embarked on the first stage of their Northern Expedition. Early in 1927, however, from his base on the lower Yangzi, Chiang rejected the political authority of the government established at Wuhan, and with the aid of conservative factions and the gangs of Shanghai he eliminated the left-wing and Communist allies of the Nationalists in the cities.

In 1928, the second stage of the Northern Expedition destroyed the power of the Nationalists' chief enemies in the north, and produced a formal reunification under the government at Nanjing, but from this time on, regardless of ideals and hopes for reform, the practical effect of the government was limited by its restricted control of national economic resources and by the problems of exercising power at a distance against strong local interest. The tension between national and local authority was expressed in rivalry between despots and, as often before in China, the regime maintained itself by dealing between various groups of influence, while the subjects of government were largely ignored.

Throughout the republican period, the intrigues and squabbles among warlords and politicians, intellectuals and men of money, had taken place within a restricted political culture. This was not a new phenomenon, for it had been the basis for the structure of all imperial China. For centuries in the past, government of the empire had depended upon a small group of gentry, intellectuals, some merchants and men of war, and the vast majority, notably the peasants, had no role but to provide food and physical service. In continuation of this pattern the political élite of China, whether they supported or opposed the Nationalist regime, were no more than a fraction of the whole population, and much of their activity was irrelevant to the interests of the mass of the people.

For an outsider, particularly from the West, it is hard to compre-

Map 3 The Republic of China

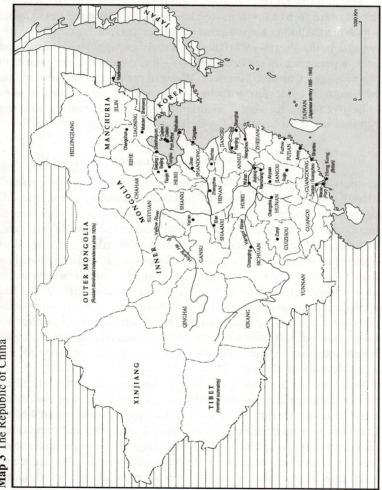

hend the strength of this traditional pattern, and much of the analysis of Chiang Kaishek's regime, both at the time and in later discussion, is flawed by superficial Western expectations. The Nationalists first received sympathy and approval because they were thought to represent a liberal reform movement; later they were criticized for their failure to live up to those alien standards. But the very points which made Chiang Kaishek and his associates, particularly his wife, popular in the West — Christianity, language, apparent culture, and even dependence on outside support — made them less acceptable to the people of China. Ultimately, Western praise and Western blame were equally irrelevant, for whether as a selfish warlord or a national hero, Chiang Kaishek was operating in Chinese terms, and the Nationalists must be judged in the context of their own tradition.

The Northern Expedition

Though the May Fourth Movement of 1919 was recognized immediately as a landmark of political activism and theory among the students and workers in the cities of China, there was wide disagreement on the programme for the future. Moderates, like Hu Shi, sympathized with the demonstrations against Versailles and fully supported involvement of the people in questions of politics, but they feared the results of widespread disorder and favoured gradual reform under proper government. With his American experience, Hu Shi approached the problems of China from a liberal pragmatic viewpoint and, just as he had opposed the doctrines of conservative Confucianism, so he wrote and spoke also against radical ideologies which sought to extend revolution in China and hoped then to build a new society in the ruins.

Chief among the new opponents of the moderates were Hu Shi's former colleagues at Beijing University, Chen Duxiu and the librarian Li Dazhao. Deeply involved with the student militants of 4 May, Chen Duxiu joined the demonstrations that followed and spent three months, badly treated, in gaol. In September 1919 he resigned from Beijing University and moved to Shanghai, and by the middle of 1920 he was the chief organizer of a Marxist Study Society and a Socialist Youth Movement. At the same time, Li Dazhao in Beijing, who had formed a Marxist research society in 1918, amalgamated it with other groups in early 1920 to form the Beijing Society for the Study of Marxist Theory. At Shanghai in July 1921, the First Congress of the Communist party of China united the Beijing and Shanghai groups, and the twelve delegates, who included young Mao Zedong from Hunan, represented a total of some fifty members.[1]

Communism in China owed its development as much to the example of the Russian revolution as to the theories of Karl Marx. In the last years of the empire and the first period of the republic there had been some translation of writings and discussions by Marx and Engels, but the main stream of revolutionary thought had been based on constitutional theories and Western-style democracy. By 1919 and 1920, however, the failure of republican ideals when faced with the reality of Yuan Shikai and the warlords had brought even Sun Yatsen to revise his plans for power, and the horrors of war in Europe made men wonder whether Western civilization was such an admirable model as it had once appeared. Some now claimed that traditional Confucianism, with all its faults, provided a better scheme for living and a more rational approach to society, and those who still hoped for progress and improvement could watch the development of the Soviet state in Russia. The Bolsheviks there, in a comparatively short time, had reunited the country and driven back foreign invasion, and their policy towards China was at first sight far more generous than any other nation. While the Western powers at Versailles had maintained their own interests and tolerated Japanese pressure against China, the new government of Lenin denounced the policies of the tsarist regime and proclaimed in 1920 that it was prepared to renegotiate the unequal treaties. In these circumstances, somewhat to the dismay of Western observers, the first Soviet envoys to Beijing were received with favour not only by the Communist Party but also by the Chinese government. At the Second Congress of the Communist Party held at Shanghai and Hangzhou in July 1922 the delegates agreed to join the Third Communist International. Under guidance and pressure from the Soviet envoy, the Dutchman Henricus Sneevliet, known by the pseudonym of Maring, it was further agreed that members of the Communist Party should join Sun Yatsen's Nationalist movement as individuals, and should work within that group to approach their final ambitions.

The Russian decision to support Sun Yatsen, and to encourage their Chinese colleagues to join him, had not been immediate and automatic. Maring and other envoys had talked in Beijing with the warlord Wu Peifu, who certainly controlled the strongest government and army in China, and they had considered negotiations with other military commanders. It soon became clear, however, that such men were no use to the Communists and would never change their ways, while

Sun Yatsen, despite his apparent weakness and his Western Christian prejudices, controlled the only significant group which offered opportunity for revolution and Communist progress, and he welcomed any foreign aid.

Already, in 1922, the Communist Party had attempted to operate on its own and had organized and supported strikes and local disturbances. There was a seamen's strike at Hong Kong and another among the mine workers of Anyuan in Hunan, but early in 1923 a strike among workers on the main north–south railroad, from Beijing to Guangzhou by way of Wuhan, was broken by soldiers of Wu Peifu's government and leaders of the union were executed. Party membership continued to grow, and was more than four hundred at the time of the Third Congress in June 1923, but the alliance with the Nationalists had become a necessity, while the amorphous nature of that organisation meant that the Communists could use it as a common front without sacrifice of principle. So Chen Duxiu and his colleagues agreed to join forces with Sun Yatsen, and the first National Congress of the reformed Nationalist Party, with Communist membership, was held at Guangzhou in January 1924.

For foreign observers in China, the alliance with the Communists and the support Sun Yatsen was receiving from Russian advisers gave the Nationalist Party a new and threatening appearance. The platform was now specifically radical, popular, and anti-imperialist, and the emotion stirred by Sun Yatsen's last journey to Beijing and his death in March 1925 was soon displayed in direct action against foreign interests. A pattern of strikes and protest marches developed and spread, and on 30 May 1925 demonstrators in Shanghai were fired upon and killed by British-led police. This Thirtieth of May Incident angered people throughout the country: trade union membership and strike activity increased sharply, there was a nationwide boycott of foreign goods, and a massive strike in Hong Kong lasting sixteen months and involving two thousand workers, from dock labourers to garbage collectors, brought trade through the colony almost to a standstill. Hundreds of people were killed or wounded in clashes with foreign or warlord Chinese police, but the demonstrations continued in every city, and even after the first wave of indignation was spent the people and their rulers remained restless and uneasy. It was only a matter of time before the revolutionaries in Guangzhou started their move to the north, and over the rest of China the last months of 1925 and the beginning of 1926 were a period of waiting.

For a time after the death of Sun Yatsen, however, the Nationalist government was distracted by rivalry and faction feuds. In August 1925 Liao Zhongkai, a close colleague of Sun Yatsen with left-wing sympathies, was assassinated, and suspicion was placed on the conservative leader Hu Hanmin. Hu went into exile, but a group of his associates, known as 'Western Hills' from their first meeting at Sun Yatsen's temporary burial place near Beijing, announced their distrust of the Communist alliance and their intention to purge the party. With the authority and influence of the Russian adviser Borodin, however, the cracks in the façade were papered over, and in January 1926, at the Second Congress of the Nationalist Party, the Western Hills group was defeated and Hu Hanmin's rival, the left-wing leader Wang Jingwei, held chief political authority. He shared his position in the government with Chiang Kaishek, superintendent of the academy at Whampoa and now commanding all the effective military forces of the Nationalists.

Wang Jingwei, a man of great personal charm, was a long-time supporter of Sun Yatsen and the revolutionaries. In 1910 he was involved in an assassination plot against the Manchu prince regent, Zaifeng, was arrested but not executed, and remained in gaol in Beijing until the revolution of 1911. After some years in France he became Sun Yatsen's close adviser, and was generally accepted as his successor to the leadership of the party and of the government. His position was weakened, however, first by the political opposition of both the right-wing politicians and the Communists, and secondly by the military authority of Chiang Kaishek.

In 1925 Chiang Kaishek was thirty-eight years old. Son of a merchant family in Zhejiang province near Shanghai, he left home as a young man to enter the army. In 1907 he went to Japan for training, and he joined Sun Yatsen's revolutionary party in Tokyo. In the first years after the revolution of 1911 he was involved with secret societies and underground opposition to Yuan Shikai in Shanghai, and became closely associated with Du Yuesheng, leader of the powerful Green Gang. His energy and promise impressed Sun Yatsen during his periods of exile in the French Concession, and he acted as his personal agent at intervals in Guangzhou. In 1924, after three months in Russia on a military programme, he was commissioned to establish the new Whampoa Military Academy and was later confirmed as head of the army. With the cadets and the graduates of Whampoa as the élite corps and leaders of the new model army, the forces of the Nationalists began to expand beyond their headquarters in Guangzhou, and in the first months of 1925 a series of victories secured control of Guangdong and Guangxi. In many towns and villages the ground had been prepared by partisans and organizers, calling the people to support the revolution and depriving the enemy of essential supplies and labour, and often enough of the loyalty of their troops as well. Despite this civil action, however, the victories were essentially military, and the run of success added to Chiang's prestige. The political leaders were weakened by disagreements and the Russian adviser Borodin accepted General Chiang as the man of real authority.

On the very eve of the northern campaign, however, in March 1926, the warship *Zhongshan*, under orders from a Communist officer, moved from her regular moorings to anchor off Whampoa. Chiang claimed this was the beginning of a Communist threat against himself and a first move to seize power, and he sent soldiers to guard the ship and to occupy workers' headquarters in Guangzhou. The Communists claimed the whole incident was a trick, but the Nationalist party passed resolutions against Communist influence. After the defeat of the Western Hills group in January, the Communists had held the majority of top administrative posts and working control of the central committee. In May, after the *Zhongshan* incident, no Communist held a chief executive position and their membership of the central committee was limited to three places out of nine. Wang Jingwei was forced into exile in France, and on 5 June 1926 Chiang Kaishek was appointed commander-in-chief of the National Revolutionary Army. On 27 June he gave his final orders for the Northern Expedition.

The Nationalists' chief and immediate opponent was the veteran warlord Wu Peifu, whose capital was now at Wuhan. To the east of Wu Peifu, with headquarters at Nanjing, the rival General Sun Chuanfang controlled the lower Yangzi basin and the provinces of Jiangxi and Fujian. The main Nationalist attack was concentrated on Wu Peifu's positions along the railway, and the first assaults brought remarkable success. In a whirlwind campaign, brushing aside the resistance of the enemy and hurtling north on commandeered trains, the Nationalists captured Changsha, capital of Hunan province, on 11 July and at the end of August, their numbers swelled by turncoats and deserters from their opponents, they came to Wuhan. Wu Peifu fled north into Henan and the flag with a white sun, badge of the revolution since Sun Yatsen's earliest days in Guangzhou, replaced the five coloured bars of the old republic.[2] In January 1927 Wuhan became the capital of the National Government.

Further to the east the campaign against Sun Chuanfang met with less spectacular success. The Nationalist army, commanded in this sector by Chiang Kaishek himself, advanced due north through Jiangxi and also along the coast from Shantou towards Shanghai. Progress on both lines was comparatively slow, partly through the difficulty of the terrain and the quality of the opposition, and partly because there was less active popular support in Jiangxi than there was in Hunan. By November, however, Chiang Kaishek had established his winter quarters at Nanchang and in December the Nationalists occupied Fuzhou.

Whatever the disagreements within the revolutionary camp, the breakout from the south was recognized by both the Chinese people and the foreigners as the beginning of a new era. In many areas, notably in Hunan and along the coast, the ground had been prepared even before the Nationalist army came to the field. Undercover agents had established contact with local groups of potential sympathizers, and the unfortunate warlords found their communications broken and their troops demoralized by small-scale guerilla action from the peasants and by strikes, riots, and demonstrations from the people of the cities. One architect of the victory was the Hunan Communist, Mao Zedong, chief of the Peasants' Movement Institute for the training of revolutionary agitators,[3] and when Wu Peifu attempted to rally support with the argument that victory for the Nationalists would be victory for Bolshevism and the Reds, Chiang Kaishek readily acknowledged Soviet and Communist support. He claimed to lead an army and a revolution which would bring independence to China, freedom to the masses, and government to the people, and he offered small comfort to the conservatives, the foreigners, and the men of military power.

The Government at Nanjing

At the end of the first stage of their march to the north, with the Nationalists established at Wuhan and Nanchang, it was inevitable that the authority Chiang Kaishek had established over his political colleagues would be weakened. From his headquarters in Jiangxi, Chiang was in no position to control events in Hunan and Hubei, and the transfer of the capital to the larger city of Wuhan was itself a defeat for his influence. The Communist and left-wing political organizers, moreover, were claiming great credit for the military success in Hunan, and on 3 January 1927 they scored another triumph: a demonstration of students and workers, led by the Communist Liu Shaoqi and other activists, entered the British Concession by force, and a few days later the small concession in Jiujiang, down river at the mouth of the Poyang Lake, was also occupied and looted. In negotiations which followed, the British gave up their rights at both places.

With Nationalist armies poised to advance against Nanjing and Shanghai, it was clear to the Western powers that any attempt to use the local forces at their disposal could only bring embarrassing defeat, and a soft answer was the most sensible policy. For the Chinese, however, the British surrender of the concessions was a major victory. In the past, as in the period of 30 May, demonstrators had faced the foreigners with courage. This time they had faced them and won. Throughout southern China, despite executions by warlord soldiers still in control, rioters under the Nationalist banner paraded their hatred and their long-felt sense of shame, and the foreign troops and missionaries and civilians were shaken by processions of men, women, and children, screaming abuse and making obscene gestures to their faces. For the first time in their history, it seemed that the ordinarily courteous, tolerant, humble Chinese people had gained a sense of power and were prepared to show their feelings for the arrogant strangers from overseas. The sight was neither pretty nor reassuring.

Yet if the Nationalist movement was now shared among the people, military power remained with Chiang Kaishek, and it was his army which held the foreigners in check while the people acted out their rage. In the first months of 1927, as the government in Wuhan threatened to expel the more moderate politicians and as the Communists in Hunan province encouraged local violence to force through agrarian reform, the businessmen of the eastern cities came to Chiang's camp to seek his assurances and to persuade him that the revolution was best served by competent government and not by extremism.[4] On 24 March, when the Nationalists entered Nanjing, there was firing from the troops against foreign warships in the river and a return bombardment against the shore, and before Chiang came to Shanghai the Chinese city was seized from the warlord troops by a Communist-led uprising. Even so, Chiang Kaishek was able to keep peace with the foreigners, and the revolutionary administration in Shanghai, on orders from Stalin, handed over to Chiang's administration.

By this stage the Communists were uncertain of their own best policy. Stalin in Moscow and the Russian advisers on the spot still felt their influence in the Nationalist movement was too weak and uncertain to risk a full take-over, and the Chinese party was divided between men like Chen Duxiu, who saw the Nationalist alliance as essential for success, and enthusiasts like Mao Zedong who sought action among the peasants and workers without regard to support from the moderates and men of substance in the towns. As Chiang Kaishek gained approval by the appearance of responsibility, the government at Wuhan was torn between its need for wide support and extremist demands for radical action. The pattern was not surprising, for any revolutionary group is subject to similar pressures, but in the confusion of the time Chiang Kaishek, with the army under his control, was the only man who knew what he wanted.

Despite the accusations of his enemies, Chiang Kaishek may indeed have been a genuine Nationalist. Like any politician, however, he was certain that his own interpretation of the creed was correct, and on a personal basis he now found himself under critical attack from the Communists and his rivals in Wuhan. For his own safety, and for the pattern of revolution in which he believed, he had every reason to act.

On 12 April 1927, in a well-planned coup with the use of his own soldiers and hired gunmen from the underworld, Chiang Kaishek disbanded the revolutionary administration of Shanghai, disarmed the workers' general union, and executed many of its members. In the next few days the process was repeated in every city under Chiang's control, from Nanjing and Hangzhou to Fuzhou and Guangzhou. On 17 April the Wuhan government proclaimed Chiang's expulsion and dismissal, but on the following day Chiang set up his own government at Nanjing, and the Wuhan regime could be certain of authority only in Hunan and Hubei. From this restricted base, the left-wing government, headed again by Wang Jingwei, attempted to renew the Northern Expedition, but their armies were forced to a halt near Zhengzhou in Henan. Further to the east, Chiang Kaishek's forces had moved towards the Huai and then, in a great battle at Longtan on the Yangzi, they destroyed a counter-attack by Sun Chuangfang.

The Wuhan government was now threatened from all sides and the Communists, in desperation and under Russian orders, attempted a coup to seize control. By August, however, they were defeated and expelled from the Nationalist alliance, and Wang Jingwei with the rump of his supporters negotiated for reunion. The new national government, purged of Communist influence, was established at Nanjing in September. Chiang Kaishek, now sure of his position, took leave of absence in Japan and Shanghai, and in December 1927 he married Meiling Soong, youngest sister of Madame Sun Yatsen. Like Sun Yatsen, Chiang Kaishek had a former wife whom he discarded for the new alliance and, also like his former leader, he accepted baptism into the Christian Methodist Church. In January 1928 he returned to office in Nanjing, resumed command of the Nationalist armies, and received orders to continue the advance to the north.

So the Communist alliance was ended and the Communist Party was broken and discarded. In August 1927 a mutiny at Nanchang was defeated by loyal troops, but further skirmishes followed in the cities of south China. On 11 December, with guidance from Moscow, there was a major rising in Guangzhou, and the Guangzhou Commune, with a soviet government, held power for three days. Then came the soldiers of the Nationalists, the rebellion was crushed, and five thousand people were executed. It is said that the rebels had worn red scarves around their necks as a badge, and took them off to escape detection when the fight was lost; but the dye in the scarves had run, and people with red stains round the neck were caught and killed.

With this final disaster, Stalin himself, whose authority in the Communist International had played a major part in the long fiasco, came under political attack and personal danger from the Trotskyists in Moscow. The attempt to follow the Russian model, with revolutionary alliance and a power base in the cities, had ended in total defeat, and the former Communist leadership in China was quite discredited. Li Dazhao had been caught and killed by the warlord regime in Beijing; Chen Duxiu was expelled from the party; the official headquarters was in hiding in Shanghai; and the main active group was now the small band of peasants and soldiers, hiding on the borders of Jiangxi and Hunan, led by Mao Zedong.

The way was thus clear for the second stage of the Northern Expedition. The Nationalists advanced into north China, with alliance and support from Feng Yuxiang and Yan Xishan, against Sun Chuanfang and Zhang Zuolin. Their main line of attack followed the railroad through Jinan, capital of Shandong province, and they captured the city on 1 May 1928. As Zhang Zuolin's forces withdrew, there were patriotic slogans against Japan, and two days later Japanese troops left their zone of occupation and forced the Nationalists from Jinan. Chiang Kaishek submitted to the outrage and resumed the advance, and on 6 June Yan Xishan's forces marched into Beijing.

The nature and course of this second stage of the Northern Expedition showed the problems which the new government would face. On the one hand was the interference of the Japanese, and on the other the alliance with the northern warlords. Apart from their hostility to Wu Peifu, Zhang Zuolin, and lesser rivals, there was no strong reason for Feng Yuxiang and Yan Xishan to support the Nationalists. They were prepared to pay lip service to the new government, but Yan had ruled Shanxi like an independent monarch since the earliest years of the republic and he did not regard his relations with Nanjing as anything but negotiations between equals. Feng Yuxiang had shown some interest in reform, but his base in the north-west was well removed from the centre of government and it would never be easy to compel his obedience.

And the problem of Japan was still more intractable. In the early days of revolution Japan had been a refuge for rebels against oppression, whether from the government of the Qing dynasty or the usurpation of Yuan Shikai, and Japanese and Chinese had talked of their common race and tradition in the face of Western imperialism. The difficulty was the question of leadership, for the Japanese, with their rapid modern development, regarded themselves as the predominant partner. Their propagandists spoke of Japan as the energetic, intelligent younger brother burdened by responsibility for his lazy, incompetent elder, and Japanese interference in China was often described as the kicks and blows which should drive her along the path of progress. It was, however, increasingly clear that Japan would not accept a rival in East Asia.

The contrast between the West and Japan is shown by the surrender of the British after the Wuhan riots in January 1927 and the aggression of the Japanese at Jinan in May 1928. Japanese concern, however, at this time lay not so much in China Proper as with their position in Manchuria. Since the Russo-Japanese War and the later collapse of Tsarist Russia, Japan had extended authority along the railroads of Manchuria, and politicians and businessmen regarded the mineral and industrial resources of Manchuria as much a part of the national economy as the rice-growing regions of Taiwan, long ago taken from China. Zhang Zuolin, warlord of Manchuria, had played a major role in north China, but he was regarded as a Japanese ally and dependent. On 4 June 1928, as his military position around Beijing and Tianjin collapsed, and as he himself fled back to Manchuria, Zhang's headquarters train was blown up by a bomb and totally destroyed. The assassination was arranged by officers of the Japanese army who had become doubtful of Zhang's loyalty and who hoped to use the disruption of his death to strengthen their position in the north-east.

In immediate effect the killing was a failure, for the government at Tokyo, which had no prior warning of the plot and found itself severely embarrassed, refused to support any further move. The assassination was an ominous sign that extremists in the Japanese army could act without any fear of punishment, but it gave the Nationalists an opportunity to extend influence into Manchuria. Zhang Xueliang, son of Zhang Zuolin and commonly known as the Young Marshal, with unexpected energy and skill established his succession through bribery and murder, and he readily accepted alliance and support from Nanjing. By the end of 1928, in theory and lip service at least, all China was united under the Nationalist regime. Beijing, 'the northern capital' was renamed Beiping, 'northern peace',[5] and the government of the republic was maintained at Nanjing.

According to the programme of Sun Yatsen, military victory should be followed by a period of political tutelage, as the people were educated for democracy. In June 1929, soon after the state funeral of Sun Yatsen in Nanjing, it was announced that political tutelage would end in 1935, but for the time being the government remained in the hands of the Nationalists. The formal institutions of Sun Yatsen's constitution were established, with a five-part government comprising executive, legislative, judicial, examination, and control councils, being the accepted divisions of Western-style democracy together with the more Chinese institutions of an examination authority to supervise education and recruitment into the civil service, and a control council to oversee the financial and general administration of all the others. The formal structure of the constitution is preserved in present-day Taiwan, but reality has never matched the theory.

In practice, chief power was held by Chiang Kaishek, president of the republic, and with the authority of that office dominating the executive council, the cabinet and the ministry. Though Cai Yuanpei, former chancellor of Beijing University, became first president of the control council, the new administration was composed largely of Chiang Kaishek's own supporters and his allies from the right wing of the Nationalist Party. Hu Hanmin was president of the legislative council, and the warlord Feng Yuxiang, whose support was too important to be ignored, was made vice-president of the executive. Wang Jingwei, who had left for France after the failure of the government at Wuhan, held no position.

Chiang Kaishek now gathered support from his old friends in Shanghai and his new relatives by marriage. Chief among his personal attendants were two nephews of Chen Qimei, a revolutionary and gangster who had held considerable influence in secret societies of Shanghai and had been Chiang Kaishek's first patron. Chen Lifu and Chen Guofu acted as Chiang's agents in the army to ensure the loyalty of his officers, and many party members and officials were quite prepared to accept their leadership. Madame Chiang, on her side, was naturally allied to the interests of her brother, the financier T. V. Soong, and her eldest sister Ailing had married Dr H. H. Kung, a merchant banker who claimed descent from Confucius. The third sister Jingling, Madame Sun Yatsen, shared her late husband's more revolutionary, left-wing ideals, and she remained isolated in exile among the communists in Russia, but the other two groups of relatives by marriage, with Chiang Kaishek himself and his friends of the Chen family, dominated the politics of Nanjing. These 'Four Families', as their enemies and rivals called them, controlled the financial operations of the government and the great corporations, and this network of interest and patronage maintained Chiang Kaishek in power and secured the funds for his army.

The failed left-wing government at Wuhan had made peace with Nanjing, but its members had small part in the new administration. Forced into permanent opposition, they were prepared to form alliances with any other faction and cause trouble where they could. Though the new government controlled the strongest army in China, it was too much to expect that squabbles would cease, and constant conflict at the top between those who were in power and those who were out made the years of the Nationalist hegemony almost as confused and often as bloody as those of the warlords who had preceded them. In February 1929 there was a rising in Wuhan and in Hunan province, and Feng Yuxiang rebelled in May. Feng was defeated and forced into temporary retirement, but it was not until December that all his subordinates and allies had been dealt with. In April 1930 Yan Xishan of Shanxi joined Feng Yuxiang and the southern warlord Li Zongren in a new uprising, and in July they were joined by Wang Jingwei, returning from overseas to proclaim an alternative government in Beijing. Fortunately for Nanjing, while Chiang Kaishek's army faced the rebels in the south, the Young Marshal, Zhang Xueliang, intervened in support from Manchuria, and a temporary peace was restored. Wang Jingwei retired for a time to Hong Kong, but in May 1931 he set up another government at Guangzhou, and at the end of that year, with growing threat from the Japanese in Manchuria, both groups of the party agreed to reconciliation. Early in 1932 Chiang Kaishek resigned as president of the republic to become chairman of the military council, and Wang Jingwei was appointed president of the executive council and effective prime minister.

Despite the apparent retreat of Chiang Kaishek, however, whether in semi-retirement at Lushan, a mountain resort town near the Yangzi, or in active command of government forces in action against the Communists and other rebels, he retained his influence, and newspapers frequently recorded how Wang Jingwei had called upon General Chiang to ask his advice. Meanwhile, Chiang Kaishek's interests in Nanjing were well guarded by such allies as the Chen brothers, who held an effective monopoly of the ministry of education and party security, and by this means supervised both official propaganda and the operations of the secret police. In similar fashion, the ministry of finance was controlled from 1928 to 1931 by Chiang's brother-in-law, T. V. Soong, and when he left office to attend to other interests he was succeeded by Chiang's other brother-in-law, H. H. Kung. With such continuity in office, the basic policies of the government showed very little change, and there was no opportunity or desire for consultation with the people.

At best, the government in Nanjing seemed a ramshackle oligarchy; at worst, it had many signs of military dictatorship. In retrospect, however, compared to the bloody struggles yet to come, the years from the late twenties to the early thirties were a period of prosperity and hope. Though the threat from Japan was clear and obvious, with steady encroachment and seizure of territory, the government had survived its disagreements and its enemies within the country and there was some expectation that the regime in Nanjing could gain tolerance, acceptance, and even support from the United States and other Western countries. The Nationalists had now broken with the Soviet Union, and Chiang Kaishek, with his wife, emphasizing their Christian principles and their admiration of United States, obtained general support in America for his defiance of European imperialism.

China maintained an independent foreign policy where it could, and during 1928–9 it successfully negotiated the right to set its own tariffs, without being bound by earlier agreements with foreigners, while Britain was induced to return the naval base at Weihaiwei and the concession at Xiamen. On the other hand, there was no more than token recognition of Chinese interests in the International Settlement at Shanghai, and the French and other powers retained their positions among the treaty ports. Similarly, despite reforms in Chinese law, no foreign government would relinquish the rights of its citizens under extraterritoriality, and the Nationalists were not strong enough to force the issue.

In these and other dealings with the Westerners, the new government had generally acted with restraint, and it gained acceptance over-

seas as long as Chiang Kaishek could muster sufficient force to suppress, or at least control, rebellion inside China. Two major and
intractable problems, however, remained: the military aggression of
Japan and the expectations of the Chinese people. As events were to
prove, no Nationalists force could hope to face Japan on equal military terms, and realization of this forced the Nanjing government
into repeated humiliation. Its army was the strongest in China, but it
had no heavy equipment and only a minimal navy and air force, so
the best policy in the face of bullying from Japan was to temporize
and hope for assistance from outside. This policy prevented immediate and disastrous defeat, but it was naturally a source of shame to
the proud party of revolution which had vowed to free China from
oppressors.

Still more important in the end was the question of the acceptance
of the government amongst the people. Behind the fragile unity which
Nanjing preserved, behind the party enthusiasts who praised General
Chiang as the leader of the revolution, and behind the reformist mor-
alizing slogans which opposed corruption and immorality while preaching modernization and a 'New Life', it was difficult to tell just what
the people of the country had gained and easy to forget how much
they might still, with reason, expect.

Government, Production and Trade

During the first part of the twentieth century, China remained a subsistence agricultural economy, and the balance between the growth
of population, about one per cent a year, and the supply of food was
maintained primarily through a low standard of living, poor public
health, and a generally high death rate. In effect, the rules of Malthus
were restraining population growth, and the common fate of the people was constant undernourishment punctuated by occasional years
of disaster and starvation.

To a considerable degree, however, the Nationalist government in
Nanjing was isolated from the generality of the people by its physical position on the eastern seaboard. The Nationalists achieved notable
gains in the development of a modernized economy, but development
was skewed toward the industries of the east, while outlying provinces
remained largely unconsidered. Politically, the Nationalists were committed to the cities rather than to the country, to the men of property
rather than to the peasants, and questions of land reform and agricultural development, most vital of all, fell into default.

Nevertheless, despite government economic planning, private capital was not encouraged to enter the market for long-term loans to new industries. In a few cases, such as railroads and some government-supported co-operative schemes, money was found from Chinese sources to finance new developments, but the greater part of local capital was put into short-term loans at high interest, or sent to chase a meaningless profit on the foreign bond-market. The shortage of capital and a lack of long-term confidence made investment cautious and borrowing expensive: in the villages of China the moneylender was entitled to charge 3 per cent per month, and his rates often went as high as 9 per cent. In the cities, government bonds paying interest on a face value of 9 per cent per annum were sold at a discount and could realize as much as 30 or 40 per cent interest on the actual cost to the purchaser. In such circumstances, neither the peasant in the countryside nor the small entrepreneur in the city could hope for real support and security.

The government did show energy in large-scale civil works, notably roads and railways. Surfaced roads trebled to more than 100,000 kilometres between 1927 and 1937, and the government also embarked on a programme of railway construction. In 1912 there had been rather less than 10,000 km of railway line in the country; during the next fifteen years 3,500 km were built, and the same amount was added by the Nationalists from 1928–37 in China proper, including completion of the line from Guangzhou to Wuhan. From 1931, however, there was ominous development in the north-east: 4,500 km were constructed in Manchuria, part of the Japanese-controlled system designed to serve as a support for the coming war against China itself. The weakness of the government can be seen from the fact that all the work in this major field was exceeded by the constructions of a foreign power operating on Chinese soil.

The growing strength of Japan in east Asia was matched by her influence on China's industry. In 1914 Great Britain held the majority of foreign capital invested in China, 37.7 per cent of the total, and Japan held 13.6 per cent. In 1931, even after the disappearance of Germany and Russia as major rivals, Britain's share was static at 36.7 per cent, while Japan's had almost trebled to 35.1 per cent, an investment of US$1,137 million. Of all foreign investment in 1931, more than a third was concentrated on Shanghai, and more than a quarter was in Japanese-controlled Manchuria.[6]

Though much of this foreign investment had taken the form of loans to the government, particularly in the days of warlord borrowings in the 1920s, or capital equipment for foreign-dominated railway systems in Manchuria, it was in the development of light industry that new investment had its greatest effect, and notably in textiles.

Since the beginning of the century the traditional exports of China, silk and tea, had shown a marked decline. During the 1870s these two products represented 90 per cent of China's export trade, but by 1913 silk represented only a quarter of total export and tea was less than a tenth, and by 1931 these figures had been halved. Both commodities were affected by foreign competition, silk by the growth of that industry in Japan, tea by the British possessions in India and Ceylon, but both industries contributed to their own decline through erratic supply and poor quality control. In 1936, after the loss of Manchuria to Japan, the single major item among China's exports was vegetable oils, one fifth of total exports, while over half the export trade was diversified among small-scale commodities such as hides and leather, eggs and pig bristles.

The import of opium, which had once been the ruin of China's foreign trade, officially ended during the early years of the republic. In 1911 the British government had at last agreed with the Manchu government to scale down the trade of opium and to stop it in 1917. This they did, and opium disappeared from the lists of imports, while successive Chinese governments launched campaigns against the drug.

In practice, however, the trade continued, and China itself became a leading producer. Though the Nationalists imposed a tax which was supposed to discourage consumption, it was a major source of government revenue, and opium, with its derivate heroin, was widely popular among those who could afford it. The greatest production of the drug was in the south-west, particularly Yunnan province, where the allegiance of the local warlords was held by the Nationalists' control of the trade route down the Yangzi. Shanghai was the centre for processing and distribution, both inside China and for export overseas, and the illegal trade was dominated by the Green Gang of Du Yuesheng and by the Nationalist Party itself, which used these illegal resources to maintain the military power upon which its government depended.

It is difficult to estimate the damage and misery which opium has caused to China. Apart from the fact that the drug is addictive and debilitating, so that people under its influence are incapable of useful work, its cultivation uses fertile ground in hill country which could otherwise produce food. In south-west China, the opium poppy may still be grown, and production is certainly maintained in Burma and the north of Indo-China. Those who have a taste for poetic justice may like to trace the history of the drug and its relatives from the days of the British and the Opium War to the American army in Vietnam and the drug peddlers of the West.

In more regular trade and industry there was important development in cotton: in 1913 raw cotton was three per cent of net exports, but made-up cotton goods were almost a third of net imports; in 1931 raw cotton was some ten per cent of net imports, and cotton yarn, often sent away for further manufacturing in Japan, was a net export. By 1936 the balance of trade in raw cotton and finished piece goods was almost even, and trade in cotton yarn showed a small profit, some two per cent of export value.[7]

Modern cotton mills were first set up in China by foreign firms under the privileges from the Treaty of Shimonoseki in 1895, and the main investors were British and Japanese. Chinese-owned firms soon joined the competition, and cotton became the fastest-growing industry in the country and a major user of modern machinery. By 1920 there were 2 million spindles for cotton yarn in China, of which 1.3 million were controlled by native companies, 500,000 by Japanese, and only 150,000 by British firms, seriously affected by the European war just ended. With British investment lagging behind, Chinese spindles doubled to 2.7 million in 1936, but investment from Japan had expanded fourfold, with more than 2 million spindles in Japanese factories on Chinese soil. Companies controlled by foreigners, moreover, were generally larger than those of native Chinese firms: in 1936, 93 Chinese mills held 53 per cent of spindles, 45 Japanese firms held 42.5 per cent of spindles, and four British firms held 4.4 per cent. In all industries, according to a manufacturing census of 1933, the average number of workers in a factory was little more than two hundred.[8]

In the years of depression after 1929, large Japanese firms were able to drive out or take over a number of their small Chinese competitors, and the energy and enterprise of the Japanese industrial invasion, with their large market and advanced industry at home, helped them to a predominant position in the slow economy of China. For their part, Chinese companies suffered from a shortage of trained administrators, accountants, and foremen. In 1933 the government established a Cotton Industry Commission to improve the quality of the raw cotton that was grown and to encourage local industry. The new organization, and a period of comparative prosperity and peace in the middle 1930s, did produce some expansion, and a number of new mills was established in the country away from the treaty ports. Even so, the city of Shanghai continued to hold more than half the cotton spindles in China and, still more disconcerting in terms of modern development, over 60 per cent of the cotton cloth produced in China during 1931 came from handicraft industry, 10 per cent was processed from imports, and less than 28 per cent was made in factories.

Here was a major limitation on the expansion of modern industry. In a rural household, labour was effectively of zero cost, and people could work at any time and season they were not needed on the land. For factories, however, the capital for equipment was expensive, and although wages were extremely low, they were still a real cost. For some processes, such as the spinning of yarn, machines had an advantage, but many, including the weaving of cloth, could be carried out by a semi-skilled domestic worker with comparable efficiency and sometimes in superior quality. As a result, even in the 1930s, it is estimated that more than three-quarters of China's industrial production came from handicraft industry.[9]

In general, moreover, apart from commodities such as wheat flour, cement, iron and steel, where popular need or government construction provided a reasonably firm market, the modern style industries of China, foreign or native owned, had limited impact upon the country as a whole. In a sense, as critics claimed, the cities of the eastern seaboard and the rivers were connected more closely to the foreign nations with which they traded overseas than they were to their own hinterland. The patterns of trade were important to the foreigners and to the government and to the businessmen, but they were of only marginal interest to the people of China, and the entire modern industrial sector of the economy, including factory output, mining, utilities, and modern transport, represented only 7.5 per cent of the estimated gross domestic product.[10]

Opinions vary as to the success of the modern economy in China during Nationalist rule. Some scholars have argued that the government encouraged progress, but others have talked of general economic stagnation, and a detailed analysis of the pre-Communist period suggests that there was a fluctuation of industrial development which reflected such factors as civil war, comparative peace, and international events, rather than the effect of any measures by the Nationalists or their various predecessors.[11]

By the 1930s the chief official imports of China were raw cotton, wheat, and rice, altogether more than a quarter of the total value, with significant quantities also of iron and steel, chemicals, kerosene, and petroleum. Far the most dangerous was the net import of food grains, for the problems of earlier years had not been solved and the growing population still pressed on the means of subsistence. China suffered a constant imbalance of imports against exports, and the balance was made up by foreign investment, payments by foreigners for goods and services in China and remittances sent from overseas Chinese to their families at home. Though the Chinese might resent the foreign forces on their soil, the money they spent was a major factor in maintaining a balance of payments.[12]

Fig.1 Reported Receipts and Expenditure of the Nationalist Government at Nanjing 1928–1937

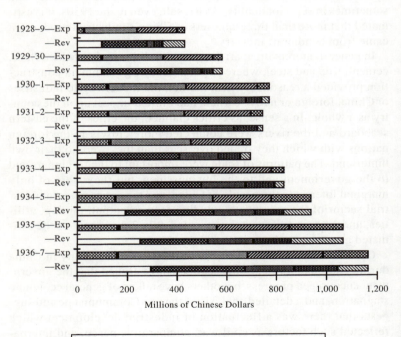

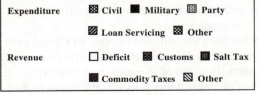

As part of the settlement which confirmed the Nationalists in power, a National Economic Conference in June 1928 allocated land tax to the provinces, and since income tax was not introduced until 1936, central-government revenues depended very largely on indirect taxation. In 1929 China achieved tariff autonomy, the right to set duties without reference to foreign powers, and when the *lijin* and other internal tolls were abolished in 1931 the loss of revenue was compensated by a steep rise in import tariffs, from a nominal 4 or 5 per cent in 1928 to an actual 30 per cent in 1936, and by a 'consolidated' tax on such staple commodities as wheat flour and cotton yarns. In theory, the freeing of internal trade and the excise barriers against imports benefited Chinese manufacturers and merchants, but the main purpose of the tariffs was to increase revenue, and much of the money went on such unproductive causes as the servicing of foreign loans and the maintenance of the army. In 1931-2, for example, government revenue was reported as 619 million Chinese dollars, of which almost 60 per cent came from customs duty, 24 per cent from the salt tax, and 15 per cent from the consolidated commodity tax. Expenditure was $749 million, of which over 40 per cent was allocated to military purposes, 36 per cent to interest payments,[13] 6 per cent transfers to the provinces, leaving only 17.5 per cent, about $130 million, for central administration and public works.[14]

This budget shows the limitations of central resources and the absence of any real tools for economic control. In particular, it may be observed that the actual revenues of the central government were almost entirely absorbed by military commitments and the service of loans, while the amount of annual deficit, made up by further borrowings, was slightly more than the amount allocated to civil expenditure. So all administration was severely restricted, the finances allowed minimal opportunity for the development of new and effective programmes, and even the collection of taxes was commonly arranged by quotas or even by tax farming, which restricted still further the government's access to funds and put profits into private hands. Except for the immediate region of Nanjing, moreover, the provincial regimes were little concerned with the central government, and besides their agreed revenue from land tax and a share of the salt tax, they added a plethora of exactions on every aspect of food, clothing, housing and trade, for a cumulative effect greater than the old *lijin*.

Allowing for their problems, it is doubtful whether Nationalist financial policies were of any advantage to the modern development they claimed to assist. Imports were restricted, but not to the benefit of industries which depended on raw materials from abroad, while the government need for money to balance its budget put quantities of discounted bonds into competition with private bids for capital and held interest rates at 18–20 per cent per annum. Modern companies, moreover, were the easiest to assess and to tax, and demands for 30 or 40 per cent of gross income could drive even large and prosperous enterprises out of business.

In effect, the Nationalist government was dependent upon commercial activity and the city-based economy, and it had no means to extend its control over the wider resources of the nation. During the early 1930s central-government expenditure varied from 2 to 6 per cent of the gross national product, and perhaps as much again among provincial administrations.[15] With this limited base, and so much of its income devoted to servicing debts and paying for soldiers, the Nationalist regime imposed excessive burdens on many of its natural supporters, yet lacked the ability to expand its activities into meaningful administration and social service.

For behind all the claims and hopes of the Nationalists, to build China's economy and develop her power, there remained the vast reality of the country itself. With feuds and factions at the capital, and loose-knit allies in the provinces still holding men under arms, the government at Nanjing could be sure of controlling only the lands about the lower Yangzi, and their orders had small effect in the far north-west and the hill country of the south. Both in numbers and in economic power, the peasants were the backbone of the country, and the Nationalists could not afford to neglect them.

Landlord and Peasant

We have observed already that the picture of agricultural self-sufficiency was based upon the limitation of population by hunger and exhaustion. It also conceals a regular pattern of local misfortune. Despite new roads and railways, the communications system was generally inadequate for rapid large-scale transport, so each region faced limited possibility of relief from a poor harvest, or from the depredations of banditry and civil war. For the vast majority of people, the constant problem was feeding themselves locally, and no farming family could feel secure from one year to the next.

There were also natural disasters. In 1920 an earthquake in Gansu buried alive 250,000 people in their cave homes among the loess.

Drought in north China during 1920–1 caused a famine which killed half a million people and left twenty million destitute, and floods in north and central China in 1924 put thousands of square miles of farmland under water, drowned hundreds of thousands of people, and drove millions from their homes. In 1928, when they came to full power, one of the first acts of the Nationalists in government was to establish public works agencies, such as the North China River Commission, to construct flood-control and transport works, dykes, and canals, which would lessen the danger from the great rivers. Inevitably, however, these were long-term projects, and in the early 1930s China continued to suffer from massive flooding. In 1931, in one of the worst floods ever recorded, the Yangzi overflowed its banks along the whole course from the Dongting Lake and Wuhan to the estuary, and the Huai River flooded the southern part of the North China Plain. More than 64,000 square kilometres were affected by the flood: cities, towns, and villages, including Wuhan itself, were covered by water, 140,000 people were drowned, and in the countryside alone 25 million lost their homes and their possessions. Three years later, in north China, the Yellow River broke its banks, with damage comparable to the great change of course in 1853. Another flood followed in 1935, and though administrators and engineers of the government were able to make some plans for controlling the damage they were faced with the massive problems of heavy rainfall, serious silting, and vast areas of open, level ground at the mercy of the water. Before they could properly begin their work, they were interrupted by the war with Japan in 1937, and to this day, after decades of peace, floods are still frequent.

Even in good conditions, the agricultural economy of China was weak. Much of the growth in output recorded through the first decades of the century was based upon the expansion of farmland in Manchuria, while in China proper, apart from some marginal increase in the Yangzi valley and the south-west, the 'internal frontier' remained closed, and there was no further room for colonization. Within the restricted area of arable land, moreover, practices were largely unchanged from the traditions of the past, and increased production was obtained either through the refinement of earlier techniques or through the addition of labour. In this respect, Chinese agriculture was a form of large-scale gardening, and the very intensity of human physical input discouraged the introduction of modern techniques: since the farming population still had to be fed, and there was practically nowhere else to go, it was more sensible to use this 'free' work-force than to incur the real costs of mechanization. In this manner, the very success of traditional techniques was a trap: estimates suggest that agriculture required almost 80 per cent of the labour force to produce 65 per cent of total domestic product, and although one could describe China as self-sufficient in providing a minimum subsistence of food, it has also been observed that the major product of a Chinese farm is more Chinese people.

Work on the land was hard and primitive. Water for irrigation was commonly obtained by simple water wheels, frequently run by manpower alone, often by two men slinging a roped bucket between them, hour after hour, to flood the ground for rice or bring water to the wheat. In the south, water buffalo dragged a wooden plough through the mud slurry of paddy-fields to prepare the ground for planting, and rice seedlings were transplanted by hand. In the drier land of the north, oxen and donkeys were used as draught animals, but the plough did little more than scratch the surface of a ground which had been sown and cropped for centuries in the past. Every house had chickens and pigs to provide a protein supplement in the heavy diet of cereal, while garbage and human waste was carried in yoked burdens to maintain the fertility of the soil. Except for planting or harvest, however, there were more hands available than there was work for them in the fields, and it was estimated that only about one-third of able men worked regularly on the land. When their labour was not required, men and women found occupation in home industries, weaving cloth, or small handicrafts.

As far as possible, the peasant family was self-supporting, and an eighth of its income came from non-agricultural production. On the other hand, while there was opportunity for diversification into cash crops such as cotton, tea, and silk, these were vulnerable to market fluctuations. In north China, wheat was often sold to buy poorer food such as sorghum or millet; in the south, cotton impinged upon the land which could be used for rice, and in the far south-west there was particular tragedy as farmers were forced by local warlords to grow opium, exploited for cash while the people starved.

By far the most popular crop was wet paddy rice, grown almost universally in the Yangzi basin and the south, but cultivated also in some areas of the north. Estimates for the period 1931–7 suggest a crop of 70 million tonnes of rice, 23 million tonnes of wheat, grown extensively in the north, together with 45 million tonnes of the less usable sorghum, millet, and barley, while sweet potatoes, the poor man's food of the south, provided the grain equivalent of 7.5 million tonnes.[16] Industrial and cash crops, including soy beans, cotton, sugar cane, tobacco, peanuts, tea and silk, totalled some 50 million tonnes.

Statistics such as these, however, given the circumstances of the time, are inevitably unreliable, and there is considerable debate whether the productivity of Chinese agriculture did in fact increase, remain static, or even decline during the years of the republic. Some scholars believe the increase in food supply largely kept pace with the growth of population, about 1 per cent each year, but others have argued there was a disaster for agriculture as the economic link between the country and the cities was broken, the price of food produced in China became greater than the cost of imports from abroad, and constant market instability, from one cause or another, brought rural misery, unemployment, debt, and the forced sale of land.[17]

Both a major survey by John Lossing Buck and government reports and statistics indicate that though ownership of land was biased towards large holdings by landlords, with a high proportion of tenancy, the situation varied throughout the country and there were fewer large estates and fewer landless men than some propagandists claimed.[18] In the north some 60 or 70 per cent of farmers owned their land, while the provinces of the Yangzi and the south, notably Sichuan, Hunan, Jiangxi, and Guangdong had the greatest number of tenants. In this region, where fertile soil and favourable climate could produce heavy crops and a good return on capital, the peasants who owned their land, the part-owners, and the full tenants were roughly equal in numbers. Over the whole country, the average family held some 15 *mu*, about 2.5 acres or one hectare. On the other hand, almost 75 per cent of families had less than that amount, while 5 per cent of families, with individual holdings over 50 *mu*, controlled 35 per cent of the land. At the higher levels, however, less than 1 per cent of families had estates larger than 150 *mu*, and they controlled only 12.5 per cent of arable land. These larger landholdings were normally rented out to tenants, for there was no widespread system of commercial agriculture based upon hired labour, and no great estates like those which had been developed in Europe or colonized in America.

Indeed, a major problem of Chinese agriculture was the small size of the average farm. Though Buck's assessment of less than two hectares is a most general statement, it can be compared with the smaller holdings of Japan, at one hectare, and the larger figures for Germany, at almost ten hectares, or England, at twenty-five. On such a small scale, farming in China was very inefficient, and the situation was made worse by the division of holdings into small parcels scattered among other properties. These tiny tracts of land were divided from one generation to another in the same family, and aerial photographs of agricultural land, showing a mass of strips of different shades, from different crops at various stages of development, illustrate not only the intensity of cultivation but also the heavy subdivision of these miniature farms.

In theory at least, there is no necessary disadvantage to tenant farming; the important question is the terms upon which the tenancies are held. By the 1930s it was estimated that more than half of all tenants in China paid a proportion of their crops as rent, a quarter paid in cash, and a very small number paid with their labour on the landlord's other fields. The balance was in share-cropping, where the landlord supplied not only the land but also the tools and seed to work it, and the farmer was not much removed from a position of a labourer.

Two matters seriously affecting the position of tenants were short-term tenure and rents paid in cash. In 1930 a special Land Law asserted that a tenant should not be expelled unless the owner proposed to farm the land himself, and like many other laws this should be interpreted rather as a sign of serious trouble than as the achievement of a solution. Most contracts had no fixed duration, and required frequent renegotiation, when the landlord could add conditions to the tenancy and increase the scale of rent. In many districts the tenant was compelled to find a cash deposit at the beginning of each new contract, and he was often in debt for this sum before he had opportunity to work the land.

The increasing proportion of cash rents, rather than payment in kind, reflected a tendency for land to be held by the residents of cities and towns, who looked upon their property as a simple investment and paid no attention to their responsibilities in the community. In the past, local gentry had taken a lead in the life and work of the peasants and villagers, but this tradition was passing rapidly, and such assistance as was received came from government officials and foreign missions, not from the wealthy men of the district. The concept of benevolence from the rich and co-operation from the less prosperous had been honoured in the past perhaps more as an ideal than a reality, but the divisions were now more harsh and clear than they had ever been before, and the landlord's agent and the money-lender were a hated feature of rural life.

In similar fashion, the obligations of taxation also forced the farmer to sell produce for cash. According to Buck's survey, in any one year some 15 per cent of the rice crop and 30 per cent of wheat were sold by the peasant community, while cash crops such as tobacco, cotton, and peanuts were sent to market annually. High transport costs meant that the peasant must sell locally, and this put him at the mercy of large traders. Often the farmer was compelled to sell his grain cheap at the glut in harvest time, and bought seed dear when it was due to be planted in spring. Almost inevitably he had to borrow, and heavy interest made it unlikely he would recover. By 1935 it was estimated that almost half of all rural families were in debt, and the borrowing had been needed to cover running expenses for the household or the farm, not for capital investment. For many people it was uneconomic to work the land.

The government made some attempt to control the situation, but with little success. Nationalist policy stated that rent payments should never exceed 37.5 per cent of the main annual crop, but this was not an enforceable limit, and there were many examples of rents as high as 50 per cent. There was some attempt to encourage the formation of farmers' co-operatives to sell produce on a larger scale and wider market, but the programme had limited application, and it would never be a quick solution.

Most serious of all, the land-tax system, one of the chief sources of revenue for the provinces, bore heaviest on the men who worked the soil: the landlord passed it on as a charge to the rents, and the local officials, commonly in collusion with the local magnates, did nothing to equalize the burden. Without proper control by the government, the highest share was paid by the poorest people, and there was no machinery to control the profiteers.

The farmers' difficulties were primarily a problem of economics in an unsettled, backward, agricultural country, but they were emphasized by the inadequacies of government. People were bound to the land by poverty, with small hope of improvement towards even a reasonable prosperity: in times of hardship many were driven to starvation and slavery, and in times of good harvest they were still oppressed by debt, rent, and taxes. No one could see a proper programme of land reform, and indeed the disruption and loss of confidence which might be caused to an economy already under stress gave the government some excuse for caution and delay.

In the 1930s, however, one observer summed up the prospects: land reform was essential to the people of China, but no government could afford to pay adequate compensation to the property owners, and the Communist Party was the only political group which proposed land reform without compensation.[19] This was a critical test, and unless the Nationalists changed their programme they would never be free of pressure from the radicals. Either by war, revolution, or peaceful politics, some change was inevitable; but both China and the Communists had a long way to go.

Shanghai and the Writers

By the end of the 1920s and the beginning of the 1930s the old traditions of literature and scholarship had been overthrown by the new writing that followed the May Fourth Movement, by the influence of Western and Soviet literature in translation, and by the acceptance of modern techniques. Archeological discoveries, notably those of the Shang dynasty at Anyang by the Western-trained Professors Li Ji and Dong Zuobin, confirmed the nationalistic view of China as an ancient culture indigenous to eastern Asia, and books and articles debated such questions as whether the civilization of China had been founded upon slavery. According to Marx, all society emerged from a slave-based economy to feudalism, capitalism, and socialism, and Marxist theorists, Communist or not, sought to prove his claim. In contrast to the past, classics and histories were now used as material for the study of society rather than for the textual criticism of philosophers, and argument depended on each protagonist's view of contemporary China. Students and professors saw themselves as an independent élite, they discussed and criticized government policy, and they were prepared to demonstrate for their ideals.

Outside the universities numbers of writers, still experimenting with vernacular literature, had wide circulation and influence. One of the earliest, and perhaps the greatest, was Zhou Shuren, a journalist and essay-writer who took the pen-name Lu Xun. Born in Zhejiang in 1881, he studied Western medicine in Japan, where he was appalled by the apparent complacency with which the people of China accepted their weakness and inferior position. His two best-known works, *Diary of a Madman* and *The True Story of Ah Q*, bitterly satirize the corruption of public life, and the futility, weakness and stupidity of the Chinese. With his picture of Ah Q, the good-for-nothing who plans big and talks loud but who can bullied and crushed by anyone who stands up to him, Lu Xun expressed scorn for his fellow countrymen, in the hope he might encourage them to resentment and action.

Lu Xun, who died in Shanghai in 1936, has been claimed by the Communists, and is honoured as chief of all writers, but he was never a member of the Party. In his later years, however, he took part in organizations to encourage writers and artists, and many of his friends and proteges were supporters of the Communist cause. Shen Yanbing, also from Zhejiang, under the literary name Mao Dun, made his reputation with a trilogy of novels about the Nationalists, and his feelings may be summarized by the titles: *Disillusionment, Dilemma,* and *Pursuit.* Shen Yanbing later became Minister of Culture in the People's Republic of China, and produced little of importance after that time, but his novels are admired not only for their political implications but also for their human understanding, their sympathy, and the reality of their characters. Of other prose writers, two of the most influential were Ba Jin, pen-name of Li Feigan, whose novel *Family* pictured the frustrations and tragedy of young people in a large, traditional household, and Lao She, pen-name of Shu Qingchun, whose *Rickshaw Boy* was widely read in China and became an American bestseller in English translation. Leading poets of the time were Wen Yiduo and Guo Moruo, and the younger Ai Qing, born in 1910, whose work expressed the sufferings of the people:

> In the land where once the Yellow River flowed,
> In numberless dry water-courses,
> The wheelbarrow
> With its single wheel
> Lets out a squeal that shakes the mournful sky,
> Piercing the wintry chill and desolation.
> From the foot of this hill
> To the foot of that,
> The sound cuts through
> The misery of the people of China.[20]

Almost inevitably, intellectuals of the republic were opposed to the governments of the day. While the warlord regimes of the first years of the republic had nothing to offer idealists and patriots, the Nationalists who succeeded them with such high hopes could never satisfy full expectations. To any thinking man, the problems of China were too great to be solved by any but radical reforms, and those outside the government were unprepared to accept excuses based on economic and political expediency. So writers formed a natural group of criticism to the regime, and cosmopolitan Shanghai became a centre of opposition. In 1930 the League of Left-Wing Writers was formed, latest of a series of literary societies which had sprung up throughout the 1920s, but now gathering together all those, Communist or not, who claimed the government was too conservative, too complacent, and was betraying the revolution.

Such a view was not entirely fair, for the government did have major problems and was never free from the threat of war and rebellion, but the reaction of Chiang Kaishek and his supporters did little to mend the breach. Persecution of civilian opponents never reached the extremes of Nazi Germany, Stalinist Russia, and imperial Japan, nor the excesses and cruelty of more recent times, but from the end of 1931 there was growing repression, with frequent arrests of writers and scholars on political grounds, and in the same pattern there appeared also a fanatic political group, the Blue Shirts, prepared to support the leadership and the policies of Chiang Kaishek and the Nationalists with parades, demonstrations, and violence. The hostility of the government did little to curb the spread of protest, but it confirmed the educated opposition in their resentment of the Nationalists, and numbers of them changed from friendly critics to outright enemies. Some joined later in the war against Japan, but many chose to go into exile or threw in their lot with the Communists.

It was, indeed, a weakness of the Nationalist Party that it was unable to produce policies and inspiration which would attract support for its programme in government. The propaganda of the 'New Life' movement, chief rallying call of the 1930s, presented little more than a jumble of Confucian and Christian platitudes, and though Chiang Kaishek's Christianity had helped him find acceptance in the West, the appeal of this faith within China was felt only by a small group of the urban middle classes. Similarly, the Blue Shirt movement, copied from the fascism then popular in much of Europe, was equally irrelevant to the mass of the Chinese people.

Again, the site of the Nationalist capital created difficulties. Formerly one of the greatest cities of the empire, Nanjing had been brought to ruins by the Taiping rebellion, and was largely by-passed by modern development and trading prosperity. Despite its rank as capital of the republic, Nanjing was little more than a provincial town, with few fine buildings and no recent cultural tradition. The government there was isolated from the peasant base of the Chinese people, but it was also difficult for the rulers to understand, to recognize or to influence the intellectuals and men of business in other cities, particularly in Shanghai.

For all the region under Nationalist control, and indeed for China as a whole, the metropolis of Shanghai was the most powerful economic entity and the most influential community. It was not surprising that Shanghai became the centre of opposition, just as it had formerly been the headquarters of the Communists and revolutionary workers, for it was to Shanghai that ideas came from the West, and it was in Shanghai that the two systems of life and thought met and clashed most bitterly.

Though the alien enclaves on Chinese soil remained a source of resentment to every patriot, the comparative security of the market and the convenience of Western-style commercial law had ensured the prosperity of Shanghai, and many refugees from China, including Sun Yatsen, had cause to be grateful for the sanctuary offered by foreign administration. By the 1930s the city had reached the zenith of its prosperity and power, and new buildings towered over the famous Bund, first constructed as a dyke to hold the left bank of the river, but soon famous as one of the great merchant boulevards of the world. Besides the traders and their captains and crews, the streets of Shanghai were crowded with people from every part of the world: missionaries from Europe and America, soldiers from Japan, White Russian refugees from Manchuria and the old empire of the tsars. On the other hand, the thoughtless benevolence which gave shelter to the exile did nothing to supply him with food or shelter, and many who came to Shanghai found only hunger and destitution in the streets. It was here that some poor whites were forced to make their living through prostitution; it was here that the legendary notice was displayed in the gardens, 'Dogs and Chinamen not allowed', and it was here that fat foreigners sat in rickshaws and compared the form of the men who hauled them: 'When you hear your boy pant really hard, you know you've almost reached the Club.'

For the Chinese of Shanghai, there were drugs and gangsters, smuggling and murder, but for those of some means there was also the opportunity to enjoy and to share in the culture of the West. Beside the books, magazines, and newspapers, Chinese and foreign, there were art exhibitions and an embryo film industry, ballet dancing in the schools, and cocktail bars in the hotels. Film actresses and dance bands had pin-up posters and a personal following, and when Nanjing launched the 'New Life' campaign against corruption and moral danger, its obvious target was the public dance-halls of Shanghai. Censorship sometimes extended also to theatres and bookstalls, but the Nationalists, no matter what they may have hoped for, could never enforce their rules to any full extent, and Shanghai, like the rest of China, remained largely a law to itself.

Ultimately, the disappointment in Nationalist government was that so much of the administration was ineffectual. Often this was due to outright corruption, and gangster influence at the highest level was well recognised, but it was frequently through inefficiency and practical weakness that the Nationalists failed to enforce the enlightened laws they placed in the statute books, or to reform the structure of the country. Above all, they seemed to have lost the driving ambition for change which had first brought them to power: they accepted the structure of power in the cities, they supported the landlords in the countryside, they accepted the alliance of warlords in the provinces, they neglected the human needs of the people, and they left radical slogans to the Communists. No longer were they men in a hurry, and they often acted as if the unity of the nation was assured, and social change, like capital construction, would come slowly and gradually. Given time, and peace, they might have been correct, but there were enemies both at home and abroad, and neither the rulers nor the people of China had time to spare.

Notes

1. Besides twelve full delegates there was also one alternate delegate from Guangzhou and two Russian advisers. The initial gathering on 1 July took place at the house of a sympathizer in the French Concession of Shanghai, with some delegates from out of town lodged in a girls' boarding school around the corner. Part way through the meeting, there was warning of an intended raid by the French police and the group scattered, renewing the meeting a few days later on a rented pleasure boat at Jiaxing, south of Shanghai towards Hangzhou.

  Neither Li Dazhao nor Chen Duxiu was present, but they were recognized as founders of the party, and Chen Duxiu was appointed head of the Central Committee.

2. The dragon flag of the Qing dynasty was replaced in 1911 by a new design of horizontal stripes, red, yellow, blue, white, and black, representing each of the five groups, Chinese, Manchu, Mongols, Tibetans, and Muslims, which now composed the people of the new republic. The badge of a white sun on a blue background had been adopted by Sun Yatsen for his revolutionary party as early as 1895, and it is said to have been designed by his friend Lu Haodong who was killed in the unsuccessful rising at Guangzhou in that year. In 1906 Sun established the flag of revolution as red, with the blue sky and white sun in one quarter. In 1912 this was adopted as the naval flag of the republic, while a red flag with a gold star outline, which had been used by the rebels at Wuhan on 10 October, was taken by the army. Later, however, Sun Yatsen reclaimed the revolutionary flag for his government in Guangzhou, and on 8 October 1928 it was adopted as the national flag of China. It is still the symbol of the government in Taiwan.

  The present flag of the People's Republic of China, red with five gold stars, one large and four small, in the corner, was proclaimed as the national flag on 21 September 1949. For most of the civil war, however, from the earliest days of the Jiangxi Soviet in the late 1920s, the Communists used a red flag with a black hammer and sickle on a single yellow star in the centre, which was the banner of the international Communist movement.

3. The Institute was established in a former Confucian temple which had been nationalized and turned into a school. During the 1970s it was maintained as a memorial to Mao's work and was a notable tourist attraction. More recently, however, it has been removed from the regular schedule.

4. The foreigners were encouraged in their negotiations by the city gangs, led by

Chiang Kaishek's former patron Du Yuesheng of the Green Gang, and their arguments were supported by vast bribes.

5. This name commonly appears in transcription as Peiping. In 1949, with Communist victory in the civil war, the city became the capital of the People's Republic of China, and the name was changed back to Beijing. To avoid unnecessary complication, in the course of this work I refer to the place consistently as Beijing.

6. Remer, *Foreign Investments in China,* and Hou, *Foreign investment and economic development,* quoted and discussed in *Cambridge China* 12, pp. 116–20 [Feuerwerker, 'Economic trends'].

7. Figures based on Chinese Customs reports, quoted by Cheng, *Foreign Trade and Industrial Development of China,* pp. 32–4, and *Cambridge China* 12, pp. 121–7 [Feuerwerker, 'Economic trends'].

8. Survey by Liu Ta-chün (D.K. Lieu), published at Nanjing in 1937, and discussed by *Cambridge China* 12, pp. 57–60 [Feuerwerker, 'Economic trends'].

9. On handicraft industries, see *Cambridge China* 12, pp. 51–7 [Feuerwerker, 'Economic trends'], discussing, *inter alia*, Liu and Yeh, *The Economy of the Chinese Mainland.*

10. Perkins, 'Growth and changing structure of China's twentieth-century economy', in Perkins (ed.), *China's Modern Economy in Historical Perspective,* pp. 116–25, discussed in *Cambridge China* 12, pp. 37–41 [Feuerwerker, 'Economic trends'], and see Sheridan, *China in Disintegration,* pp. 224.

11. For example, Eastman, *The Abortive Revolution*, who finds stagnation, and Rawski, *Economic Growth in Prewar China,* who speaks of useful progress. Chang, *Industrial Development in Pre-Communist China,* presents an index of industrial production which is widely accepted: e.g. *Cambridge China* 12, pp. 49–50 [Feuerwerker, 'Economic trends'] and 824 [Bergère, 'Chinese bourgeoisie'].

12. In 1930, for example, imports were valued at over 1,300 million Customs Taels and exports, allowing for under-valuation, at less than 900 million Taels. Remittances from overseas Chinese were estimated at 210.9 million Taels and expenditures by foreigners, for legations and embassies, gunboats and garrisons, merchant ships and missionaries, totalled 145.4 million Taels. Remer, *Foreign Investments in China,* pp. 221–2, quoted by Cheng, *Foreign Trade and Industrial Development of China,* pp. 259–60, and discussed by *Cambridge China* 12, pp. 121–5 [Feuerwerker, 'Economic trends'].

Figures such as these are based on customs statistics, though there was also a flourishing smuggling trade importing dutiable goods and exporting bullion. Estimates of this activity naturally vary, but the value was perhaps 5 per cent of the official trade.

On the Customs Tael see note 15 to Chapter 3. It remained the official unit for overseas trade valuation until 1930, with value at par with the new silver Chinese Standard dollar of $1.50.

13. Most of these loans were owed overseas, but despite the natural resentment which any Chinese might feel against such exploitation, the Nanjing government was in no position to repudiate the debts. For future development, China relied on international credit and further loans, while there was well-founded anxiety that some powers would enforce their rights, if necessary by war. The government negotiated to improve the terms of tariff agreements and similar matters, but there were limits to the tolerance which foreigners would show for a wholesale rearrangement of their sources of profit.

14. Figures from the annual *China Year Book*, and other tabulations, cited in *Cambridge China* 12, pp. 106–7 [Feuerwerker, 'Economic trends']. The Nationalist government at this time was still in the process of currency reform, but the exchange rate of the new Chinese Standard dollar was approximately 1s. 3d. sterling or 22 cents US.

15. *Cambridge China* 12, pp. 99–100 [Feuerwerker, 'Economic trends']. The comparable figure in Australia at that time was 30 per cent, and in America, under Roosevelt's New Deal administration, it was almost 20 per cent.

16. From Perkins, *Agricultural development,* pp. 266–89, discussed in *Cambridge China* 12, p. 66 [Feuerwerker, 'Economic trends'].
17. Compare, for example, *Cambridge China* 12, pp. 64–5 [Feuerwerker, 'Economic trends'], citing Perkins, *Agricultural development,* with *Cambridge China* 13, pp. 256-60 [Myers, 'Agrarian system'].
18. Land holding and tenure is discussed by *Cambridge China* 12, pp. 76–87 [Feuerwerker, 'Economic trends], with citations from John Lossing Buck and other more recent works. Buck's research is still a chief source of information for this period. Chinese scholars have argued recently, however, that his investigations were distorted by being limited for the most part to territory under Nationalist control, and also because many of his field-workers were university students drawn from the landlord classes.
19. Fitzgerald, *The Tower of Five Glories,* p. 219.
20. 'Wheelbarrow', in *Selected Poems*, pp. 47 and 273.

Further Reading

On government:

Eastman, Lloyd C. *The Abortive Revolution: China under Nationalist rule, 1927–1937*, Harvard University Press, 1974.
Jordan, Donald A., *The Northern Expedition: China's national revolution of 1926–1928,* Honolulu, Hawaii University Press, 1976.
Seagrave, Sterling, *The Soong Dynasty*, New York, Harper and Row, 1985.
Sheridan, James E., *China in Disintegration: the republican era in Chinese history,* New York, Free Press, 1975.

On the general economy:

British Naval Intelligence Division, B.R. 530, *China Proper*, 3 vols, 1944–45.
Feuerwerker, Albert, 'Economic trends, 1912–49', in *Cambridge China* 12.
~~odhead, H.G.W., and others (eds,), *The China Year Book* [annual from
No. 1 (1912) to No. 20 (1939)], Kraus Reprint.

~~ulture:

Lossing, *Chinese Farm Economy*, Nanjing, 1930, and *Land
~~n in China*, Nanjing, 1937.
H., *China: Land of Famine*, New York, International Famine
~~ission, 1926.
'The agrarian system', in *Cambridge China* 13.
~~ricultural development in China*, 1368–1968, Chicago,

~~elopment in Pre-Communist China: a quan-
!dine, 1969.

Cheng, Yu-Kwei, *Foreign Trade and Industrial Development of China,* Seattle, University of Washington Press, 1956.

Coble, Parks M. "The Kuomintang Regime and the Shanghai Capitalists," in *China Quarterly* 77 (London, March 1979).

Hou, Chi-ming, *Foreign investment and economic development in China 1840–1937,* Harvard University Press, 1965.

Liu Ta-chung and Yeh Kung-chia, *The Economy of the Chinese Mainland: national income and economic development, 1933–1949,* Princeton, Princeton University Press, 1965.

Perkins, Dwight H. (ed.), *China's Modern Economy in Historical Perspective,* Stanford, Stanford University Press, 1975.

Rawski, Thomas W., *Economic Growth in Prewar China,* Los Angeles, California University Press, 1989.

Remer, C.F., *Foreign Investments in China,* New York, Macmillan, 1933.

On foreign experience and reminiscences:

Crow, Carl, *Foreign Devils in the Flowery Kingdom,* London, Hamish Hamilton, 1941.

Fitzgerald, C.P., *The Tower of Five Glories,* London, Cresset, 1941.

Fitzgerald, C.P., *Why China? recollections of China 1923–1950,* Melbourne, Melbourne University Press, 1985.

Spence, Jonathan, *To Change China: Western advisers in China 1620–1960,* [first edition 1969], Penguin, 1980.

Wei, Betty Peh-t'i, *Shanghai: crucible of modern China,* Hong Kong, Oxford University Press, 1987.

Two Western works of fiction which present an atmosphere of the time:

Bridge, Anne, *Peking Picnic,* London, Chatto and Windus, 1932, reprinted 1962.

McKenney, Richard, *The Sand Pebbles,* New York, Gollancz, 1963.

On society and literature:

Ba Jin, *Jia* [Family], Beijing, Foreign Languages Press, 1978.

Chiang Yee, *A Chinese Childhood* [third edition reprinted] London, Methuen, 1953.

Ai Qing, *Selected Poems,* edited by Eugene Chen Ouyang, Beijing, Foreign Languages Press, 1982.

Selected Stories of Lu Xun, translated by Yang Hsien-yi and Gladys Yang, Beijing, Foreign Languages Press, 1980.

C.T. Hsia, *A History of Modern Chinese Fiction,* Yale University Press, 1961

Lee, Leo Ou-fan, *The Romantic Generation of Modern Chinese Writers,* Harvard University Press, 1973.

On society:

Bergère, Marie-Claire, 'The Chinese bourgeoisie, 1911–37', in *Cambridge China* 12.

Fei Hsiao-t'ung [edited by M.Park Redfield, *China's Gentry*, Chicago University Press, 1953 [Professor Fei, trained in the United States as a sociologist, later became a leading scholar under the Communists. This early work describes the society at the end of the republic, and includes sample life- histories compiled by Yung-teh Chou].

Holmgren, J., "Myth, Fantasy or Scholarship: Images of the Status of Women in Traditional China", in *Australian Journal of Chinese Affairs* 6 (Canberra, 1981).

Lin Yueh-hwa, *The Golden Wing*, London, Kegan Paul, 1984 [anthropology, in fictional form, on a traditional family under the republic].

Wolf, M. and Witke, R., [eds.], *Women in Chinese Society*, Stanford, Stanford University Press, 1975.

5

The Enemies of the Nationalists

THE defeat of 1895 established the military weakness of China against Japan, and the imbalance of power continued for fifty years. This known superiority had enabled the Japanese to humiliate and destroy Yuan Shikai, it had brought them approval at Paris for the seizure of former German possessions in Shandong, and it allowed them to gain effective control of Manchuria. There were frequent protests and boycotts within China against Japanese economic influence and imperialist aggression, but the Nanjing regime was in no better case than its predecessors to face the threat. In 1928 the local Japanese army contemplated the full take-over of Manchuria. In 1931 they realized that ambition, in the following year they established a puppet government, and in 1933 they forced their way into north China. On each occasion the Nationalists were also attacked elsewhere, first at Jinan in Shandong and then twice, most bloodily, at Shanghai.

Chiang Kaishek, however, was primarily concerned with the rebellion inside China. Holding the Japanese as best he could, he sought most particularly to remove the Communist regime established by Mao Zedong in the south of Jiangxi province. After campaigns of increasing severity, the Communists were driven from their base, but in the Long March from 1934 to 1935 they travelled to refuge in the north-west. From that isolated territory they held off the local Nationalist armies, and called for a united front against Japan. In a mutiny at Xi'an in December 1936, Chiang Kaishek was captured and compelled to negotiate, and despite his justified anxiety for the outcome, he committed the Nationalists to open defiance of Japan.

The Communists in Jiangxi

When Chiang Kaishek came to open hostilities with the Communists in 1927, the failure of the Guangzhou Commune in December marked the end of the urban-based party as a real threat to the new government. The fighting was bitter, but the Communists had only limited support, and groups of workers frequently aided government troops in the destruction of local insurrections. Though the Nationalists had many faults, and though much of their policy was ineffective, they were perceived as allies by the city-dwellers of China, and they still held their approval.

Throughout the countryside, however, the pressures, of debt, rent, and taxation remained, and there was small evidence of government concern. During the days of the Northern Expedition, Mao Zedong had argued there was wide resentment and desire for change, and in September 1927, as the official leadership of his party attempted rebellion in the cities, Mao led the Autumn Harvest Rising in Hunan. Again, however, support was limited and uncertain, the military reaction of the government was swift and effective, and the rising was as complete a failure as those of the cities. Mao was expelled from the Politburo, governing body of the Chinese Communist Party,[1] for his 'military opportunism', and he fled with a few followers to the Jinggang Mountains on the Hunan-Jiangxi border.

Early in 1928, he was joined by Zhu De, a leader of the ill-fated Nanchang mutiny, and they combined forces to form the Fourth Red Army, with Zhu De as commander and Mao as political commissar. With some vicissitudes, they established a base in the border region, and in July they moved to Ruijin, in the south of Jiangxi near Fujian, and established a soviet government. They maintained links with other Communist groups in similarly inaccessible territory, and for the time being these small-scale fighting units operated independently of the central organization in the cities.

In the countryside which they controlled, Mao Zedong and his comrades gained remarkable success. At first the Communists led attacks on all men of property in the villages they dealt with, and reallocations of land stripped even middle-rank peasants of their possessions and frequently entailed mass execution. Soon, however, policies became more moderate: absentee landlords were dispossessed as a matter of course, but for those who farmed their own land the Communists enforced only a serious redistribution, not total confiscation. Since the very presence of the Communists meant a state of civil war against local administrators and supporters of the Nationalist government, there was fighting, bloodshed, and murder on both sides, but the Communist rule in Mao Zedong's area was so restrained that he was criticized by main-line leaders in the cities as a right-wing deviationist.

During the first years in Jiangxi, Mao's chief problem was not the disapproval of Communist theorists but survival against attack from the Nationalist army. Through 1929 and most of 1930 he was able to remain independent of the politics of the centre, and he was faced only by local resistance. In November 1930, however, having crushed the Communists in the cities and having also, by compromise with Wang Jingwei, obtained brief respite from the disputes of politicians and warlords, Chiang Kaishek launched the first of a series of campaigns against the rebel strongholds. Attacked on their home territory, the Communists used mobility and knowledge of the country to defeat the first two campaigns in very short order. In July 1931, however, in the Third Campaign of Encirclement and Extermination, Chiang Kaishek took personal command of 130,000 men, well armed and trained by German advisers, and although they suffered some defeats they maintained steady progress. In September, however, the Japanese attack on Manchuria compelled a halt, and the Communists remained in possession of the field. Preoccupied with the north and east, and also with a rebel regime in Guangzhou, the Nationalist forces did not return to Jiangxi for more than a year.

Mao Zedong and his people continued to expand their territory and consolidate control, and, as the Nationalists withdrew their offensive, Mao Zedong proclaimed his regime as a Soviet Republic, an alternative government for the whole of China. On 7 November 1931, he opened the First All-China Soviet Congress, and from his position of strength he invited the leaders in the cities to attend.

Despite the disasters of 1927 the rulers of the Communist Party and their advisers in the Comintern at Moscow had continued emphasis on the industrial proletariat in Shanghai, Wuhan, and other centres. In this policy they were encouraged by Stalin, and under his authority the new party leader in China, Li Lisan, stirred up strikes and sabotage. In July 1930, during a civil war between factions of the Nationalists and their warlord allies, Li Lisan sent units of the Red Army under Peng Dehuai to seize Changsha, capital of Hunan province. Once again, however, Chiang Kaishek's forces were well-prepared to deal with an enemy who faced them so foolishly in the open, and the Communists were driven out with heavy losses.[2]

Li Lisan was dismissed, and he was succeeded by a new group of young men in their early twenties, Wang Ming, Bo Gu,[3] and others, who had lately returned from training in Moscow. Commonly described as the Twenty-eight Bolsheviks, these newcomers were far better attuned to the wishes and policy of Stalin and the Communist International, but they were quite inexperienced in the realities of China. They made no headway against the government, they were seriously harassed by the secret police, and in June 1931, after a series of successful raids, the Nationalists captured and executed the secretary-general of the party, Xiang Zhongfa.

It was at this low point in their fortunes that the Communists of the cities were invited to the Congress of Soviets in Jiangxi, and the Twenty-eight Bolsheviks found themselves outvoted and almost disregarded. With complete support from his own people, and with allies among those out of office in the cities, Mao became Chairman of the Central Executive Committee of the new Soviet Government. The Politburo leaders were almost totally excluded, but they refused to recognize Mao's authority. As the man of practical power, he was head of the new government, but he had no status in the national Communist Party.

Potentially, this was a danger, for Mao's position was based on personal control in Jiangxi, and no matter how well they were weighted in his favour the committees established by the Congress circumscribed his authority. Even in earlier years there had been discontent among the Communists in Jiangxi, and in the Futian Incident of December 1930 there was open rebellion within the Red Army which was put down only at the cost of two or three thousand lives. Success against the Nationalists in 1931 restored Mao's position, but rivalry remained, and during 1932 there was a gradual erosion of his influence. During a conference at Ningdu in August, Mao was criticized on both practical and theoretical grounds, and some of his authority over the army was transferred to Zhou Enlai, who had come to Jiangxi from Shanghai with the Twenty-eight Bolsheviks, and who supported their aggressive military policy. In May 1933 Mao Zedong's vital post, Commissar to the Red Army, was taken from him and Zhou Enlai took his place. From the point of view of internal politics, Mao had lost his link with his military supporters; and at the same time, in external policy, the Soviet was set on a course of rigid resistance and strict pursuit of Comintern theory.

The change could hardly have come at a worse time. In January 1933 a fourth Nationalist campaign was interrupted by further Japanese invasion in the north, but in May the Nanjing government patched up a peace and turned its full attentions once more against the Soviet. The new line committed the Communists to stiff and inflexible military strategy, and politically, though Mao Zedong had offered to join the Nationalists in a united front against Japan, the new leadership rejected such compromise and so weakened their appeal to the rest of China. In Jiangxi itself the Communists now enforced strict redistribution of land: where well-to-do peasants had formerly been permitted to retain some of their former property, the new leadership insisted they should be attacked and dispossessed, and poor peasants and landless men were encouraged to join in class warfare. Apart from questions of ideological purity, the funds obtained from confiscations went to strengthen the Red Army but, far more dangerous for the future of the Soviet state, the new policy caused bitter disturbances in territory already under control, and alienated many who had been prepared to accept the expansion of Communist rule. When the Nationalists came in force to Jiangxi in 1933, people who had been passive in the earlier campaigns welcomed them as liberators.

In fairness to the new leadership, it is doubtful whether any administration or military defence could have survived the onslaught of the Fifth Campaign of Encirclement and Extermination. The preliminary moves began in October 1933, but where Nationalist forces in Jiangxi had formerly numbered only some 150,000 men, the new attack was brought with 700,000, generally well trained and supported by heavy weapons and a small, but unchallenged, air force. The Nationalist armies were the most powerful military force within China, and this time they suffered few distractions to their operations against the south. Learning from past mistakes, and benefiting from the Communists' loss of local popularity, Chiang Kaishek and his officers consolidated each advance with military fortifications and improved lines of supply, and they organized the people into small-scale political units, *baojia*, with mutual responsibility between families and villages both to encourage their own defence as militia and to ensure a network of informers and hostages.

At the beginning of the campaign, the Jiangxi Communists had one offer of alliance which might have distracted their enemy: in September 1933 a small rebel government was established in neighbouring Fujian, drawing support from the disaffected Nationalist army which had faced the Japanese in Shanghai earlier in the year, and inspired by protest at the authoritarianism and corruption of the regime in Nanjing. The leaders of this gallant band of idealists offered alliance against Chiang Kaishek and against Japan, and they sent letters to leaders in the West, statesmen and philosophers, asking their support for democracy in China. Predictably, the practical men overseas paid no attention, and the Communists in Jiangxi, after some hesitation, reverted to their strict line and refused a united front. The People's Revolutionary Government at Fuzhou, formally established in November 1933, was extinguished by the Nationalist armies in January

1934. Short-lived and impractical though it appeared at the time, it was the last attempt to achieve a political movement in China which would be neither militaristic, authoritarian, nor Communist.

In the meantime, as the Nationalists pressed against the territory of the Jiangxi Soviet, Mao Zedong was in political trouble. In January 1934, though he still remained Chairman of the government, the central executive committee was taken over by the Twenty-eight Bolsheviks, and one of them, Zhang Wentian, became chief of the commissars and practical head of the administration. For some time, Mao Zedong remained as a figurehead, but in July the pretence of respect was ended. By orders of the Comintern in Moscow, he was criticized and accused; he was demoted from his offices and barred from party meetings, and for the next few months he was held alternately under house arrest or in close confinement.

From some of his reported sayings, it is possible that Mao's quarrel with the new regime was based rather upon natural rivalry than on any major disagreement in policy, and he is said to have claimed that the Communists were strong enough to withstand the Nationalist assault. With his enemies in control, however, he was in no position to say much else, and in fact the policies of the Communists in the first months of 1934 were quite disastrous. The army, now largely controlled by Zhou Enlai, exhausted its strength in unsuccessful attacks on Nationalist strong points, and these defeats were accompanied by ruinous losses of men and material. The Nationalists enforced a strict blockade of the Communist areas, and in the summer of 1934 there was serious shortage of salt and an enormous rise in the price of food. By the middle of the year it was clear the only hope of survival lay in retreat and flight and in the middle of October, after some months of preparation, the Long March of escape began.

The Long March

The main body of the refugee force numbered about 100,000: 85,000 soldiers and 15,000 government and party officials. Several smaller groups were ordered to join the escape, and a large party, many of them chosen because they were out of favour with the ruling clique, was left as rearguard. On 15 November the Nationalists captured Ruijin, and in March 1935 Mao Zedong's brother Mao Zetan was killed in a mopping-up operation.

Though the only route of escape lay to the west, the final destination was uncertain and there seemed small hope of survival. Some

Map 4 The Long March

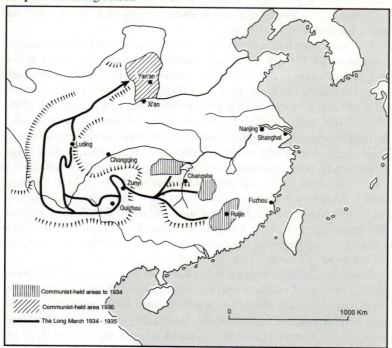

Communist-held areas to 1934

Communist-held area 1936

The Long March 1934 - 1935

0   1000 Km

have described the Communists slipping away unseen through a ring of Nationalist armies, but many units sought to break through and were smashed to pieces in the process. For several weeks, though the Nationalists were unable to swing all their armies in pursuit, the main body of the Communists, slow and clumsy, continued to suffer heavy loss as they fled across the mountains into southern Hunan. Crossing the Xiang River is said to have cost the Red Army half its effective strength, and the marchers fled west again into the wild terrain of Guizhou. The Wu River was crossed only through the gallantry of volunteers who fought to hold a bridgehead and silence the enemy forts which opposed them, and the market town of Zunyi, first resting place for the refugees, was seized by trickery and false flags.

In January 1935, in conference at Zunyi, the failure of Mao Zedong's rivals was confirmed. Zhou Enlai is said to have denounced his own errors; the secretary-general Bo Gu was dismissed and replaced by Zhang Wentian; and Mao Zedong was restored to the Politburo, and became chairman of the command team for future operations.

For more than half a century, the Long March has been acclaimed as a triumph for the Communists and, unlike many other great themes of history, the story of the march is equal to the emotions which it arouses in memory and retrospect. Beginning with a series of disasters, it ended with a grim struggle against odds, with feats of heroism, hardship, and ultimate survival similar to those the British celebrated at Dunkirk. Marching through deep forest, among humid mountains and valleys, following small tracks away from main roads, and crossing deep gorges and rushing streams with rafts and small boats and bridges of rope and wood, Mao Zedong and his followers twisted away from their pursuers. Skirting the wild uplands of west China and the borders of Tibet, they passed the upper reaches of the Yangzi and struggled over the great ridges of the Qinling Range to the comparative haven of Shaanxi province in north-west China.

Some incidents of the march appear more like a pilgrim's progress than a reality of military and political history. The soldiers of the Red Army crossed mountain peaks covered with snow; they made peace with non-Chinese tribes of Yunnan by oaths of brotherhood and animal sacrifice; and they gained respect from the people by their honest dealing and their ready payment for goods with silver dollars. In some areas, however, the invasion was bitterly opposed. In northwestern Sichuan the army passed through trackless marsh known as the Grasslands, where the men were surrounded by quicksand and bog and many, with one false step, disappeared forever. In that region too, the mountain tribesmen hurled rocks as they came through the passes, or shot at them from hiding along the way. On only one occasion did they contend with the main force of Nationalists, but in May 1935, at the crossing of the Dadu River near Luding in central Sichuan, the whole Communist force faced destruction, and Chiang Kaishek set headquarters at Chongqing to see the battle through. After a forced march by one army of eighty miles in twenty-four hours, and a front assault along a cable bridge in the face of machine-gun fire, the crossing was seized and the army escaped the trap.

Soon after this, one large group under the veteran leader Zhang Guotao, who had quarrelled with Mao, turned west in the hope of establishing a base on the edges of Tibet, but Mao Zedong and his forces continued north to southern Gansu. In October 1935, as they crossed the ridges of the Liupan Mountains and came into Shaanxi from the west, Mao Zedong composed one of his most famous poems:

> Lofty the sky
>   and pale the clouds
> We watch the wild geese
>   fly south till they vanish.
> We count the thousand
>   leagues already travelled.
> If we do not reach
>   the Great Wall we are no true men.
> High on the crest
>   of Liupan Mountain
> Our banners billow
>   in the west wind.
> Today we hold
>   the long rope in our hands.
> When shall we put bonds
>   upon the grey dragon?[4]

Mao had with him at this time no more than some seven or eight thousand men, all exhausted and many sick from starvation, dysentery, and exposure. Of the one hundred thousand that began the Long March twelve months before, less than a tenth reached the northwest. For some time there had been a small Communist base in Gansu and Shaanxi, but the territory was not secure, and the newcomers were alien to the local people. On the other hand, though the forces of the Nationalists could ultimately have destroyed them, the advantage of time was now on the Communist side. The 'grey dragon' which Mao referred to was the empire of Japan, and by 1936 the threat had become so great that Chiang Kaishek was compelled to change policy. Mao Zedong was a genuine patriot and a consistent supporter of alliance among all Chinese against Japan, but he achieved survival and triumph through Japanese aggression against his enemies in Nationalist Nanjing.

The Japanese at Mukden and Shanghai

We have noted earlier how the armies of Chiang Kaishek were faced and defeated at Jinan in 1928.[5] The confrontation was ordered by the elected government at Tokyo, who were with reason afraid that united China might threaten Japanese interests in Asia, and their officers carried out the task with enthusiasm. Not only did they attack the Nationalists and drive them from the city, but when the Chinese sent a party of senior officials under a flag of truce the envoys were seized by Japanese soldiers, and every man was shot.

The Jinan Incident set a pattern for the years that followed: whether its government was democratic or militarist, the policy of Japan towards China was one of deliberate intimidation coupled with total ruthlessness and indifference to the people. In the modern history of Japan, the years of the 1930s are the 'Dark Valley' when economic depression and political weakness brought aggressive extremists, formerly just tolerated by the government, into real power. For the Chinese people, who suffered the full onslaught of the invaders' brutality, there may be some justice in the massive destruction that their persecutors brought down upon themselves, but for more than ten years, until war in the Pacific consumed the empire of Japan, China faced that heavily-armed neighbour alone in a long one-sided fight.

By the end of the 1920s there was no comparison between the might of Japan and the weakness of China. China still achieved most production by traditional techniques, and had no more than minimal capital and modern technology to develop her economy. Japan was already one of the most powerful nations of the world, with an advanced industrial base, effective financial machinery, and a highly skilled labour force. The depression of 1929 was a blow to the country's prosperity and confidence, but it strengthened the hands of nationalists and lowered the prestige of constitutional government. With world decline in trade, and particularly the fall of silk exports to America, there came great hardship to many small farmers, and much of Japan's regular army was recruited from these rural groups. Under pressure from ministers of the army and navy, anxious to keep their juniors reasonably content lest their frustrations break into mutiny, the government continually expanded its armaments and military strength.

It was clear to all observers, and later events proved their calculations correct, that the army, navy, and air forces of the Nanjing government were no match for the Japanese, but during the first years of Nationalist rule the main centre of conflict lay not in China proper but in Manchuria and the north. In 1895 Japan had defeated the Chinese empire in this region and claimed a sphere of influence, and ten years later, in the war with Russia, the Japanese confirmed their interests. To many Japanese, Manchuria was the land they had fought for twice in living memory, and since Japanese investment had inspired its development it was only right the Japanese people should control the country's future.

Indeed, from the end of the Russo-Japanese War, the dominance of Japan over Manchuria grew steadily more effective. An occupation force was established to guard Port Arthur and the South Manchurian Railway. Its first headquarters were at Dairen, but they transferred to Mukden, capital of the region, in 1928, and from there, almost independent of control by the government in Tokyo, the officers of the Kwantung Army planned the expansion of the Japanese empire.[6]

Their immediate opponent was Zhang Xueliang, the Young Marshal, who had taken over from his father after Zhang Zuolin was killed in the bomb explosion on the Beijing–Mukden railway in 1928.[7] Zhang Zuolin had held power with Japanese support, but the manner of his death, widely and accurately attributed to Japanese agents, made it clear that the alliance was ended. Zhang Xueliang turned to the Nationalists, and in the time at his disposal he attempted to establish a position in Manchuria which might counter-balance Japanese power.

Although he was an opium addict, and notorious for the number and variety of his concubines, Zhang Xueliang possessed real political talent. Given the military situation, however, he had no hope of success. The Japanese soldiers dominated the region from their posts in the cities and along the railway, and their equipment and training was well above the standard of any warlord army. For a few months in 1929 Zhang Xueliang and the Nationalists attempted to extend their authority by taking over the Chinese Eastern Railway, in the northern half of Manchuria, but the strong Soviet reaction, with fighting on the frontiers and around the city of Harbin, compelled them to back down, and the only result of the exercise was to remove any chance of support from Russia and to encourage the Japanese by signs of weakness.

On the other hand, in the years since the Russo-Japanese War and the fall of the Qing, Manchuria had received a vast flow of immigrants from China, and progress owed as much to the new settlers as to the influx of Japanese capital and enterprise. From 1900 to 1930 the population doubled from 17 million to 34 million: there were 250,000 Japanese, 750,000 Koreans, many thousands of Russians, Mongols, and others, but the remainder were all Chinese, generally from the neighbouring provinces of Hebei and Shandong, and all regarding themselves as citizens of the unified republic.

For his own part, Zhang Xueliang attempted to divert some of the profitable overseas trade from the main South Manchurian Railway and the Japanese-controlled Port Arthur to secondary Chinese-held lines and smaller alternative ports, and he also established a military airfield and a naval base. In 1931, however, when the Chinese captured and shot a Japanese intelligence officer as he was travelling in disguise, and accusations were made that the Japanese were sending agents to cause trouble in Mongolia, the Kwantung Army decided to attack.

On 18 September 1931 there was a bomb explosion near the railway station at Mukden. The damage was minimal, but Japanese troops accused Chinese soldiers of firing upon them, and they forced their way into the city. The attacks continued, and by 21 September their army controlled all the cities of southern Manchuria. In November the Japanese moved north, and by the end of the year they had eliminated all but the last remnants of Chinese resistance. In March 1932, with the whole area under Japanese control, the victors proclaimed the new state of Manchukuo.

Inside China, as Zhang Xueliang collected his scattered forces at Beijing, there was natural fury, with demonstrations, boycotts of Japanese goods, and demands that the government should fight. Prudently, Nanjing took no such action: much of its military force was involved with the campaign against the Jiangxi Communists and with a rival government of Wang Jingwei in Guangzhou, and years of feuding and civil war had put the country in no better shape to fight the Japanese. Instead, the Chinese turned to the League of Nations, and the council of the League resolved that Japan should withdraw her troops to their original lines of occupation. Predictably, Japan refused.

In fact, the government at Tokyo was in serious straits. Even before the Mukden Incident and the campaign of conquest that followed, the emperor himself had given orders against it. The instructions were sent to Manchuria, but the message was delayed by accident or design, and the government was presented with an accomplished fact. Prime Minister Wakatsuki and his civilian colleagues, with mounting criticism from foreign powers, attempted to explain things away and even tried to cut off funds from the Kwantung Army. In this they failed. The government itself fell in December 1931, and the nationalist extremists of Japan, buoyed by such easy victory, confirmed their country in the policy of aggression. In May

1932 the last elected premier, Inukai Tsuyoshi, was murdered in his official residence by a group of army and navy officers and cadets, and Japanese politics began a thirteen-year period of military-dominated government, interrupted by occasional assassinations, and ended only by defeat in war.

There have been claims that Chiang Kaishek was so anxious to avoid conflict with Japan that he ordered Zhang Xueliang to offer no resistance. Certainly, his policy during these years emphasized the suppression of rebellion at home, and overseas he sought only to show China as the victim of aggression. It might have been better for his country and his party if he had fought an all-out war and rallied the people, as he was to do in later years, to the cause of the nation. But that policy would have brought immediate defeat, the loss of all the Nationalists had gained such a short time before, and the likely return of China to full-scale civil war tempered only by foreign occupation. As it was, war came very close in January 1932, and the people of China and the world had an excellent view of Japanese military style.

As a foreign power which had a treaty relationship with China, Japan possessed a share in the administration and defence of the International Settlement at Shanghai, and in the first days of 1932, as Chinese indignation mounted against the attack on Manchuria and demonstrators appeared to threaten all foreigners indiscriminately, the authorities in Shanghai made preparations for defence. By previous agreement, the northern part of the Settlement was in the charge of a Japanese garrison, and their lines lay very close to the crowded Chinese quarter known as Zhabei. At one point, a tongue of Chinese territory cut into the foreign area, and the international command appear to have agreed that the Japanese should occupy the salient in order to strengthen their defensive position. They gave, however, no explanation to the Chinese and on the night of 28 January 1932, when officers and men of the Chinese Nineteenth Route Army, the main force stationed at Shanghai, saw Japanese troops advancing against their positions, they stood their ground and fought.

Whatever misunderstandings may have taken place, there was no question that the Japanese were waiting for some such chance. Besides the soldiers of the garrison there was also a naval force, including an aircraft carrier, and the fighting at Zhabei was joined by aircraft bombing and strafing the crowded civilian district, and by shelling from the sea, first against the forts which guarded the harbour from the Yangzi, then against Chinese positions in the villages and fields outside the city. While Western troops held their lines in an embarrassed neutrality, the armed forces of Japan, from the sanctuary of the International Settlement, displayed their methods of war against Chinese civilians and soldiers alike.

For more than a month, the Nineteenth Route Army held out against the Japanese attack, but the power of the invaders settled the issue. Within days of the outbreak of fighting, even as the Chinese government attempted to negotiate an armistice, Japanese warships sailed up the Yangzi to bombard Nanjing and the Nationalist capital was shifted to refuge at Luoyang in central China. Not until 5 May, when they had driven their opponents completely from the field, did the Japanese agree to a ceasefire, feeling that they had well demonstrated their strength.

In this they were correct. Despite their early difficulty in overcoming resistance in Shanghai, the Japanese had shown they could defeat any force the Nationalists sent against them, and no realistic adviser could encourage the Chinese government to embark on a major war. On the other hand, whatever they may have proved, the Japanese had lost all sympathy in the outside world. For a time, perhaps, they could ignore the opinion of Europe and America, and their publicists and politicians in later years tried hard to gain support for the code of *bushido*, the way of the samurai warrior. But for those who had seen and heard of the battle at Shanghai — the courage and gallantry of the Chinese soldiers against overwhelming force, the red crosses bombed and shot up, the destruction of homes, the murder and forced labour of civilians and prisoners — there were no ideals or excuses which could pardon the Japanese war in east Asia. It may seem strange that the British and French and others whose countries in the past had fought the Chinese with equal cruelty should now take exception to the Japanese, and certainly the Japanese found it hard to understand. But the age of imperialism in Asia was passing, and the Japanese were behind the times. Whatever their feelings, one legacy came from Shanghai: Englishmen had sometimes spoken of the Japanese, part in jest, part in respect, as 'little yellow gentlemen'; few now used the word 'gentlemen'.

Japan and Manchukuo

In contrast to the grandstand display at Shanghai, few people in the outside world had any opportunity to observe events in Manchuria.

Many in the West and even some in China were prepared to regard the territory as a 'semi-detachable' part of the Republic, for the regime of Zhang Zuolin and his son had encouraged no sense of unity, and others, anxious about Bolshevism and the power of the Soviet Union, found reassurance in a strong Japan face-to-face against Russian power. But despite these considerations, China's appeal to the League of Nations was accepted over Japanese protests, and in December 1931 a special commission led by the British Lord Lytton was sent to observe the situation.

On 9 March 1932, however, the Japanese proclaimed a new government of Manchukuo, headed by the Manchu prince Puyi. Now twenty-seven years old, Puyi had been the infant last emperor of the Qing. He formally abdicated in 1912, but was permitted to maintain a court at the old imperial palace until the Christian General Feng Yuxiang seized Beijing in 1924. He was then driven out and took refuge under Japanese protection at Tianjin and he was brought from there to this puppet state in the Japanese empire.

The League of Nations proved quite ineffective. Lord Lytton's report, which was not presented until September 1932, stated clearly that Manchukuo was merely an invention of Japan, but offered no proposals for action. By refusing to recognize the new state, the League showed disapproval of Japanese policy, but its failure to move against aggression was an early indication of the League's weakness and ultimate futility.

In the meantime, although an armistice was effective at Shanghai, and Chiang Kaishek had leisure for one more campaign against the Communists in Jiangxi, the invasion of the north-east was still not finished. After a few months pause to consolidate their position, the Kwantung Army turned against the province of Rehe, to the east of Inner Mongolia, which controlled the Great Wall frontier north of Beijing. After a preliminary skirmish in August 1932, and a forced occupation of Shanhaiguan, chief defence point on the railroad to Beijing, the Japanese launched their main attack in January 1933. The Nanjing government had expected Zhang Xueliang would be able to hold his defence lines, but his troops were forced swiftly back in disorder to Beijing. The Japanese came in pursuit, and by April their soldiers had crossed the Great Wall and entered China Proper. With their available armies in consistent retreat and with no chance of counter-attack, the Nanjing government was compelled to ask for peace, and on 31 May 1933, by the Treaty of Tanggu, the Japanese agreed to halt.

The terms of the treaty were quite as harsh as any Chinese might expect. Not only was the Japanese occupation of Rehe confirmed, it was also recognized that they possessed a special and privileged position in north China. To maintain this position, it was further agreed that the whole of eastern Hebei, including Beijing and Tianjin, should become a demilitarized zone: so the treaty rewarded Japan for her aggression, established the new frontier at the line of the Great Wall, and also stripped China of her defences in the north.

Whatever debate might have continued at the League of Nations on the question of Manchuria, the Japanese attack in Rehe and north China put an end to tolerance at Geneva. In the face of bitter criticism, as her armies were still advancing against Beijing, the delegates of Japan withdrew from the League in March 1933, and they cancelled all agreements on arms control. In December 1936 Japan joined Nazi Germany in the Anti-Comintern Pact aimed at the Soviet Union. The enmity for Britain and France was already declared, and the German threat from Europe should dissuade the Soviets from intervening when Japanese armies turned once more against China. As if to confirm their disregard of the League and other outside opinion, on 1 March 1934 the Japanese enthroned their puppet ruler Puyi as Emperor of Manchukuo.

Public opinion in the West could find no reason in this pattern of aggression by Japan except simple lust for power and shameless militarism, and this seems to be mostly true. In the context of the time, however, there were other reasons for Japan's concern about China, and for her disregard of Western sympathies.

Though the economic power of China was small, and her industrialization and foreign trade no more than a fraction of Japan's, they were a matter for growing concern. The bulk of Japanese trade was built upon small-scale, low-cost industry, notably in the field of textiles, and though Japanese investments had gained control of many factories in China, the cheap labour and growing production of Shanghai and other cities were beginning to afford serious competition to Japan itself. Other countries, such as the United States and the European powers, could view China's trade with equanimity and approval: they obtained cheap piecegoods from China, and they found an expanding market for heavy equipment and advanced machinery. Japan, however, already suffering badly from the 1929 depression, did not have heavy industry to sell abroad and the Chinese competition was threatening Japan's established markets.

In these circumstances, the invasion of Manchuria and the dominance of north China held clear advantages for the Japanese economy. The conquest enlarged Japan's domestic base and gave access to coal, oil, and iron from known fields about Mukden, while the neutralization of north China prevented the government at Nanjing from exploitation of the coalfields in Shanxi, Hebei, and Shandong, and also deprived the Nationalists of market and industrial potential in the northern cities. So the aggression of the Kwantung Army served the interests of the industrialists and the community at home, weakening the position of an economic rival and offering new prosperity based upon a wealth of mineral resources.

Japan had never felt fully accepted among the great powers of the West, and she found small sympathy for her internal economic problems or for her political difficulties with China. The United States of America and the imperialist powers, chief among them Britain and France, were prepared to support the Nationalist government as a regime which offered peaceful conditions for trade, which showed some tolerance for their financial interests and their rights under the old treaties, and which might eventually provide a counter to the growing military power of Japan. Perhaps more important, Chiang Kaishek's war against the Communists in Jiangxi put him clearly in the Western camp. Sympathy and support, however, in this world of *Realpolitik*, stopped short of military commitment: the Western powers could accept the dismemberment of China so long as their own interests and privileges were preserved; and by a touch of irony, the apparent success of Chiang's anti-Communist campaign in Jiangxi in 1934 deprived him of that leverage in seeking foreign aid. In fact, on the eve of the outbreak of war in 1937, with the short-sightedness that characterized so much of the diplomacy of the 1930s, the British were negotiating with Japan for an understanding that might divide China into spheres of influence, just like the old days before the fall of the Qing.

For the time being, immediately after the Treaty of Tanggu, the Japanese were content to consolidate their gains in Manchuria and Rehe, and to extend their influence in north China without open war. Politically this was not very difficult, for the northern provinces had never been closely involved with the Nanjing government, and there were many people who held small affection for the Nationalists. By 1935 there were plans for an 'autonomy movement' over five provinces, from Shanxi and Suiyuan east to Shandong. This proved too ambitious, and the Japanese were compelled to accept a compromise, but the Hebei–Chahar Political Council, nominally responsible to Nanjing, was largely controlled by men opposed to the Nationalists. Many of the leaders had risen to prominence in the days of the Manchus, several of them had close ties with Japan, and none was a friend of the central government. Aided by armies across the border and garrisons within China, but also applying considerable political skill, the Japanese took advantage of the Nanjing regime's loose administration and one-party rule to raise disturbances from which they might hope to profit.

On the other side, however, though the Nationalists had not eliminated all their enemies within China, they found it increasingly difficult to justify the continuation of civil war without regard to the threat from Japan. As early as 1933, when the leaders of the Nineteenth Route Army formed their ill-fated People's Government in Fujian, the centre of their policy had been a united China which would include Nationalist and Communist forces in alliance against Japan. By 1935, when the Communists had been driven from Jiangxi and had staggered to their refuge in the north-west, there was real debate whether the government should make a final attempt to exterminate them, or whether the threat from Japan was now so close that only a united front could save the country. Chiang and his colleagues, with their experience of Communist influence on the Nationalist Party in earlier days, and with their long commitment to war, saw no way to deal with the Communists but by a fight to the death, and they regarded the call for a united front merely as a device to gain the Communists time to regroup their forces and return to the attack. In this they were surely correct, but the danger from Japan was too great, and the soldiers and people of China were losing sympathy with an autocratic government which thought only of its individual advantage and appeared to ignore the safety and defence of the country.

The United Front

Driven from Manchuria by the invasion of Japan in 1933, the Young Marshal Zhang Xueliang and his remaining troops had taken refuge in the eastern part of the North China Plain. As the Communists arrived in Shaanxi, however, after their long march of escape from the south, Zhang and his army found themselves again in the front line of fire. In November 1935, under the orders of Chiang Kaishek, the soldiers of the North-eastern Army under Zhang Xueliang and of the local North-western Army under General Yang Hucheng were ordered into a final campaign to suppress the Communists.

Neither the soldiers nor their officers were anxious for the fight. The Communists, well on the defensive, avoided open battle and concentrated propaganda against this civil war, urging the alliance of all patriotic forces against Japanese aggression. Not surprisingly, they found a sympathetic hearing from Zhang Xueliang and his troops, for these men had lost their land and their homes at the hands of the Japanese and they could see no reason why they should exhaust their strength against fellow-Chinese for the profit of a government in distant Nanjing. By the middle of 1936 both Zhang and Yang Hucheng were in correspondence with the Communist leaders, and the offensive was at a standstill.

The reaction of Nanjing was delayed by a minor rebellion in Guangxi, but at the beginning of December 1936 Chiang Kaishek flew to Xi'an, headquarters of the anti-Communist armies, to consult with his recalcitrant generals and give them personal encouragement in the war. In the early morning of 12 December, however, while Chiang was still asleep in his quarters, he was aroused by shots as mutineers from Zhang Xueliang's personal guard and the North-eastern 105th Division gunned down his bodyguard and came to seize him. Chiang Kaishek fled but was captured and brought back, and in the forced negotiations which followed he was compelled to agree to a united front with the Communists and to full opposition against Japan.

As commentators and historians have noted, the Xi'an Incident is one of the more remarkable events in all the turbulent history of modern China.[8] For the Communist negotiators, led by Zhou Enlai, the objective of a united front was achieved, and their area of refuge, with its capital at Yan'an in Shaanxi, was made secure. For the Nationalists, one of the largest questions is why Chiang Kaishek ever felt he could trust himself to Zhang Xueliang and his allies, who were clearly half-hearted in their support of his policy. Once the disaster had taken place, however, and Chiang was in the hands of his enemies, negotiations appear to have passed off with courtesy and goodwill. It would not have been surprising if Chiang had been killed by his captors, but even his enemies recognized that the Nationalist leader was essential to the rallying of non-Communist China against the outside enemy. Zhang Xueliang, protesting that the kidnap was rather the result of misplaced enthusiasm than a deliberate plot, accompanied Chiang Kaishek on the plane back to

Nanjing. The party, including Madame Chiang, who had travelled to Xi'an to share her husband's brief imprisonment, returned to Nanjing on 25 December. Zhang Xueliang was immediately put under house arrest; he never held office or rank again, and he accompanied the Nationalists on their exile to Taiwan.[9]

Within the government in Nanjing, despite the humiliation he had suffered and the fact that there was now no hope for a speedy destruction of his Communist enemies, Chiang Kaishek gained immense prestige. No one else could have negotiated with the Communists nor been accepted by them, and he was now indeed the leader of the nation, with authority far above his associates and rivals. For a time, at least, the Communists would keep their side of the agreement, and in February 1937 a plenary session of the Nationalist Party confirmed the truce.[10] For all their hesitation in the past, Chiang and the Nationalists had always feared that war with Japan was inescapable and now, with the crisis only weeks away, they faced the threat with a fair determination and the best military array that they could manage.

The real beneficiaries of the accord, however, were the Communists. On the one hand, they had halted the attacks which were being sent against them, and they had compelled their enemies to acknowledge them as patriots, with a legitimate role to play in the nation at large. At the same time, moreover, they had forced Nanjing into direct confrontation with the military might of Japan, and the Nationalists would be in the front line of that conflict. Often in Communist history, the slogan of a 'united front' may best be interpreted by the cynical proposal of 'Let's you and him fight', and in this case it was extraordinarily successful. The very existence of the agreement ensured that war would come, and quickly, for the alliance was clearly aimed at Japan, and the Japanese had no cause to wait until China was better prepared. Within a few months of that concord in the far north-west, the miseries of war had engulfed the heart of Nationalist China.

Notes

1. The term Politburo is an abbreviation of Political Bureau, traditionally the controlling body of a Communist organization.

2. In a massacre which followed this debacle, Mao Zedong's wife and his younger sister were killed by the Nationalists.

3. In accordance with common and necessary practice among revolutionaries, both these names were pseudonyms: Wang Ming, taken from a well-known Chinese novel, was the name adopted by Chen Shaoyu, and Bo Gu was the pseudonym of Qin Bangxian.

4. *Mount Liupan,* scanned to the tune *Qing bing luo,* translated by Michael Bullock in Ch'en, *Mao and the Chinese Revolution,* p. 337; also by Stuart Schram in *Mao Tse-tung,* p. 188.

5. See p. 96–7 above.

6. The Kwantung [= *guandong*] Army took its name from the territory where it was stationed: Manchuria is east (*dong*) of the passes (*guan*) which lead from Korea and Manchuria to Beijing and China Proper. The term had been used in earlier times, and in 1936, when the Japanese took over the Liaodong peninsula from Russia, they named their holdings there the Kwantung Leased Territory.

7. See p. 97–98 above.

8. Based upon the old transcription of the name of the city, it is frequently referred to as the Sian Incident.

9. Fifty-four years later, in June 1990, Zhang Xueliang attended a public reception in Taiwan to celebrate his Chinese ninetieth birthday (eighty-ninth by Western count), and he later visited some of his family in the United States.

10. Apart from the ceasefire, the main points of agreement were that the Communists recognized the government in Na..jing and accepted the Three Principles of the People as the ideological base of the Republic. The Nationalists agreed to turn their energies against Japan, and in token of the new unity the Communist Red Army was renamed the Eighth Route Army (Route Army being the term for an army group), with an approved and limited recruitment and subject to theoretical control from the Nationalist high command under Chiang Kaishek.

Further Reading

On the Communists:

Chen, Jerome, *Mao and the Chinese Revolution,* New York, Oxford University Press, 1965.

Guillermaz, Jacques, *A History of the Chinese Communist Party, 1921–1949,* [translated by Anne Destenay], London, Methuen, and New York, Random House, 1972.

Kagan, Richard C., 'Ch'en Tu-hsiu's Unfinished Autobiography' in *China Quarterly* 50 (London, 1972).

Rue, John E., *Mao Tse-ung in Opposition, 1927–1935,* Stanford, Stanford University Press, 1966.

Snow, Edgar, *Red Star over China* [revised edition], New York, Grove, 1968.

Schram, Stuart, *Mao Tse-tung,* Penguin, 1966.

Wilson, Dick, *The Long March, 1935: the epic of Chinese Communism's survival,* New York, Viking and London, Hamilton, 1971.

The Rise of the Chinese Communist Party: the autobiography of Chang Kuo-t'ao, 2 vols., Kansas University Press, 1971–2.

On Japan:

Coox, Alvin D., *Nomonhan: Japan against Russia, 1939,* 2 vols, Stanford, Stanford University Press, 1985 [including a history of the Kwantung Army in Manchuria].

Duus, Peter, Ramon H. Meyers and Mark Peattie (eds.), *The Japanese Informal Empire in China, 1895–1937*, Princeton, Princeton University Press, 1989.

Hata, Ikuhito [translated by Alvin D. Coox], 'Continental expansion 1905–1941,' in *The Cambridge History of Japan*, volume 6.

Iriye, Akira, 'Japanese Aggression and China's international position' in *Cambridge China* 13.

Thorne, Christopher, *The Limits of Foreign Policy: The West, the League and the Far Eastern Crisis of 1931–33*, London, Hamilton, 1972.

6

The War Against Japan, 1937–1945

WHATEVER their patriotic motives, the policy of the united front served the Communists well. When Japan attacked China in 1937, the brunt of the fighting was borne by the Nationalists. Within eighteen months they had been driven from Nanjing to Chongqing in the far west. There, deprived of the sources of former prosperity and power, the government of Free China could offer no riposte to the Japanese invaders, and did little more than wait for American aid.

When that aid did come, after the attack on Pearl Harbour in December 1941, the heart of the Nationalists had been broken, and the regime of Chiang Kaishek was little more than that of a warlord, chiefly concerned with personal interests, riddled with corruption, and all but irrelevant to the conduct of the war and the fortunes of the nation.

At the same time, however, the Communists in the north-west, suffering no such direct attack or losses to the Japanese, were able to take the opportunity to establish a network which guided guerilla activity across the countryside of north China. Not only did they fight the Japanese and the puppet troops of the Nanjing government under Wang Jingwei, but they also aroused the peasants to redistribute the farmland and remove their landlords.

Besides this political and military initiative in the field, Mao Zedong at Yan'an developed the discipline that his movement required. The principles had been indicated earlier, but the Rectification Campaign of 1942 may be seen as decisive, and the Yan'an Forum, with its slogan that literature and art must serve the people, was designed to ensure that intellectuals should apply their energies not to personal ideals but to work within a narrow framework for the revolutionary cause. The problem was not to stir up the peasants — the resentment and potential for violence was always there — it was to maintain control of the party and its agents, so that each followed orders, even in the absence of direct contact with headquarters. In coming years, the ordered co-operation of the Communist movement, at every level, was the decisive factor in its ultimate victory.

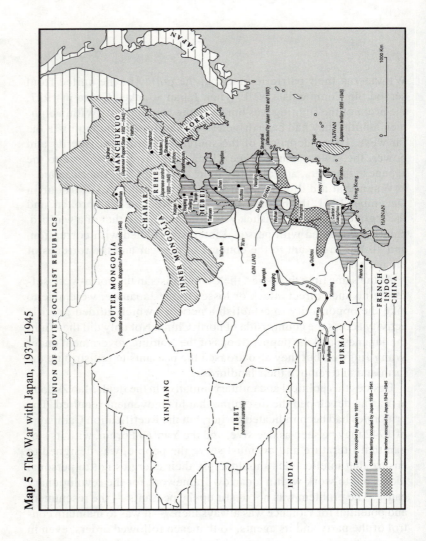

Map 5 The War with Japan, 1937–1945

The Japanese Offensive

In the second half of 1936, as Chiang Kaishek's government planned the final destruction of the Communists in Shaanxi, there were negotiations with the government of Japan under Prime Minister Hirota, a former diplomat. The Japanese asked for an end to boycotts and demonstrations against their interests in China, for a recognition of their special interests in the north and for joint action against Soviet influence in Mongolia, and they called also for economic co-operation and for the acceptance of Japanese advisers in China. It is possible they genuinely believed these terms to be no more than appropriate recognition of their dominance in east Asia, but the Chinese, reasonably enough, found them unacceptable. They required that Japan should withdraw her troops from north China and should cease her encouragement of separatist movements. The argument dragged on to the end of the year and then, to no one's surprise, as Chiang Kaishek concluded his agreement with the Communists at Xi'an, negotiations were abandoned.

Hirota was a civilian, and the Diet had influence on his government, but Japan's foreign policy now rested with the army. The cabinet was brought down soon afterwards by the resignation of the Minister of War, General Terauchi, and although members of the Diet distrusted the Kwantung Army, no government could be formed to establish a peaceful settlement. At the same time, the new united front in China appeared a clear threat to the ambitions of the Japanese militarists and in July 1937 a local skirmish at Beijing was expanded into full-scale war with China, and ultimately with the world.

On 7 July Japanese troops in north China held exercises outside Beijing. They claimed authority from the Boxer Protocol of 1901, but the Chinese protested that the number of soldiers was four times that permitted by the agreement, and the region they were using for their manoeuvres had never been approved. The technicalities, however, became secondary when the Japanese claimed that one of their soldiers was missing and demanded entry to the city of Wanping in order to look for him. General Song, commander of the garrison Twenty-ninth Army, refused permission, and the Japanese attacked him.

The first clash took place at Lugouqiao, known to Westerners as Marco Polo Bridge. For a short time there was a chance that the Japanese would attempt to smooth things over, for the new Prime Minister, Prince Konoye, and the navy and foreign ministers in Tokyo were all anxious for restraint. The Minister of War, however, General Sugiyama, threatened again to bring the government down with his resignation, and the cabinet gave in. Within a few days, men and munitions were pouring south from Manchuria, and on 13 August a second front was opened against Shanghai.

The Chinese armies were already at a disadvantage in the north, for the Japanese were established within the country. On 27 July the enemy attacked Beijing, and the Chinese, rather than accept destruction of the ancient capital, abandoned it the following day. Brushing local defences aside, the Japanese thrust south across the plain, east into the Shanxi hills towards Taiyuan, which fell to the invaders early in November, and north-east across Inner Mongolia. The Chinese held their position near Xuzhou on the Huai River and achieved one notable success at Taierzhuang, but in the long campaign from December 1937 to May 1938 the Japanese destroyed the last regular Chinese forces north of the Yangzi.

In the north, the Chinese had been forced by circumstances to give ground quickly, but when the Japanese attacked Shanghai the armies there were ordered to hold their positions. The policy was questionable, for the Japanese had overwhelming air and naval superiority, and some of the best Chinese troops were committed to restrictive defence against a well-prepared and mobile enemy. For almost three months they kept their lines, to the admiration of foreigners and the inspiration of their own people. At last, after years of retreat and conciliation, the Nationalists had shown they were prepared to fight, and they did so with gallantry, but also with heavy casualties. Then, however, the Japanese turned the flank with a landing on the north side of Hangzhou Bay, the armies at Shanghai retreated up the Yangzi in confusion, and Nanjing fell on 13 December.

By the summer of 1938, though not without difficulty, the Japanese had broken the back of Chinese resistance. They had defeated the best armies under Chiang Kaishek's command and they had seized all the industrial centres which had given China hope for modernization and development. Besides Tianjin and Shanghai, the port of Qingdao in Shandong was taken in January 1938, Xiamen fell to seaborne invasion in March, and Guangzhou in October. Moving up the Yangzi from Nanjing the invaders also captured Wuhan in October. By the end of the year the Nationalist government was penned into its western capital of Chongqing in Sichuan, while Japan controlled all the communications systems of the North China Plain and the Yangzi valley, and all the ports along the coastline.

The Nationalists had suffered desperate blows, and in military terms the war was lost. Only a few ragged armies escaped the destruc-

tion in the east. The Japanese themselves believed that the fall of
Nanjing would compel Chiang Kaishek to surrender, and after the
capture of that city in December 1937 the German ambassador, whose
country was both an ally of the Japanese and also a long associate
of the Nationalist military regime, was asked to act as go-between.
The terms included payment of an indemnity, acceptance of Japanese-
influenced autonomous movements, and alliance with Japan and
Manchukuo in an extension of the anti-Comintern pact. Chiang
Kaishek's German advisers urged him to accept the terms, but Chiang
held firm. With the slogan that he would 'use space to buy time', he
ordered the withdrawal of the government and as many of the people as possible from the occupied and threatened east to the security of the west. Throughout 1938, as Japanese armies spread across
the north and up the Yangzi, the people of China adopted a scorched-
earth policy, evacuating what they could and destroying much of
what remained.

If nothing else, the great retreat to the west showed the authority
of the Nationalist government, and its acceptance amongst the people, even in defeat. Sometimes the destruction was disastrous, as
when the Chinese armies broke the Yellow River dykes to halt the
Japanese advance into Henan in June 1938: the floods engulfed
homes, towns, and farmlands in a swathe of devastation, and the gap
remained unrepaired until the end of the war almost ten years later.
With reason, modern Communist histories have criticized the disregard for life which the sabotage showed, and the flooding of the
Yellow River in 1938 must rank as one of the greatest man-made
misfortunes of history.

The exodus to the west, ill-planned and often confused, yet saw
remarkable ingenuity. As the people in their hundreds of thousands
fled before the Japanese advance, industrial machinery and equipment was packed up and transported into Sichuan, and foreign
observers described with awe how whole factories were brought by
rail, steamer, and junk to start work again near the new capital. In
much the same way, teachers and students from Beijing, Qinghua,
and the other great universities of the north, as well as those from
Nanjing and Shanghai, moved west across the mountains. In three
main centres, at Chongqing, Chengdu, and National South-west
United University at Kunming in Yunnan, the transplanted universities, which actually claimed more students than in the years just
before the war, maintained the scholarship, the research, and the
pride of Nationalist China.

For the Japanese, military success brought national triumph and international shame, and both emotions were strong. The army and navy were proud of their achievements, but some of the greatest victories had been marred by excesses which cast a shadow on the chivalry of *bushido*. The attack on Shanghai in 1932 had blackened Japan in the eyes of the West, but when Nanjing was captured in December 1937 tens of thousands of civilians were slaughtered out of hand, women were raped, and houses were burned ar.d looted. As horrified foreign witnesses sent the story throughout the world, even the Japanese authorities realized their men had gone too far: in a sense, the whole war was a defiance of the government in Tokyo, but the discipline of the army itself was now in danger. No newspaper in Japan was permitted even to hint at the stories of atrocities, but the local commander and two other generals were withdrawn and whole units of reservists called up at the beginning of the campaign were hastily bundled home.

Besides this, in the enthusiasm of victory there were a number of attacks on neutral Western powers. Best-known of the victims was the United States gunboat *Panay*, bombed and sunk in the Yangzi near Nanjing, and for a short time it seemed the Americans might join the war. Tokyo, however, apologized and paid compensation, and Roosevelt's government was not yet prepared to break the policy of isolationism.[1] On the other hand, the attack on the American flag, coupled with descriptions of the massacre and rape at Nanjing, removed all vestiges of American sympathy for Japan's mission in east Asia.

As the American ambassador in Tokyo, Joseph Grew, remarked about this time:

In dealing with Japan we are, in effect, dealing with two distinct authorities who are sometimes very far apart in their respective conceptions of foreign policy. The home government, which alone is able to appraise the international aspects of this situation, is for the moment very nearly powerless to compel the military authorities in the field to implement its assurances.[2]

Both the military and the government, however, discriminated between countries. Confident of their power, the militarists despised the American will to fight, but they recognized the United States as a potential danger, and they were willing to accept mediation. In dealing with the British, however, committed to conciliation in both Europe and the East, far less care was taken. One gunboat was shelled, and the British ambassador was strafed by Japanese aircraft in his car between Nanjing and Shanghai. In contrast to the *Panay* affair, the Japanese authorities showed little concern. Britain was an old associate of the Chinese government, and her empire was a rival to Japanese ambition, while protests from the League of Nations, of which Britain was a leading member, had confirmed Japan's sense of international loneliness and lack of understanding.

The moral support of the League for China, while it irritated the Japanese, had no effect on their policy, and no foreign power gave practical help to the Nationalists. In a strange, short war, almost ignored in the West, the armies of Japan did clash with those of the Soviet Union, first in the summer of 1938, near the borders of Korea, Manchuria, and the Soviet Maritime Province, and then a year later in a bloody campaign about Nomonhan in Outer Mongolia. The results were remarkable, for the hitherto invincible Japanese were sharply and decisively defeated, and throughout the Pacific war they continued to keep a respectful eye on the Russians in Siberia.

In August 1937, after the Japanese attack on Shanghai, the Nationalists confirmed their agreement with the Communists by entering a treaty of non-aggression with the Soviet Union, and in the months and years that followed they gained some hope from their ally's success in the north and from the sympathy they received in America and Europe. Though the Japanese had torn the best of the country from Chinese control, Chiang Kaishek had still no thought of surrender. From his refuge capital at Chongqing, he claimed that the Japanese were bogged down in the vastness of the territory they had to conquer; he pointed, with some accuracy, to the strain that the war had placed on their economy, and he hoped that military aid would eventually come from foreign powers who could not afford to see Japan grow strong at China's expense.

It was a brave stand, but not all the party would accept it. At Beijing, Nanjing, and in Inner Mongolia, the Japanese established puppet governments, and although few would accept their legitimacy it was clear that these new regimes, with the support of the invaders' army, controlled most of the former Nationalist territory. In December 1938, when the leaders at Chongqing resolved again to reject Japanese proposals of peace, Chiang Kaishek's old rival Wang Jingwei refused to accept their decision. Flying from Chongqing to Hanoi, he returned to Shanghai with Japanese support, and in March 1940 he established a 'reformed' national government at Nanjing.

There is room for discussion on Wang Jingwei's motives for this betrayal. He was, after all, one of Sun Yatsen's oldest associates, and he had held high power in Nanjing before the war. In part, at least, it was simple jealousy of Chiang Kaishek, distrust of his authoritarian power, and fear for his own safety.3 He claimed that alliance with Japan was a fulfilment of Sun Yatsen's wishes, but the Japanese, with their wider concerns throughout the Pacific and South-east Asia, had little interest in window-dressing, and the people of eastern China, more than half the nation, were condemned to a fruitless period of controlled occupation. Wang Jingwei's puppet government had no wide acceptance, and his troops served only on local garrison duty, with occasional, generally unsuccessful, offensives against Communist-held areas in the north and north-west.

Regardless of jealousy and distrust, however, Wang Jingwei had prophesied defeat for Chang Kaishek and the Nationalists, and he sought to throw in his lot with the winning side. So his defection was more than the treachery of a turncoat: it cast serious doubt over the future of the Nationalists and on their programmes. From his stronghold in Sichuan, Chiang Kaishek continued to broadcast defiance, but he had no plans for counter-attack, and there were few indications that any Western power would ever give active support.

By 1939 the Japanese had achieved their immediate aims in China, and their chief interest was to consolidate these gains. The one major route they did not yet hold was the railway from Wuhan through Changsha to Guangzhou, and between 1939 and early 1942 there were three great campaigns in this region, all of which resulted in Japanese withdrawal. Further north in Shanxi, the armies of the Nationalists and Yan Xishan, with the Communists to the north of them, held a defence line separated from Japanese-controlled territory by a no man's land of devastation. And though the Japanese sent daylight air-raids against Chongqing, they never made a serious attack.

Economically as well as militarily, however, the position of the Nationalists was weak. Despite the transport of factories and equipment to the west, the heart of Nationalist China had been cut away by the Japanese invaders and the rump of territory in the south-west supplied few of the requirements for modern industry and munitions. This region of China produced no oil and very little coal, and other essential items such as salt were always in short supply. The military equipment lost in the east could never be properly replaced; medical supplies were seldom available; the transport system, formerly to some extent motorized, declined rapidly to wagon and pack-horse and then to coolie-carriage on foot. Eyewitnesses told of outposts armed with nothing more than an outdated machine-gun, and of whole armies with no heavier artillery than a pack-howitzer. Often enough, the Chinese outnumbered their enemies in the field, but there was no way they could attack Japanese fortifications against the armament that was brought to bear against them.

In the last years of Nationalist power in Nanjing, the government budget, allowing for foreign loans and increased tariffs, had almost balanced. With the retreat to Chongqing, however, the cost of war swamped all official finance, and it was impossible to extract adequate revenue from the limited and backward territory which remained under its control. To solve the immediate problem, the government printed banknotes, and inflation inevitably followed. In 1935 the Chinese dollar had been worth about 1s.8d. sterling, but by the end of 1941 the value on international exchange was down to a little more than 3d. In 1942 the situation became worse, for the Nanjing government under Japanese control began a currency war against the Chongqing finance system. Both sides refused to recognize and exchange the other's currencies and each attempted to replace the other's notes by issues of their own. The Chongqing government pegged its international exchange rate to 3d sterling, and supported this fictitious value by subsidies to remittances overseas, but the inflation continued. By 1944, the official exchange rate for the US dollar was twenty National Chinese dollars, but US dollars were widely circulated about Chongqing, and the black market value was three hundred Chinese dollars. In Shanghai, now under Japanese/Nanjing control, the cost of living index had risen from 100 in 1936 to almost 2,000 in 1942 and to 30,000 in the middle of 1944.[4] Eventually, the Chongqing dollar had slightly the better of the exchange, but by the end of the war it was clear the currency was in desperate straits, and it would require great skill to restore it.

Until the end of 1941, although the Japanese had conquered eastern China, and their puppet regime was established at Nanjing, they still recognized the treaty rights of foreigners. The International Settlement at Shanghai remained a centre of banking and finance, with quotations for both National Chongqing dollars and their rivals issued by Wang Jingwei, not to mention a variety of Japanese-sponsored and occupation currencies which circulated in north China and elsewhere. When war came in Europe between Japan's ally, Nazi Germany, and Britain and France, the situation in the Far East was naturally tense, and the British were compelled to make a series of humiliating compromises with Japan: in July 1940, isolated after the fall of France, they agreed to close the Burma Road from Chongqing to India for three months, and by doing so they cut Nationalist China's chief link with the outside world. At the low level of the economy, the damage was not severe, and in October the British recovered their confidence and allowed the route to be reopened.[5]

The foreign trade of Nationalist China had indeed almost disappeared, but Britain and the United States extended credits to Chongqing, which were used for food, for cotton and other textiles, minerals, machinery, and weapons of war. Exports were tong-wood oil, minerals such as tin, tungsten, and manganese, all of which are found in south-west China, and, very importantly, silk. Silk was needed urgently by the allies, particularly for parachutes, and the whole industry was nationalized in 1941. Until the Japanese attack on the Western powers at the end of that year, the main channels for trade were the Burma Road, the highway north into Soviet Russia, and a variety of tracks leading to eventual market in Hong Kong. When these were interrupted by fighting in South-east Asia, communications were maintained with India and with Russia by air, and silk continued to be exported.

There was some exchange between the Nationalists at Chongqing and the territories of the east. The Chinese Post Office continued to serve both regions, travellers could move around the battlefields from one area to the other, and though both sides maintained a blockade it was sometimes possible to buy even Japanese-made goods in Chongqing. On the other hand, the production of eastern China was integrated with the war economy of Japan, and in many regions there was danger from famine because two-thirds of the rice crop was requisitioned for the army. Industries could no longer compete with Japanese manufactures, and by 1941 almost the whole of occupied China's overseas trade was directed to Japan.

The disruption of commerce caused by war, and the rapid inflation during these years of division, were serious for the whole development of China, but their immediate effect was as much political and social as economic. The inflation hit hard on investors, capitalists, and landlords, and more fiercely still against salaried and professional employees. In the nature of things, so long as the landlord was still in contact with the peasants who owed him money, and was able to maintain a modicum of control in the village, he could pass on his difficulties to those who worked his land or owed him money, while the weakness of the economy provided one more hardship for the poor farmers who must pay high rents, exorbitant interests, and increasing land tax to the government. For salaried men, however, there was no such escape, and the income of government servants in particular, both military officers and civilian administrators, became worthless.

In a general sense, the inflation destroyed the prosperity of the middle classes, the one group with generally liberal ideals which had consistently supported the Nationalists. More directly, it fostered the spread of corruption: with impossibly low salaries, any man who cared for his own survival and for that of his family was driven to graft. Army officers, no matter how well and how bravely they might fight the enemy, supplemented their income by false reports of the numbers under their command, by holding back a percentage of pay and supplies, and by regular trade on the black market. In civilian government, every official demanded personal 'squeeze' for all services and certificates, and tax and customs administration, where the really big opportunities lay, was notorious for bribery and false returns.

Above all this weakness and growing disillusion stood the figure of Chiang Kaishek. Chiang had made himself the symbol of Chinese defiance against Japan, and he was respected even by enemies and rivals for his courage and determination. At the same time, however, he dominated the government at Chongqing as he had never been able to manage the politicians of Nanjing. His chief rival, Wang Jingwei, had shamed himself and his cause by surrender to the Japanese, and in the tight community of south-west China there was no one of civil authority to defy the 'Generalissimo'.[6] The only possible opponents within Chiang's ranks were the semi-independent generals of his scattered armies, but he handled these with confidence, playing one off against another, ignoring many of their misdemeanours, but insisting always on respect and obedience.

Chiang Kaishek held power in the Nationalist regime through personal influence and control of the army. In these dark years, his strengths and weaknesses became very clear. As no other man, he held the imagination of the people of China and of sympathizers in the world outside, but even among his greatest admirers few would claim he had any ability as an administrator. His orders were often contradictory, he paid inordinate attention to minor details, and he largely ignored the major problems of corruption, military recruitment, and the supply of armies in the field.

Most serious of all, the government at Chongqing gradually took the appearance of another warlord regime: dominated by one arbitrary chieftain, with the centre of power in the army, and with minimal, discredited, civil and political influence. All governments, even among advanced democracies, have a tendency towards centralized authority and military power in time of war, but the restraints maintained in Britain and the United States were notably absent at

Chongqing. For Nationalist China at war, the only questions were those of survival, and ideals of reform had to wait until the government was restored to power.

In these terms, the administration at Chongqing concentrated its attention on military organization and strategy. Even though the army had suffered massive defeat, was seriously under-equipped, riddled with corruption and nepotism, and depended rather on personal loyalties than allegiance to the republic, chief emphasis was placed on forced recruitment and expenditure for arms, and all questions of social change and land reform were neglected. Most notably, despite the patriotism of their appeal to the nation, the Nationalists made small use of guerilla forces in occupied areas. For some time after the first Japanese invasion, isolated units which had been overrun by the enemy maintained a gallant resistance behind the lines, but the retreating Chinese armies had neither the flexibility nor the administrative capacity to maintain contact with these groups or provide them support. In time, these lost units were either mopped up by the Japanese or rescued and recruited into another network of command, spread throughout northern and central China, and guided by the Nationalist's present allies, constant rivals, and future enemies, the Chinese Communist Party.

The Communists in the North-west

The territory to which the Communists came in October 1935, at the end of their long march from Jiangxi, was of greatest advantage to them simply because of its distance from the government in Nanjing and the armies of Chiang Kaishek. It was a poor region, with constant threat of crop failure and famine, and although both Shaanxi and Shanxi contain major reserves of coal and oil, they were backward and isolated from the rest of the country.

On the other hand, the hills and rivers which separate this region from the south and east had often served as a line of defence for armies in the north-west. Here the Communists could find some security and a breathing-space to consolidate their position against attack. The local leaders Gao Gang and Liu Zhidan had maintained a soviet regime in the Shaanxi–Gansu border area since 1928, but despite the natural suspicion with which the men of the north-west regarded the newcomers from the south, Mao Zedong had established authority in the Communist movement, and his government expanded the area under its control and its support amongst the people.

To some extent, the fighting in the north-west during 1936 resembled the campaigns of the old Jiangxi Soviet, with government troops relying on the *baojia* system of mutual responsibility, and with disorder widespread along an uncertain front line. But the local Nationalist troops and their allies from Manchuria under Zhang Xueliang showed little energy, and the Communists developed their position without great difficulty. They suffered some reverses, notably in the first months of 1936, when an invasion of Shanxi, territory of the warlord Yen Xishan, was driven back with heavy loss and Liu Zhidan was killed, but by the end of the year the Communists controlled the greater part of northern Shaanxi. In December, after their forces captured the county city of Yan'an, Mao Zedong moved his headquarters there, and in January 1937 Yan'an became the capital of the soviet government.

There had been some change in policy. Besides the principles of land reform and freedom from local oppression, Mao Zedong sought support from the Muslim minority, from the Mongols of the north, and from the secret society *Gelao Hui*, a peasant movement influential in many provinces, including Mao's native Hunan. And in all his statements and writings there is constant appeal to patriotism and opposition to the threat from Japan.

The Communists had long argued for a united front, but the current situation, with Japanese encroachment across the north and with refugees from Manchuria in the armies opposed to them, made the policy far more relevant. During the first months in the north-west the official line still described Chiang Kaishek as a secret ally of the imperialists and a traitor to the nation, but from the middle of 1936, as the possibility grew of some *rapprochement* with the Nationalist armies which faced them, this theme was dropped. In December 1936, when Chiang Kaishek was trapped in the Xi'an mutiny, Mao and his emissary Zhou Enlai were prepared to seek a genuine, though perhaps temporary, united front.

Through the latter part of 1937 and the bulk of 1938, however, as the Japanese armies forced their way into eastern China, the Communists could offer little support. The Red Army, now the 'Eighth Route Army' of Nationalist command, was in action against the Japanese in September 1937, immediately after the final agreement with Nanjing, and, under the command of Zhu De and Peng Dehuai, they fought beside Yan Xishan and the Nationalists against the invasion of northern Shanxi. Like the other armies, they gained small success in open fighting against the Japanese, and the fall of Taiyuan in

November ended that phase of operations. Henceforward the tactics and strategy of the Eighth Route Army were based on guerilla activity behind enemy lines, with occasional raids by the main force, and they gave only limited co-operation to their allies. With their experience and training, and most notably their ideology and genuine patriotism, the Communists made successful sorties against Japanese territory and they encouraged the people to maintain resistance and to sabotage Japanese operations.

It is difficult to assess the military significance of Communist operations in north China. At first, the Japanese were concerned rather with the maintenance of supply lines, chiefly railroad, to support major fighting in the south and, later, the expansion into South-east Asia. Some commanders claimed that the guerilla attacks in the north were no more than pinpricks, and indeed almost three-quarters of their soldiers in China were concentrated against the Nationalists. On the other hand, no regular officer will readily concede success to guerillas or saboteurs, and the Japanese used numbers of men under the nominal command of Wang Jingwei or the other puppet governments to maintain their position and protect their major posts. So north China was divided into zones held firmly by one side or another, with large disputed areas subject to raids and reprisals.

From the point of view of the Communists, the situation had many advantages. Organizing groups behind the enemy lines, and sending cadres to maintain communication and transfer instructions, they enlisted supporters throughout north China. Many men, local patriots or isolated troops of defeated Nationalist armies, joined the Communists as allies in underground struggle against Japan, and the assistance they received from the base at Yan'an confirmed them in their new allegiance. For their own part, the Communists emphasized the national patriotic struggle, and in areas where they held power they acted with consideration for those who had joined them. In marginal territories, it was laid down that authority should be shared three ways, between Communist cadres, leaders of other political organizations including local Nationalists or similar groups, and those people without party affiliation. In accordance with the agreement with the Nationalists there was no confiscation of land, but rents were limited to no more than one-third the value of the crop. Both landlords and peasants were prepared to accept the compromise, for taxation was reasonable and the government was honest. At Yan'an itself the Communist regime renounced the name 'soviet' and styled itself simply as the administration of the 'Border Region'.

The care with which the Communists proceeded, and the skill with which they united all classes in the struggle against Japan, did not, however, bring the Nationalists at Chongqing to trust their good intentions. By the agreement of the united front, the Eighth Route Army was limited to 45,000 men; but the organization of guerilla units, each of which might vary in strength from one to two thousand, gave ample opportunity to expand numbers beyond the limits agreed, and the Chongqing government had no way to control the Communists or to compete with them in the north. The one alternative was the warlord Yan Xishan, who had been driven from his capital at Taiyuan by the Japanese advance in November 1937, but who still controlled the greater part of southern Shanxi. In an early attempt to rally support to his defeated troops, Yan Xishan encouraged the popular *Ximeng Hui* or 'Sacrifice Society', and this patriotic, non-Communist group received considerable support. Yan, however, later realized that the society's ideals of reform and fair dealing were a threat to his regime, and in 1939 he turned against these former allies, killed their leaders, confiscated their equipment, and recruited their men by force into his own command. Returning thus to his old style of warlord rule, he left the field of action and propaganda against the Japanese in north China open for the Communists.

Further south the situation was more difficult. Throughout the Yangzi valley there still survived some pockets of Communist guerillas who had remained behind when the Long March left Jiangxi in 1934, and when the Japanese broke the Nationalist defence in 1938 these groups emerged to join the resistance. As in the north, Communist units allied themselves with local forces and with isolated Nationalist troops. Linked by radio and secret communications with Yan'an, a New Fourth Army was established in Anhui, with contacts into south China. For this type of underground movement against Japan, Communist guidance was consistently more effective than that of the Nationalists, who still concentrated on formal military arrangements, and it appeared for a time that the Communists and their allies would pre-empt the role of popular resistance in south China as they already had in the north.

In January 1941, however, the Nationalists struck back. On orders from Chongqing, the New Fourth Army was deprived of recognition and disbanded, and in a renewal of civil war behind the Japanese lines the local Nationalist commander turned against his fellow countrymen and destroyed their organization and their leadership. From the Nationalists' point of view, the encroachment of the Communists had gone too far: the heartland around Nanjing, although it might be occu-pied by Japan, must still be preserved from other rivals. The Communists protested against the Nationalist aggression, they boy-cotted the meetings of the political council which formed the king-pin of their alliance with Chongqing, and they made it clear that the united front was ended. From this time forward, the two parties fought separate wars against the invader, and both looked forward to ulti-mate victory not only over Japan but also over their opponents with-in China.

In other respects, too, Yan'an was now a rival to Chongqing. The territory controlled by the Communists was smaller than that of Free China to the south, and Shaanxi province was certainly less pros-perous than Sichuan and Yunnan, but the years of consolidation in the north-west and the experience of war against Japan gave the Communists a confidence and sense of achievement which the refugee Chongqing government found hard to match. As early as February 1936, a few months after the end of the Long March, Mao Zedong, Zhou Enlai, and Peng Dehuai had established a Red Army University to train recruits as military officers, guerilla leaders, technicians, and political cadres for work against Japan. There was no class distinction in enrolment, the courses were intensive in three or four months, and an entrant needed only to have completed his education at primary level. In following years, particularly after the Japanese invasion, young men from every part of the country joined the Red Army for their training, and whatever their political beliefs may have been at the beginning they came to accept the discipline and the ideals of the Communists.

In contrast to the Nationalists, who paid lip-service to scholar-ship and freedom of thought, but then left intellectuals to official neglect, occasional persecution, and political futility, the Communists at Yan'an emphasized that all training must be devot-ed to immediate practical use. The so-called university was essen-tially a technical college, but even in the lowest ranks of the army soldiers were taught to read and write, and officers and men were encouraged to see themselves as genuine comrades in a struggle for specific goals and a meaningful programme. In the higher grades of the party and the army the same pattern held true: of all the strands that maintained the coherence of the Communist move-ment, surely the most important was the sense that everyone had something to contribute and that the common cause deserved effort, attention, and loyalty.

Mao Zedong and the Spirit of Yan'an

By the beginning of the 1940s Mao Zedong was not only the political head of the Communist movement, he was also the dominant intellectual figure. His political authority had been confirmed by the elimination of his two major rivals in the army and the party, Zhang Guotao and Wang Ming. Zhang Guotao, an experienced soldier and a second leader of the Long March, had taken a large contingent on a separate expedition into western Sichuan and the foothills of Tibet. He was later compelled to rejoin the Shaanxi group, but in 1938 he deserted and took service with the Nationalists. Wang Ming, former leader of the Twenty-eight Bolsheviks, who had spent several years as Stalin's protégé in Moscow, returned to China in 1937 to urge, like Zhang Guotao, a stronger alliance with the Nationalists against Japan. He was posted for liaison to Wuhan, however, where he was isolated from the leadership at Yan'an, and his patriotic work was lost when that city fell to the Japanese. By 1939 Mao and his supporters held predominant authority.

Besides his political and military responsibilities Mao Zedong found opportunity to write and publish his own thoughts on the future of the war and the role of the Communist Party in China. In his work on *Basic Tactics*, he discussed the importance of guerilla combat and the final development of irregular units into a conventional force strong enough to defeat the enemy in open battle; though at this time he did not believe Japan could be defeated without major support from overseas and the full co-operation of regular Nationalist armies. The major ideological work of this period is the essay on *New Democracy*, which gave a Chinese slant to the established Marxist-Leninist interpretation of revolutionary development. Mao claimed that the Communist Party could play a leading progressive role even in a pre-industrial society such as China, and that the revolution should be led and supported not just by the national capitalists and intellectual patriots nor, in its later stages, simply by the industrial proletariat, but by all revolutionary classes, including soldiers, workers, and peasants. The Nationalists might continue as official leaders for a time, but it was essential they should maintain their principles of reform and should not falter in their task. In due course, the Communists would succeed to leadership, and they were at all times the guardians and watchdogs of full revolution.

In February 1942, Mao embarked on the Rectification Campaign, forcing all members of the Party to understand and accept the correct relationship between theoretical and practical work. Just as the Red Army University was planned largely for technical training of political and military cadres, so it was essential that a balance be struck between the extremes of simple empiricism, or trial and error, and the opposite mistake of pure theorizing. The ultimate goal was a fully Communist society, and everything should be directed towards that end. For this, the minority must always give way to the interests of the majority, and the individual, no matter how high his rank, remained the servant of politics. Art and literature, the subjects of the special Yan'an Forum in May 1942, were given no other functions but to encourage political consciousness among the people, to educate them for their work, and to inspire them towards Communist liberation. For all cadres and for many of the people, basic texts of Marx, Lenin, Stalin, and Mao himself were prescribed for study, and those who failed to put theory into effective practice were criticized by their fellows and encouraged to analyse their own faults. With such training and indoctrination, each party member had a definite role to play and a proper pattern of conduct.

In earlier writings at Yan'an, Mao Zedong had recognized the importance of the alliance and the united front, but the New Fourth Army incident of 1941 changed the situation. Besides withdrawing from liaison committees the Communists re-established the New Fourth Army, under the command of General Chen Yi, and encouraged its expansion into Nationalist spheres of influence. On their side, Chiang Kaishek and his government cut off all supplies and assistance, and it was alleged they made agreement for Wang Jingwei's puppet troops to attack the Communists in this time of weakness. Certainly, in July 1941, within six months of the breakdown of the united front, the Japanese and their associates began a major campaign in the northwest, pressing against the Communist position with fixed blockhouses, and subjecting the people of unsettled and guerilla areas to the 'Three-all' campaign: burn all, kill all, loot all. Bitter at the lack of help they received, the Communists announced that by 1942 the numbers of the Eighth Route Army had been reduced from 400,000 to 300,000.

Within Communist-controlled territory, the end of assistance from Chongqing faced the administration at Yan'an with serious difficulties. The winter of 1941 was hard, and there were shortages of food and clothing as well as military equipment. Careful rationing, however, overcame the immediate danger, and thereafter the encouragement of co-operatives for industry and agriculture, and of local militia and soldier-farmers, strengthened the meagre economic base of the region. Ultimately the fact that the economy of the north-west was independent from the finances of Chongqing and Nanjing proved a blessing in disguise, for by 1943 the inflation in Communist territory was largely under control, and supplies of locally produced food and clothing were sufficient for both the soldiers and the people who supported them.

To all observers, the contrast between Nationalist and Communist China was impressive. In the government at Chongqing, corruption was widespread, open, and uncontrolled. In the region controlled from Yan'an, civilian officials and military men were seen to work with the people, and during the great production drive of the early 1940s men and women of all ranks shared the labour. In Nationalist armies, corruption at the top was matched by brutal treatment of private soldiers, often led to battle with ropes about one another to hold them from running away; and the pressures of war had produced every form of vice and exploitation, from black markets and prostitution to murder, rape, and senseless destruction. In Communist territories, although there were certainly scandals of dishonesty, though illiteracy was still high, and though many people still followed traditional customs and superstitions, the basic social pattern seemed secure and accepted, men and women dealt together on equal terms without the former distinctions of class and occupation, and those who worked were fairly rewarded with food, clothing, and a sense of purpose.[7]

In the history of the Chinese Communist Party, the 'Yan'an Period', from the end of the Long March in 1935 to the eve of final victory in 1948,[8] is rightly regarded as a turning point in its fortune and the formative period of the future. From the caves and houses of this small town in the hills of loessland, the Communists established a viable administration and defended themselves with success against both the Nationalists and the Japanese. Partly by accident, partly by design, their leader Mao Zedong became the focus of respect and admiration. People throughout China recognized the strength of his authority and the appeal of his programme, and they found a hopeful tolerance in his adaptation of Marxist theory to Chinese conditions and the egalitarian, reformist nature of his government.

Foreign correspondents and other overseas visitors played an important role in the acceptance of the Communists throughout China and the outside world. Even before the capture of Yan'an the American journalist Edgar Snow had visited Mao at his headquarters and his book *Red Star over China,* with detailed accounts of long interviews with Mao, Zhou Enlai, and other leaders, was speedily translated into

Chinese and very widely circulated.[9] Later correspondents, including James Bertram, who visited Yan'an in 1937, and Agnes Smedley, who wrote the story of the Long March as well as a report of conversations with Mao Zedong, presented a man of self-confidence and wide understanding, with a straightforward, effective government and way of life. Some men from the West, impressed by what they heard, came to join the Communists, and one of Mao Zedong's best-known essays is the piece written in memory of Norman Bethune, a Canadian doctor who served with his armies and died in the north-west in 1939.

Despite the publicity and the stability of the Yan'an regime, with its growing influence throughout China, the centre of resistance was still at Chongqing, and Chiang Kaishek was seen by enemies and allies alike as leader of the Chinese people. As the threat from Japan to the region south of China grew more serious, Western support for Chiang and Free China became gradually more practical and immediate, and at the end of 1941 Chiang Kaishek found the allies he had waited for so long.

The Yanks are Coming

By mid-1941, though no war had been declared, China had been fighting Japan for four years, but the refugee government in Chongqing had not received direct assistance from any foreign power, and there was no reason to believe the Nationalist or Communist armies could present an effective counter-attack against the vast territories of their country which were held by the invaders.

In 1941, however, the sympathy of Western governments, and particularly that expressed so often by the Roosevelt administration in the United States, began to take a more tangible form. In part, this was a reaction to the aggression of Japan's allies in Europe, Germany, and Italy, and support for Free China in the East was a corollary to the military and economic aid which America was passing across the Atlantic to Britain.[10] In part too, there was growing fear of Japanese ambitions in the Pacific. Many foreigners were still prepared to do business with Japanese interests in Tokyo or Shanghai, but increasing numbers of people foresaw the threat against the European colonies and dependencies of South-east Asia. With the defeat of France in 1940, the Vichy government was compelled to accept Japanese troops in northern Indochina around Hanoi, ostensibly to cut supply lines to the Chongqing government, but in July 1941, when the Japanese occupied the region around Saigon, it became clear their interests lay also

in the oil and rubber of Malaya and the Netherlands East Indies, present-day Indonesia. From that point on, the outbreak of war was only
a matter of time.

The Americans, moreover, had already pushed neutrality to its limits. In April 1941 the Lend-lease agreement with Britain was extended to Nationalist China, and US$45 million was sent as a first instal-
ment of military aid. In August 1941 an American Volunteer Group
of the Chinese Air Force, later famous as the Flying Tigers, was established under the command of US General Claire L. Chennault. Most
important of all, however, in reaction to the Japanese move on Saigon,
Britain and the United States imposed crippling economic sanctions,
and demanded that Japan withdraw not only from Indochina but also
from China. A period of desultory negotiations ended on 7 December
1941 with Japan's attack on the American Pacific Fleet at Pearl Harbour
in Hawaii, and her belated declaration of war.[11]

For Chiang Kaishek, the outbreak of full hostilities in the Pacific
was a triumph of strategy and diplomacy, and his prestige as the
national leader gained him international fame. Officially, China
declared war against Japan, Germany, and Italy on 9 December 1941,
and in January 1942 Chiang Kaishek was made Supreme Military
Commander in the China Theatre. As a leading ally against the Axis,
Chiang took part in the Cairo conference of November 1943, and in
October 1944 Nationalist China joined negotiations to draft the char-
ter of the future United Nations. On 5 March 1945 China joined in
issuing invitations for the first United Nations conference at San
Francisco, and at that meeting, which began on 25 April, the government of China, with the other four 'Great Powers', America,
Britain, France, and the Soviet Union, was made a permanent member of the Security Council.

Despite this international approval, however, and despite the aid
which the United States offered his government, the position of
Chiang Kaishek and his Nationalists improved very little, and in the
first year of the extended war the Japanese continued their record of
success. In 1942 their armies captured Singapore, the Philippines,
and the Netherlands East Indies, and their navy swept the seas from
the western Pacific to the Indian Ocean. For China, one early result
of alliance with the West was the despatch of troops in an expedi-
tionary force to relieve the retreat of the British in Burma. The
Japanese cut the Burma Road, but some supplies were maintained
by airlift over the 'Hump' of mountains between Assam in India and
Kunming in Yunnan, and Chinese forces continued to operate in the

north of Burma, supporting the British in their defence along the frontier with India, and joining the final attacks in 1945 which broke the Japanese armies.

Sadly, however, although individual Chinese units achieved some success in these campaigns, they were seldom regarded with high favour or great confidence by their allies, and the judgement of the foreigners was largely justified. The years of defeat, the lack of equipment, and the poor quality of discipline and training had rendered the armies of the Nationalists a very uncertain quantity, and the brunt of the fighting in Burma was borne by the British and the Indians. The Chongqing government was reluctant to commit its forces at a distance, but even within China itself, there was no energy for an offensive. As in all the years since 1938, the front line in China remained a wavering, devastated no man's land, with Japanese punitive expeditions raiding the peasants' fields at harvest time, while the Chinese watched impotently from the hills, skirmished occasionally on the fringes of the enemy's advance, and claimed victory when he retired according to plan.

Neither the military men of America nor the people of China were impressed by the Chongqing regime and its army, but the political situation, under jealous dominance by Chiang Kaishek, provided little opportunity for improvement. Early in 1942, as an attempt to coordinate operations against Japan, the American government sent General Joseph W. Stilwell to take command of all United States forces in China, Burma, and India, and to act as chief of staff in the China Theatre under Chiang Kaishek as Supreme Commander. Tough and effective as a leader, deservedly popular among his troops, both American and Chinese, Stilwell had experience as an interpreter to military missions in China, he spoke Chinese well, and he had great sympathy for the people. In every other respect, however, and certainly in the light of events, it is hard to imagine a more unsuitable choice for the chief liaison officer at Chongqing. Adequately described by his nickname of 'Vinegar Joe', Stilwell was an aggressive, plainspoken soldier, with an unfortunate distrust of all foreigners and all bureaucrats. He despised the weaknesses and failures of Chiang Kaishek's regime and criticized them unmercifully; he showed an almost equal dislike and distrust of the British who were fighting as his allies in Burma, and he was always jealous of his personal authority in the chain of command and the control of operations.[12]

From this last point of view, it was Stilwell's policy to develop the Chinese army under his own guidance, so that it could operate with some stiffening of American forces and take the offensive on land against Japan. In order to achieve this programme he sought direct control over Chinese units, and he protested Chiang Kaishek's blockade of the Communists at Yan'an. For an American, it was dangerous and absurd that two parties with a common enemy should fail to join forces, and Stilwell made his feelings very clear.

It was impossible, however, that such proposals could be put into effect: Chiang Kaishek would never relinquish the army on which his power depended, and neither the Nationalists nor the Communists were prepared to trust one another. Stilwell and other Americans were irritated and impatient with the corruption and disorder which they saw under the Chongqing regime, and they were impressed by the achievements of the Communists and by the apparent moderation of their land reform and other programmes. They found it difficult to effect any meaningful changes in the Nationalist administration or in the organization of their armies, and the defensive, do-nothing strategy of Chongqing remained largely unaffected.

In mid-1944, under Stilwell's personal command, several divisions of Chinese troops fought their way through stiff Japanese opposition to a clear victory in Burma and the capture of the railroad towns Myitkyina and Mogaung, but success on this front was balanced by severe defeat in the east. In July 1944 the Japanese, under increasing pressure from the Americans in the Pacific, made their last great offensive in China, driving south from Wuhan and north from Guangzhou to capture the railroad through Hunan and secure the inland route between north and south. In the face of their advance, Chinese defences collapsed, Changsha was taken with ease, and the Japanese moved west to threaten Guizhou, the province south-east of Chongqing. Faced with this crisis, the Americans asked Chiang Kaishek to transfer authority over all Chinese troops to General Stilwell. Chiang refused point-blank; President Roosevelt sent a special envoy, General Patrick Hurley, to mediate the quarrel, and in October 1944 Stilwell was recalled. His successor, Albert Wedemeyer, was made commander of all American forces in China but could give no orders to Chinese troops, and his policy was far more agreeable to Chiang Kaishek. Hurley himself returned to Chongqing as American ambassador in November 1944, and in the following month the Japanese advance was halted and the enemy withdrew to the line of the railway at Changsha.

In fact, the Japanese had gained everything they wanted in China, and their concern lay now with the American offensives in the Pacific.

In the remaining months of the war, all attention was concentrated on the islands and sea-lanes which led towards the heart of their crumbling empire, and their position in China was static. For their part, the Nationalist armies had never taken a full offensive, and the end of the war saw no change in their position. To the allies, the military significance of both Nationalist and Communist China remained just what it had always been: a distraction which served to occupy a number of Japanese troops,[13] and a region for air bases and landing fields which could serve for bombing raids against the Japanese mainland. Both Chiang Kaishek and General Stilwell had hoped, each in his own way, that China might play a leading military role in the final campaigns of the war, and there was at one time a plan that the invasion of Japan should be launched from the preliminary reconquest of the Shandong peninsula. The armies of Chongqing, however, were far too ineffective to play any notable part in such moves, and the whole question was settled, suddenly and decisively, by the dropping of two atomic bombs on the Japanese cities of Hiroshima and Nagasaki on 6 and 8 August 1945. On 14 August the government of Japan agreed to unconditional surrender.

For the people of China, eight years of conflict ending in a total defeat of their enemy had brought obvious changes which dulled the edge of their triumph and prepared the ground for a last, bitter civil war. Apologists for the leaders of Japan in the 1930s have claimed, with some justice, that the government in Tokyo sought only that the Chinese should recognize their supremacy in the East and should co-operate with them against the West. The terms they considered reasonable, however, were more than any Chinese patriot could accept, and both Chiang Kaishek and the Communists, with courage and determination, had fought them to the limit of their resources. The war had given the Communists an opportunity to claim a leading part in the national struggle, and had also, by distracting their Nationalist enemies, gained them breathing space to build their government and to train a new society. For the Nationalists, however, swift defeat had been followed by years of weakness and frustration, and it seemed now that they had lost not only all the progress achieved in the late 1920s and early 1930s, but also, far more serious, the inspiration which might give them virtue in the future. While the Communists could claim local victories, and had taken advantage of opportunities to build guerilla communities across north China, it was the fate of the Nationalists that they returned to power by others' work, with no glory or prestige of their own. In the eyes of the people of China

Chiang Kaishek and his supporters owed their survival and success only to the support of outsiders, and there were many observers, both foreigners and Chinese, who doubted their ability to hold and maintain their government.

Notes

1. Two stories are told of this incident. One is that the Americans had acquired a Japanese aircraft, one of the few casualties of the fighting over Nanjing, and they were shipping it back for study. The Japanese, anxious their secrets should not be revealed, attacked and sank the ship.

   The other story is that after compensation for the *Panay* and other vessels had been agreed and paid, the Japanese asked to confirm that they now had salvage rights; they had, after all, paid for the ships. They also suggested that if the Americans wished to replace the *Panay* then Japanese shipyards might tender for the contract. The authorities in Washington were not amused.

2. Quoted in Coox, *Year of the Tiger*, p. 93.

3. After Wang Jingwei left Chongqing, he was pursued by agents of Chiang Kaishek, and an assassination attempt in Hanoi killed his personal secretary and protégé. From this experience, Wang turned to join the Japanese.

4. Figures given in *China Proper* vol. 3, p. 217.

5. The situation of the French was more equivocal. The fall of France in 1940 meant that French possessions in the Far East came under the authority of the Vichy government, now an ally of Germany, and the Japanese gained access to French Indochina for troops and air bases, threatening the Nationalists from the south and assisting also in their forced alliance with Thailand and their later attack on Singapore. Moreover, even after the Japanese declared war against Britain and the United States in December 1941 they continued to recognize the French Concessions at Shanghai, Wuhan, and other places, and these retained some form of autonomy when all the rest had disappeared. They were not taken over until the fall of Vichy in 1944.

6. Though Chiang was frequently described in English-language works as the 'Generalissimo' (and in America even as the 'G-mo'), it is hard to determine which Chinese title this refers to. In 1917 Sun Yatsen took for a time the title *dayuanshuai*, which basically means 'commander-in-chief'; though it is a general term rather than a military rank, it was translated as 'Marshal' or 'Generalissimo'. In 1926, after Sun's death, when Chiang gained authority as commander-in-chief (*dayuanshuai*), journalists again rendered the title as Generalissimo and the style has remained with Chiang ever since (Payne, *Chiang Kai-shek*, pp. 1-2). From 1932 Chiang was Chairman of the National Military Council, and in the government at Chongqing, although he was not the formal head of state, he controlled all practical power as *zongcai* (director-general) of the Nationalist Party and Chairman of the Supreme National Defense Council, the highest organ of civilian government. He was also patron or head of so many other organizations that even his secretaries lost count (White and Jacoby, *Thunder out of China*, quoted in Schurmann and Schell, *Republican China*, p. 245). In common reference, however, Chiang was the Generalissimo, and at least one American adviser is said to have remarked during a particularly bleak period of the war against Japan, 'I just wish the Generalissimo would do a little generalissimo-ing'.

7. One aspect of Communist life may be noted here: the freedom and official equality of relations between the sexes. Under Communist auspices, the traditional 'feudal' marriage and concubinage were abolished, and marriage and divorce needed no more than a simple registration with the government. It appeared that Sun Yatsen's ideal of female emancipation had come to effect, and visitors to Yan'an remarked on the austerity and almost sexless restraint of the energetic, youthful population.

Mao Zedong also made some changes in his private life. He was first married in 1907, by arrangement of his parents, but he repudiated that wife early in his career. His second wife, Yang Kaihui, whom he married in 1921, bore him a son, Mao Anying, but she was killed by the Nationalists after the failure of the Communist attack against Changsha in 1930. He mourned her deeply, and one of his finest poems 'The Immortals', written in 1957, is dedicated to her memory (Ch'en, *Mao and the Chinese Revolution*, pp. 86 and 347, and Schram, *Mao Tse-tung*, pp. 352-3).

During the time of the Jiangxi Soviet, Mao Zedong lived with and married He Zichen, who accompanied him, although she was pregnant, through all the Long March and was wounded in a bombing raid. She bore him five children, but in 1939 she went to Moscow for medical treatment and during her absence Mao Zedong divorced her and married a film actress from Shanghai, Lanping, now better known by her style among the Communists, Jiang Qing. Jiang Qing bore two daughters to Mao Zedong, and later took a leading position in the Cultural Revolution (see Chapter 10).

8. The Communists did not capture Yan'an until December 1936, and it did not become the capital of their administration until the following year. The whole period in the northwest, however, commonly and naturally takes its name from this place.

9. The autobiographical account of Mao Zedong, presented in Snow's work as 'Genesis of a Communist', was the first statement of his life and thoughts, and it was soon translated into Chinese.

10. The Japanese, however, were not particularly interested in giving military support to Germany. To keep Manchuria secure while operations continued in China and the south, Tokyo signed a neutrality pact with Stalin in April 1941, and Hitler's attack on the Soviet Union in July was welcomed only as a means to ensure that the northern frontier would remain quiet.

11. Because of the effect of the International Date Line, the day of Pearl Harbour was 8 December in east Asia.

12. In *The Stilwell Papers*, Chiang is commonly referred to as 'Peanut', at first perhaps with some affection, but later with nothing but hatred and contempt. The British were 'Limeys', and he was frequently touchy and suspicious in his dealings with them.

13. And this was not an insignificant contribution. Even in 1945, at the last stages of the war, there were one million Japanese soldiers on the Chinese mainland, with another 650,000 awaiting the Soviet threat in Manchuria. Though the quality of the troops varied, and the ability to deploy them would have depended upon circumstances, notably sea power, it is hard to believe some of those divisions would not have made a difference to the campaigns of the Pacific theatre or in south-east Asia.

Further Reading

Boyle, John Hunter, *China and Japan at War, 1937–1945: the politics of collaboration,* Stanford, Stanford University Press, 1972 [with a discussion of Wang Jingwei and the puppet government at Nanjing].

Ch'i Hsi-sheng, *Nationalist China at War: military defeats and political collapse 1937–1945,* Ann Arbor, Michigan University Press, 1982.

Coox, Alvin D., *Year of the Tiger* [1938], *Orient/West,* Tokyo, 1964.

Esherick, Joseph W. (ed.), *Lost Chance in China: the World War II dispatches of John S. Service,* New York, Random House, 1974.

Eastman, Lloyd E., 'Nationalist China during the Sino-Japanese War 1937–1945' in *Cambridge China* 13.

Kataoka, Tetsuya, *Resistance and Revolutionary China: the Communists and the Second United Front*, Los Angeles, California University Press, 1974.

Payne, Robert, *Chiang Kai-shek,* New York, Weybridge and Talley, 1969.

Selden, Mark, *The Yenan Way in Revolutionary China*, Cambridge Mass., Harvard University Press, 1971.

Stilwell, Joseph, (arranged and edited by T.H. White), *The Stilwell Papers,* New York, Schoken, 1948.

Tuchmann, Barbara, *Stilwell and the American Experience in China 1911–54*, New York, Macmillan, 1970.

Van Slyke, Lyman, 'The Chinese Communist movement during the Sino-Japanese War 1937–1945' in *Cambridge China* 13.

White, Theodore H., and Jacoby, Annalee W., *Thunder out of China*, London, Gollancz, 1947.

7

Communist Victory, 1945–1949

B Y the end of the ruinous war against Japan, and despite the victory which the Americans had brought, the Republic of China had gone through the same litany of failure, frustration, and misery as the Qing dynasty before it. As the dreams of liberal democracy gave way to the brutality of warlords and the corrupted ideals of the Nationalists, many of the élite were in despair and looked with some hope to the Communists, who offered at least an alternative, and surely an end to civil conflict.

On the other hand, despite the losses in war, Chiang Kaishek had re-established a traditional structure of authority. Frightened of the Communist programme, landholders throughout the country supported the regime, industrialists, merchants, and the gangs of the cities either accepted the Nationalists' power or joined in the share of it, and the government in Nanjing controlled the chief military force of the nation.

Tradition itself, however, now appeared inadequate to the problems of the modern day, and the very forces which supported the Republic contained the seeds of their own destruction. In the countryside, the local power of the landlords and their agents was limited and fragmented, no match for a determined peasant rebellion under the coherent guidance of the Communists, while the very threat of such opposition made the conservative forces more aggressive and violent, and bound the government ever more tightly into a losing struggle against reform. Within the cities, the inflation first induced by war was quite out of control, destroying the basis for any secure prosperity and encouraging the corruption already endemic among those who wielded power.

The Return of the Nationalists

On 5 September 1945, three days after the formal surrender of Japan, the first units of the Chinese army returned by air to Nanjing. Grants of aid, money, and goods, came by air to Chongqing and by sea to

Shanghai, and as the Nationalist government returned to the east there seemed some hope of reconstruction and future security. On 25 October, after fifty years under the Japanese empire, the island of Taiwan was officially returned to the government of the Republic of China, and the British and American governments now confirmed their renunciation of extraterritoriality and other rights of foreign settlement which they had promised in 1942.

Not all omens, however, were so favourable, and the losses and damage of eight years of war would not be recovered so easily. Wang Jingwei, leader of the puppet government at Nanjing, had died in December 1944, but his ministers and associates were imprisoned and executed, and accusations of collaboration allowed the returning Nationalists to confiscate the property of many wealthy men who had remained in the east. It was also announced that major capital enterprises would be controlled by state monopolies, a move which not only alienated business interests but also created enormous opportunities for corruption. Throughout the territory under Nationalist rule, there were well-attested accusations that supplies and relief aid were being stolen by government officials, and all the old scandals concerning the Four Families were revived. The years of solitary power at Chongqing had increased the dependence of the Nationalists upon Chiang and his immediate supporters, and no one could believe that they were not gaining profit from their power.

With such suspicion of personal gain at the top, there was little restraint on lower levels in the party, the government, and the army, while desperate inflation, making normal wage and salary levels meaningless, strengthened the trend. Many honest men of patriotism and goodwill still served the Nationalist cause, but they found themselves overwhelmed by the pressure of events and by the moral and economic weakness of the state they sought to serve. The currency battle between the governments of Wang Jingwei and Chiang Kaishek had already disrupted the monetary system of the country, but postwar inflation continued the trend. In Shanghai, for example, from 1937 to 1945 the price of ten litres of wheat rose from $1 (Chinese National currency) to $665, but over the next twelve months, through 1946, the price of the same quantity of wheat increased ten times, to more than $6,700. In such circumstances as these nothing could maintain the rate of foreign exchange, and the only hope of financial survival for an individual lay in the ownership of property, no matter how it might be obtained, rather than in the earning of a salary or the entitlement to a debt. Everywhere the uncontrolled inflation confirmed people's distrust of the government and, by the same token, rendered the government incapable of constructive action.

On the political scene, in the territory now controlled by Nanjing, the position was little more impressive. The Nationalists had held the leading role in the republic since 1928, but many people, notable among them the students and professors of South-west United University returning from their exile in Yunnan, had high expectations of liberal reform now the pressures of war had eased. For a time these non-Nationalist civilian political groups operated in alliance as the Democratic League, and the scholar and poet Wen Yiduo was among their leading spokesmen. Chiang Kaishek, however, had no wish to relinquish any powers he had acquired, and he protected his regime with all the apparatus of a police state. In the summer of 1946 Wen Yiduo was assassinated by members of the secret police, and in 1947 the League itself was disbanded by government order. On 15 November 1946, as an elected National Assembly held its opening meeting, Chiang Kaishek announced the end of the Nationalists' period of political tutelage; but since the two major alternative parties, the Communists and the Democratic League, had found it meaningless to take part in the assembly, his statement had a hollow, cynical ring.

At the same time, although the regime at Nanjing was both autocratic and corrupt, the pressures it was facing would have taxed the abilities of any government. The Western powers, notably America, regarded the Nationalists with some benevolence, and supplied quantities of aid; but the Communists still maintained their armies in the north-west, great parts of the countryside were under their control, and in the territory of Manchuria the situation was seriously confused by the intervention of Soviet Russia.

Supporters of Chiang Kaishek have paid great attention to his conference at Cairo with Roosevelt and Churchill in 1943, but they often fail to mention that the meeting was immediately followed by another conference at Tehran, this time between the two Western leaders and Stalin. Chiang Kaishek, suspicious and embittered by Soviet support of the Chinese Communists and by lack of Russian aid in the war against Japan, had refused any meeting with Stalin, and for the time being the American and British governments were prepared to tolerate his foibles. At the beginning of 1945, however, as the end of the war against Germany came in sight, Roosevelt, Churchill, and Stalin met again at Yalta on the Black Sea, and it was agreed that Russia would enter the war against Japan in exchange for control over the Manchurian railway systems, the renewed lease of Dairen and Port Arthur as a naval base, and, of most permanent significance, the recognition of Outer Mongolia, already controlled by a Soviet government subordinate to Moscow, as a nation independent of China. So although Chiang Kaishek's allies had won the war against Japan and had regained Taiwan for China, they had also agreed to the cession of one great territory and rendered control of the other both difficult and doubtful.

In August 1945, when Japan agreed to surrender, the Soviet Union had been at war with her for less than a week, but Russian armies continued to advance until they controlled both Manchuria and Korea. In the rest of China, Japanese commanders were ordered to hold their ground and surrender their positions only to Nationalist forces, and even in the north, where great areas of the North China Plain were controlled by Communist troops, the Americans supplied aircraft so that the Nationalists could take over the Japanese garrisons without interference. The American General Wedemeyer described the ferrying of Nationalist units to strategic points throughout China proper as the greatest air and sea movement in history, and in immediate terms it was remarkably successful, but it supported the Communist and other observers' view of the Nationalists as dependents of a foreign power, it spread government troops into isolated posts across the country, and it confirmed Nationalists and Communists in their mutual distrust.

For the time being, the Russians remained in Manchuria. As protests raged throughout China, and the Nanjing government attempted to gain some agreement for their withdrawal, the armies and technicians of the Soviet Union concerned themselves primarily with looting the country they occupied. They claimed they were taking equipment as war reparations against Japan, but it is hard to justify their rights against those of China, which had suffered Japanese aggression so much longer. Within a few months, the Soviet Union acquired some US$2,000 million worth of machinery and heavy equipment, and the factories, railway yards, and mines of Manchuria were left gutted and bare. Neither the Nationalists nor the Communists who succeeded them have forgotten or forgiven this plunder of China's most important industrial territory.

Gradually, however, the Russians agreed that they would withdraw, first by the end of 1945, and then by May 1946. At the same time they secretly aided the growing strength of the Communists. Unlike the Americans, who justified their airlift of Nationalist troops as aid to the recognized government of an ally, Stalin was compelled to take a more devious line, but by one means or another strong-points in the countryside and quantities of captured Japanese equipment were transferred to the Communist forces in Manchuria now commanded by General Lin Biao. Sometimes Soviet garrisons of supply dumps and other positions of value were deliberately weakened and then withdrawn as soon as the Communists attacked. On other occasions, Nationalist troops attempting to occupy a town would be delayed or held off with some excuse, and by the time they arrived the Communists had taken possession.

After considerable hesitation, the government in Moscow decided to accept the Communists under Mao Zedong as a real contender for power in China and to render it a modicum of help. There was no question of military intervention: the Soviet Union, all but exhausted by the war with Germany, had still to consolidate its gains in Eastern Europe. There is no evidence that even the operations in Manchuria were planned for anything more than short-term gain. By the middle of 1946, however, as the Russians withdrew the last of their men from the northern towns of Harbin and Qiqihar, these places were taken over directly by units of the Communist army. The Nationalists were left with only a few garrisons in the cities further south, and the countryside was dominated by the Communists.

In Manchuria, therefore, the Nanjing government was faced with a serious threat. The territory surrendered by the Japanese had been at last relinquished by the Russians, but it was now almost entirely controlled by the Communists; and while the loss of these wealthy industrial provinces would weaken the central government, they also supplied the enemies of the Nationalists with a new base for expansion. No one had any doubt that civil war would come again; the only questions at Nanjing were the time they should open hostilities and the precise objectives they should aim for.

For the Communists, the programme was equally clear, but Mao Zedong was convinced that time was on their side, and he did not expect a quick victory in straight military terms. The Communist armed forces amounted to perhaps one million men, but the Nationalists and their allies such as Yan Xishan claimed three or four times that number. On the other hand, besides their old base in the north-west around Yan'an and their new authority in Manchuria, the Communists had maintained influence throughout north China. It is estimated that even in the early 1940s, during the war against Japan, the Communists had more fighting men in the North China Plain than in their base area around Yan'an. By the end of 1945, though the Nationalists held the cities and other main points, they were surrounded by an uncertain, hostile countryside, and they were threatened by guerilla bands of various sizes. In his writings on the People's War, Mao Zedong had said that the revolutionary army must move among the people like a fish in water, and find among them its natural support. One year after the destruction of Japan, the Communist position among 'liberated areas' in northern China was approaching that ideal.

In the years of fighting against Japan the leaders at Yan'an had made a fair show of co-operation with non-Communist allies, and the only Chinese they proclaimed as enemies were those who followed the puppet government of Wang Jingwei. As the Japanese gave place to the Nationalists, however, the patriotic struggle changed to class warfare: everywhere the Communists established their influence the common people were urged to see capitalists, landlords, and agents of the government as their enemies, and to take the law into their own hands. With support from political cadres and military men, and with growing confidence in their own united power, in one village after another the people carried out this programme.[1]

In a graphic description, the American correspondent Jack Belden told how the people of Stone Wall village united under Communist leadership to overthrow the power of their oppressive local landlord, try him in a people's court, and kill him.[2] This was basic revolution, and the people who destroyed their oppressors in this way committed themselves fully to the cause of rebellion. Elsewhere, conditions naturally varied: absentee landlords could not be dealt with so summarily, and government officials with soldiers supporting them could often overawe the local people for a time. On the other hand, the very attempts of terrified landlords and their agents to frighten the local people created divisions in rural society which the Communists were swift to exploit. As in Vietnam in more recent years, terror bred terror and violence created violence, and there came a stage when the basis of order was gone, when government troops controlled villages only as long as their patrols were on the spot, and at all other times these places served as havens for Communist guerillas. The people themselves, faced with high, uncontrolled rents, and heavy taxation for an unconcerned and distant government, obeyed official demands only when they had to, and local headmen, if they valued their lives, kept well out of the way. Sometimes slowly, sometimes almost overnight, territory was removed from government control, taxes and military recruits were gathered now for the Communist cause, and whole areas of the country that were theoretically under Nationalist rule were actually in the hands of their enemies. With their programme of land reform, the low level of their demands for taxation and other support, and the impressive authority and honesty of their cadres, the Communists at grassroots level had already brought revolution to the people, and the Nationalists were neither willing nor able to offer any viable alternative.

The Americans in China

Despite the support and advice they had given to Chiang Kaishek, and the evident importance which he and his colleagues placed upon their alliance, neither the government nor the people of the United States was prepared to intervene actively in the struggle between the Nationalists and the Communists. On the spot, in the closing stages of the war against Japan, opinions were divided: many private observers and well-informed officials, notably the career diplomat John S. Service, had reported accurately on Communist achievements and on their popularity among the people; while at higher levels, despite the embittered criticism from General Stilwell, men such as Ambassador Hurley, a close political associate of President Roosevelt, had urged that the United States should commit itself firmly to the unification in China under the leadership of the Nationalists.

The major achievement of Hurley's diplomacy was the visit of Mao Zedong to negotiations at Chongqing on 28 August 1945. Naturally enough, Mao suspected the good faith of Chiang Kaishek, but Hurley travelled to Yan'an to give personal assurances, and the Communist leader agreed to attempt a solution. The exercise was largely meaningless, for neither side was prepared to trust the other. Politically, Mao Zedong appeared ready to compromise, and he accepted the principle of a national convention which would discuss the introduction of a new constitution and the calling of a national assembly. In practical terms, however, he refused to accept Nationalist claims of control over Communist-held territories, and he refused to reduce the strength of his army from its existing 48 divisions unless the Nationalists made a similar reduction in their own forces, from an existing strength of more than 250 to a limit of 120 divisions. Considering past history, the Communists were disinclined to attempt survival without significant military backing, and the later fate of the Democratic League at Nanjing both justified and confirmed their distrust of Nationalist power.

1 The Empress-dowager: an offical photograph of the early 1900s

2 Popular resistance to the foreigners: the Black Flags of Vietnam (from the Shanghai broadsheet *Tianshizhai huabao*)

3 An offical photograph of Yuan Shikai as President of the new Republic

4 Sun Yatsen (Government of the Republic of China)

5 Wang Jingwei (L) with Chiang Kaishek (Government of the Republic of China)

6 Shanghai: the Bund in the 1930s (from *The Living China,* an offical publication of the Republic)

7 Street scene, Beijing (courtesy Hedda Morrison; from *A Photographer in Old Peking,* Hong Kong, Oxford University Press) 1987

8 Shanghai 1932: civilians in flight from the Japanese (Collection of the New China News Agency [NCNA])

9 Nanjing 1937: Japanese tanks enter the city (NCNA)

10 Communist cavalry among hills of loess (NCNA)

1 Mao Zedong with peasants at Yan'an (NCNA)

12 Cultural Revolution: Lin Biao, Zhou Enlai, and Mao Zedong (*China Pictorial*)

13 Cultural Revolution: students denounce Liu Shaoqi (*China Pictorial*)

4 In the late 1960's, young people sent down to the country were encouraged to open up new ground in the north west, guided by local people with traditional methods of farming (*China Pictorial*)

5 Meanwhile, on Taiwan, more mechanized techniques are being used: crop-dusting a rice field (Courtesy the Government of the Republic of China)

16 Deng Xiaoping (NCNA)

17 Demonstrators at Tiananmen Square in June 1989

Six weeks later the conference broke up, with a communiqué calling for a Political Consultative Conference to settle unsolved problems. In the meantime it was agreed there should be a truce, and as both parties jockeyed for position, outbreaks of fighting were at least restricted. For the Nationalists, however, one great change had already taken place: in November 1945 the new government of President Truman in Washington announced a policy of continuing American support for the government in Nanjing so long as United States arms were not employed in any civil war, and so long as a serious attempt was made to come to terms with the Communists. Ambassador Hurley, who had consistently sought a clear commitment to Chiang Kaishek, resigned in disgust, and the President appointed General George C. Marshall, one of the chief American commanders in the war just ended, as special ambassador to China with instructions to act as mediator between the Nationalists and the Communists.

For Truman, one point was clear: there would be no American military involvement in any Chinese civil war. With the end of the struggle against Japan, United States forces were being demobilized as fast as reasonably possible, and no one had either the will or the tolerance for another great war in Asia. Some politicians and officials described the Chinese Communists as agents of Soviet power, but in these early months of peace the Americans were not yet seriously concerned by such a threat, and many saw the Communists primarily as agrarian reformers. As in the time of the airlift, the United States government was prepared to give non-combatant aid to Chiang Kaishek, but the tenor of Marshall's advice was that the Nationalists should set their own house in order before they attempted to bring the Communists under full control.

Despite this attempt at disinterest, however, the Americans did lend general support to the Nanjing government, and American arms supplied to the Nationalists for the war against Japan could obviously be turned against the Communists in the north. For their part, the Communists made propaganda from Nationalist reliance on the United States, and when American private soldiers acted in China as arrogantly and as stupidly as any private soldier may in any occupied country, the news was spread as widely as possible. For the Chinese themselves, free at last from the aggression of Japan, there remained a natural fear and hatred of further foreign intervention, and it was unfortunate for the Americans, regardless of their government's policy, that both the Communists and the Nationalists believed they were committed to the Nanjing regime.

By the beginning of 1946, although the Political Consultative Conference did achieve a measure of agreement on paper, there was still no reality in the discussions and the coming of war was clearly no more than a matter of time. In the summer of 1946 Chiang Kaishek determined to attack the Communist positions in Manchuria. In April Communist troops had seized the city of Changchun from the Nationalist garrison that had taken over from the Russians. On 23 May Nationalist troops recaptured it and from that time on, despite American mediations, the war for control of mainland China was fought through to a finish.

The Civil War

Both in Manchuria and in north China the first eighteen months of the civil war were fought at two levels: direct military conflict, and the Communist 'liberation' movement in the countryside. Estimates for the numbers of troops involved on either side vary considerably, but most authorities agree that at the beginning of open conflict the Communist forces numbered little more than one million, and the soldiers of the Nationalists were three times as many.[3] On the other hand, though Nationalist troops held the cities and railway lines of northern China and southern Manchuria, they were tied down in these defensive positions, and almost all the countryside was dominated by the Communists.

In the first period of the war the Nationalists achieved remarkable success. In the north-east they secured control of southern Manchuria, and in March 1947 a major drive in the north-west captured Yan'an and drove Mao Zedong and his headquarters into the hills to play hide-and-seek with the government troops. By the middle of the year, twelve months after the start of open war, the Communists themselves estimated that they had lost a considerable area of their original base region, and some eighteen million of their former subjects were now under Nationalist control.[4]

The situation, however, was not nearly as satisfactory for the Nationalists as it may have appeared. The very occupation of enemy territory meant that still more troops were now tied down in extended positions with little room for manoeuvre. The Communists in the meantime had recruited great numbers from the regions they controlled, and it was now estimated that the effective fighting strength of the Nationalists had been reduced by casualties, desertions, and the responsibilities of garrison duty to rather less than three million men, while the People's Liberation Army had increased to perhaps one and a half million. Despite the run of success, the year of fighting had weakened the Nanjing government and put severe strain on its ill-disciplined and poorly-led army, and the American refusal to maintain supplies of military aid for a civil war meant that equipment once lost was not readily replaced. In July 1947 the Communists began their counter-attack.

In two separate drives south across the Yellow River, a field army under Liu Bocheng and Deng Xiaoping occupied the hill country of the Dabie Shan in northern Hubei and eastern Anhui, and another corps seized the high ground of western Henan. These forces now threatened all routes in the region north of the Yangzi, and at the same time Communist troops in Shandong commanded by Chen Yi came into the open to harass and attack the Nationalist positions on the east of the plain. Within the space of a few months, Chiang Kaishek's advance into north China degenerated into a desperate holding action to guard lines of retreat along the railways.

In the far north-east the situation was still more serious, for although the Nationalists had committed more than half a million troops to the Manchurian campaign, they were outnumbered by the Communists against them. In the last months of 1947 the attacks of Lin Biao destroyed the Nationalist outposts and drove them back into a defensive triangle around Mukden/Shenyang. At the end of the long winter, in the spring of 1948, the Communists turned against the railway line through Jinzhou, the one line of retreat towards Beijing. Demoralized and almost immobile, the Nationalist commanders quarrelled among themselves, Jinzhou was captured on 14 October 1948, and the bulk of the remaining troops surrendered at Shenyang two weeks later. In the two-year campaign in Manchuria the Nationalists had lost almost every man they sent there.[5]

Further south, the situation was equally disastrous for Chiang Kaishek and his soldiers. In September 1948 Chen Yi's army captured Jinan and cut the main north–south line from Beijing and Tianjin to Nanjing. At the same time, the Communist forces moved from the west against the city of Xuzhou, junction of the Beijing–Nanjing railway and the Longhai line which runs east from Shaanxi to the sea. Here, in October and November 1948 took place the last great battle of the civil war, with half a million men on either side.[6] On the open plain, under Chiang Kaishek's direct command but with contradictory and mistaken orders, the remaining mechanized units of the Nationalists stood and fought in fixed positions while the Communists pressed against their flanks, broke through their lines, and at last combined to destroy them. Many Nationalist units defected to the enemy; others surrendered in swift despair; and the remainder, outnumbered and out-manoeuvred, held their ground and fought. On 15 December, when the Communists entered Xuzhou, the Nationalists had lost two good generals and two hundred thousand men, and their soldiers in the north were cut off from all support.

From the middle of 1947 to the beginning of 1949, the Nationalist armies had lost 1,500,000 men, and great numbers of these, either by defection or in the aftermath of defeat, had joined the Communist side. It was now claimed that the People's Liberation Army, three million men, outnumbered the Nationalists, and the victories that followed made it clear that there was no further chance of opposition. In January 1949 the Nationalist commanders in Beijing and Tianjin surrendered, and all north China was in Communist hands. The remaining Nationalists, totally demoralized by one of the greatest series of military disasters that any army has suffered at any time in history, could hope only for some miracle of aid or an agreement which might halt the Communist advance across the Yangzi.

In April 1948 the Nationalist-dominated First National Assembly had appointed Chiang Kaishek as President of the Republic. In January 1949, faced with the collapse of his armies, Chiang Kaishek resigned the presidency and authorized his Vice-President Li Zongren to act in his place and attempt to make terms with the Communists.[7] But the Nationalists had nothing to negotiate with. American observers, including General Marshall, now Secretary of State, and General Wedemeyer, who had been sent on a special mission as adviser in 1947, reported in gloomy terms of their corruption, incompetence, and brutality. Equally unfortunate, in the American presidential elections of 1948, the so-called 'China Lobby', friends of the Nationalists, had aligned themselves with the Republican candidate, Thomas E. Dewey, whom they and many others expected to win; when President Truman was returned he had small sympathy to spare for the problems of Chiang Kaishek and his allies. Even the most hopeful supporter of Chiang Kaishek could see no relief without millions of dollars in aid and over a hundred thousand American combat troops, and though the Soviet takeover of Czechoslovakia in 1948 marked the beginning of the Cold War between the West and the Communist powers, Truman's administration was no more prepared than it had been in the past to commit soldiers to the mainland of Asia. To any observer of the later conflict in Vietnam, that decision must seem very wise.

For their own part, the Communists found negotiations meaningless and refused to accept anything less than complete surrender, including the punishment of Nationalist war criminals. Li Zongren was prepared to negotiate even on these terms, and a peace delegation was actually sent to Beijing, but on 21 April 1949 the talks broke down and the Communists crossed the Yangzi River. Two days later they had captured Nanjing; on 15 May the Nationalists evacuated Wuhan, and on 27 May they abandoned Shanghai. On 1 October 1949, even as his armies still thrust south in the last campaigns of the war, Mao Zedong, now Chairman of the People's Republic of China, proclaimed the establishment of the new Communist state in a great ceremony in Beijing.[8]

The Flight to Taiwan

As the armies of the Communists spread across south China, almost without opposition, the majority of the people — peasants in the countryside, workers in the cities, and even soldiers in the abandoned Nationalist armies — had no choice but to accept the new government. A few units of the Nationalists retreated south-west to the hill country of the border between Yunnan province and Burma, and they or their descendants, operating now as independent bandits, still remain a trial to the government at Rangoon and a minor embarrassment to the Chinese of both parties. Other local leaders in the south had hoped to maintain some military power in isolated regions of Guangxi and Guizhou, but the speed of the Communist advance and the spread of their political influence removed all strength from these positions and the warlords who had remained in China turned hastily to foreign refuges in Taiwan, Hong Kong, South-east Asia or the United States. Among them was Li Zongren the Acting President, who left Hong Kong for America in December 1949, and who played no further part in Chinese politics until he returned to an amnesty and a welcome in Beijing in 1965.

For many Nationalists, however, the policies of the Communists, no matter how well administered, remained distasteful, and for the leaders and chief supporters of the fallen regime the future was obviously dangerous. As the Communist armies moved against the great south-eastern cities of Nanjing, Shanghai, Xiamen, Fuzhou, and Guangzhou, refugees in their hundreds of thousands struggled to board ships which might take them to some place of safety, and for the great majority this was the island of Taiwan. On 10 December 1949, after two months of travel by air through south and south-west China, from Guangzhou to Chongqing and Chengdu in Sichuan, never more than a few days ahead of the Communist advance, Chiang Kaishek himself flew from Chengdu to Taipei, and in Taipei on 1 March 1950 he again took the title President of the Republic of China.[9]

It was a curious twist of fate that set the Nationalist Chinese on the island province of Taiwan, for the territory had been governed by Japan for fifty years until 1945, and it had been returned to China only as a spoil of war. In the short time after their return to Nanjing the Nationalists had shown little interest in their acquisition, and the provincial government had followed a recognized mixture of arrogance and corruption. The native-born Taiwanese had no affection for their new rulers, and there was a strong movement for independence. In particular, during February and March 1947 widespread riots broke out in Taipei, and they were only put down by harsh use of armed police and with several civilian deaths.

In January 1949, however, after the fall of Xuzhou to the Communists, the Nationalist General Chen Cheng was appointed governor of Taiwan province, and in the months that followed some serious attempt was made, with land-reform programmes and new currency issues, to establish a favourable climate of opinion for the otherwise discredited Nationalist government. By the end of 1949, moreover, with far more immediate effect, an estimated two million refugees, many of them soldiers trained and armed, had arrived from the mainland. The native-born Taiwanese, whose population at that time amounted to little more than six million, were dominated by these newcomers, and they were now placed in a new and unexpected role as bastion of anti-Communist China.

Whatever else might be said of the Japanese occupation, Taiwan, like Manchuria, had benefited from an administration generally superior to anything on the mainland. The public health services had eliminated major outbreaks of cholera, typhoid, smallpox, and malaria, and although the island was used primarily as a source of primary production for the Japanese economy, the farming of rice, sugar, and tea was well established, and remained for several years the backbone of Taiwanese foreign trade. Japanese officials were forcibly repatriated at the end of the war, but the work they had done provided a basis for prosperity in the future, if only the Nationalists could find the time, energy, honesty, and competence to manage it.

Certainly it was clear that little else would remain under Chiang

Kaishek's control. During October 1949, in a surprising local rally, some Nationalist troops which had evacuated the city of Xiamen maintained possession of the small offshore islands of Jinmen, or Quemoy, and another rearguard action held on to the Mazu (Matsu) group outside Fuzhou. On the other hand, though there had been hopes that a stand might be made on Hainan, the large island south of Guangdong, this territory was already partly controlled by local Communists, and it was evacuated almost without a fight in May 1950. At the same time, the Nationalists abandoned the Zhoushan Archipelago, and they now held only the island of Taiwan, the Pescadores Islands in the Taiwan strait, and Jinmen and Mazu off the Fujian coast.

Whatever plans the Communists may have had, however, for pursuing their rivals across the sea and compelling a final settlement, came to a halt in June 1950. On 25 June the Communist government of North Korea invaded the territory of the Republic of Korea in the south, and two days later, taking the excuse to halt another Communist threat, President Truman ordered the United States Seventh fleet to patrol the strait of Taiwan and keep the two Chinese enemies apart. From that time on, open war came to an end, and military operations by either side were restricted to propaganda, spying, and sabotage, with very occasional shooting.

The End of Old China?

One of the controversies which has occupied historical scholarship in the People's Republic is the attempt to fit China into a pattern of Marxism. In that theory, the history of mankind should be marked by socio-economic development from the primitive to a slave society, then to feudalism, to industrial capitalism, to socialism, and finally to communism. From the perspective of Europe and America in the nineteenth century, when Marx was writing, the line can be identified, and the influence of economics on society and politics is an important consideration in any modern study of history.

For China, however, Marx's theory is difficult to apply. There are two major problems. Firstly, the concept of feudalism, as it is understood and was practiced in the West, fits rather poorly with the imperial system as it operated over two thousand years in China: as a result, there has been much debate, of limited usefulness, whether the 'feudal' period in China began with the unification of the empire by the First Emperor of Qin at the end of the third century BC, or rather at the fall of the Qing dynasty in 1911. There must be something wrong with a model for historical analysis which produces such room for disagreement.

The second problem, not unrelated to the first, is that socialism and then communism are supposed to emerge from the industrial proletariat. In practical terms, this was the reason the youthful Chinese Communist Party was advised and ordered from Moscow to concentrate effort on cities and among workers of the mines and railroads. The policy accorded with Marxist theory, and also with the experience of the Soviet Union, where support among the industrial proletariat provided the basis for Bolshevik victory. In China, however, as we have seen, the results were different: Communists in the cities were hunted down by their Nationalist enemies; the movement survived only by taking refuge in the backward, isolated north-west; and the party rose to power through the development of guerilla activity based upon peasants in the countryside. And even on the eve of Communist victory, the urban proletariat was neither powerful enough nor sufficiently interested to raise successful revolt against its masters — at the beginning of their government the Communists found difficulty bringing such cities as Tianjin under proper control, for most of the workers' organisations were dominated by gangsters, not by left-wing unions.

The achievement of Mao Zedong and his comrades, however, defied the models not only of the Russian revolution but also those of traditional China. In Chinese terms, the Nationalist Party, even after their losses at the hands of the Japanese, had established and were in a position to maintain the regular alliance of power which had dominated the former imperial state: support from the landowners and the leading merchants, and control of the major military forces. There were, as we have seen, major weaknesses in each of the components of that alliance, but in traditional circumstances those weaknesses would produce ineffective government, not one which had been destroyed.

It was the power of Mao Zedong's vision, however, and the essence of his revolution, that brought a new force into the Chinese political scene. The common people of the countryside had hitherto been the objects of action, outside and below the arguments and struggles of the small political élite, but they now came to centre stage. The achievement of the Communists was to take advantage of the energy and resentment of the peasants, inspire them, control them, and lead them to overthrow the old establishment of power. In this respect they truly created a new China.

Modern Communist historians express sympathy and some admiration for the popular rebellions of the past, the line which leads to the Taipings, the Nian, and the Boxers of the nineteenth century, but they criticize such movements, rightly, for their inadequate leadership and lack of a sustained programme. In the past, government forces had certainly been unattractive, but rebels such as the Taipings and the Boxers always appeared worse. This time, however, Chinese and Western witnesses told how conscripted Nationalist soldiers were replaced by the courteous men of the People's Liberation Army, who paid for the goods they received and repaired the damage of war. The Chinese people had never seen anything like it, and it was an extraordinary reversal of history, that the peasant forces of the rebels were looked upon as model soldiers; the rank and file Communists, guided by their genuine, dedicated cadres, confirmed their victories on the battlefield by winning also the admiration and respect of the people they came to govern.

The success of Mao Zedong, however, did not come easily: it required great discipline to maintain a line of command from the centre through the party cadres to the troops in the field. For its immediate purpose, Mao's model of Communist-led, peasant-based revolution was brilliantly successful. In the longer term, however, as the new regime took over the whole of China, there were two major questions: could the totalitarian model, so suitable for revolutionary war, be applied with equal success to such a vast and varied nation? And, even on a more limited scale, could the discipline of the cadres be maintained so tightly now that the tension was released and victory achieved?

Notes

1. 'Cadre', the general English translation of the Chinese expression *ganbu*, is a comparatively vague term for a person in authority in the Communist system. In the early years of the Communist state it was normal for a cadre to be a member and local representative of the Chinese Communist Party among the people. The term has now acquired a wider connotation, and 'cadre' may describe any person who holds an administrative and salaried post under the government, whether or not he is a member of the party.

2. Belden, *China Shakes the World*, quoted in Schurmann and Schell, *Republican China*, pp. 316–26.

3. See the various interpretations and estimates listed in Appendix E of Ch'en, *Mao and the Chinese Revolution*.

4. *Ibid.*, pp. 299–300.

5. Casualties were often heavy, but many Nationalist troops changed sides. After the victory in Manchuria, Lin Biao was able to bring 800,000 men through the passes to attack Beijing, and a number of them were former Nationalists.

Hereafter, in referring to the major city of Manchuria, I use its Chinese name Shenyang rather than the Manchu Mukden: see also note 17 to Chapter 1.

6. This battle is commonly known as the battle of Huaihai (from the Huai River and the Longhai railroad); compare with the battle for Xuzhou early in 1938 between the Japanese and the Nationalists (p. 148 above).

7. Li Zongren, a former warlord from Guangxi province, had thrown in his lot with the Nationalists at the time of their Northern Expedition. In 1936 he was one of the leaders of a revolt supporting Wang Jingwei, but he made his peace with Chiang Kaishek and accompanied the Nationalists in wartime exile to the south-west and their return to Nanjing.

8. In 1928, after the Nationalists established their capital at Nanjing and had successfully occupied the former northern capital, they changed its name to Beiping, also transcribed as Peiping, 'northern peace' (see note 5 to Chapter 4). When the Communists captured the city and proclaimed the People's Republic, Beijing again became the capital, and its former name was restored. The Nationalist government on the island of Taiwan still calls the city Peiping.

9. Though Pinyin transcribes the name of the capital of Taiwan as Taibei, the Nationalist government there calls it Taipei, and I follow that convention.

Further Reading

Barber, Noel, *The Fall of Shanghai,* New York, Coward, McCann and Geoghegan, 1979.

Belden, Jack, *China Shakes the World*, New York, Harper, 1949.

Chesnaux, Jean, *Peasant Revolts in China* [translated by C.A. Curwen], London, Thames and Hudson, 1973.

Chou, Shun-hsin, *The Chinese Inflation, 1937–1949*, New York, Columbia University Press, 1963.

Fitzgerald, C.P., *The Birth of Communist China*, Penguin [first published as *Revolution in China*, London, Cresset, 1952].

Loh, Pichan P.Y. (ed.), *The Kuomintang Debacle of 1949: collapse or conquest?*, Boston, Heath, 1965.

Melby, John F., *The Mandate of Heaven — Record of a Civil War: China 1945–49*, London, Chatto and Windus, 1968.

Pepper, Suzanne, 'The KMT-CCP Conflict 1945-1949' in *Cambridge China* 13.

8

New Age, New Outlook, 1950–1957

IN the first years of the new order, the Communist government was remarkably successful in establishing its power and restoring a measure of prosperity. Land reform brought some equality to the countryside and destroyed the traditional authority of the landlord gentry; the Marriage Law offered rights to women, while weakening the position of the individual and the clan against the government. And overall, the ending of civil war and the reunification of China brought an immediate improvement in the economy and a new sense of national pride and achievement.

The new regime advertised its support for the Soviet Union, and gained prestige and self-confidence by defiance of America and its allies in the Korean War. As a result of that conflict, however, the People's Republic found itself barred from the United Nations, faced with implacable hostility from the United States, unable to force a settlement with the Nationalists on Taiwan, and all the more dependent upon the Soviet Union. The government was accepted as a leading power in Asia, but many of the new nations distrusted the continuing Communist interest in world revolution.

Inside China the First Five-Year Plan, based on the Russian model, emphasized the development of heavy industry, while the government sought also to apply central planning and collective labour to peasant agriculture. Politically, power was firmly in the hands of the party leadership, and the people were urged towards reform and development with campaigns and slogans. For a short time, as an echo of the post-Stalinist era in the Soviet Union, there was a move towards collective leadership and a more open debate on the work of the party, but the experiment of the Hundred Flowers was too provoking, and free speech was halted with critical persecution. Ultimately, in Communist China, there can be only one true line at a time — and that is official.

China Reconstructs

It was clear that the victory of the Communists was the beginning of new revolution, and the political and social structure which had main-

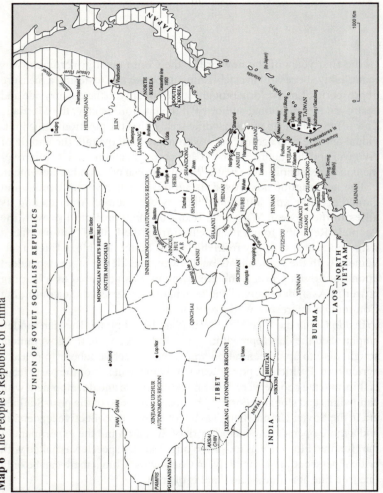

Map 6 The People's Republic of China

tained the Nationalist state was discredited and doomed. As conquerors who proposed to establish permanent power, the Communists
ensured from the very beginning that their policies and propaganda
should be widespread and without rival in the territory they controlled,
and those who failed to give them full support had no opportunity to
show their doubts. Independent newspapers and journals disappeared
immediately from the cities; despite military preoccupations and their
lack of genuine democracy, the Nationalists had tolerated many publications which opposed their policies, but the Communists ended all
pretence at freedom of the press, and nothing was published now that
did not reflect the official party line.

Throughout the country, mass propaganda and personal persuasion
were used to ensure acceptance of the new regime. Among the people, even those who had never shown any concern with politics were
required to join demonstrations, giving them for the first time a sense
of participation, and binding them also to the cause they had now publicly endorsed. In kindergartens and schools, swiftly and efficiently
organized even in the poorest communities, children were taught slogans and songs supporting the government and the party, and they
were urged to tell their parents and elders of the new morality and to
criticize or report those who failed to live up to it. In all circumstances,
the Communists taught the people, each individual should consider
the good of the whole, and no single person was entitled to excess
advantage or privilege by reason of his place in society. Most important of all, members of the party maintained this doctrine by their own
conduct, and the new administrators were honest and courteous in
their dealings with the people.

There were, of course, exceptions to the amnesty. In many regions,
as the Communists arrived, leading Nationalists had fled, but those
of middle or lesser rank often made their peace with the newcomers,
and some officials continued in office under the new administration.
Obviously this situation could not continue, and no one who had held
a position of leadership under the old system escaped investigation
and control by the new rulers. There were many cases where genuine
oppressors and local bullies were hauled to people's courts and dealt
with summarily by their victims and former neighbours, but the
Communist concern was not merely to organize revenge and destroy
the old system; they were equally anxious to ensure that no local prestige or power should be held by anyone outside the party. In his account
of a south Chinese village under the early years of Communist rule,
C.K. Yang tells the story of Li Feng, a moderate liberal leader in the

community, with considerable ability, who had for a short time held membership of the Nationalist Party, and had also led a guerilla force against the Japanese. It was not so much his political background as his personal acceptance in the village which appeared a threat to the local cadres, and in due course, two months after the takeover, he was arrested at night by soldiers, held on trumped-up charges and sent away for re-education in Guangzhou. When he returned, three months later, his health was broken by torture and ill-treatment, he was under constant and obvious surveillance by the police and the local council, and where he had once been popular there were few who now cared to talk to him or to visit his home.[1]

In due course, as general authority was established and accepted, the attention of the people was turned from political to economic and social inequities, and in June 1950 the Agrarian Law of the People's Republic was proclaimed from Beijing. The party's land reform in this early pattern did not appear designed for major change, for in theory the new legislation provided only for a return of 'land to the tiller' so that each family should own enough to support itself, and even rich peasants and landlords were allowed to keep a share of their former possessions.

Land reform was said to be largely completed by the end of 1952, when government figures claimed that some 700 million *mu* (a *mu* being one fifteenth of a hectare) had been redistributed among more than 390 million peasants, with an allocation between one and two *mu* per person. In absolute terms of land changing hands this was not particularly significant, and numbers of 'middle peasants' remained largely unaffected. Behind the official terminology, however, there was the reality of its application in the countryside. In many regions of China, redistribution of land gave opportunity for a mass meeting where old scores were remembered and repaid, and where local cadres often instituted violence against wealthy men and their families. Many landlords and rich peasants did purge their wrongs before an assembly of their fellow-villagers, and were then permitted to find a place in the new, equal society, but in other areas the excitement rose higher, and atrocities were reported where men were shot out of hand or hanged by a furious, vindictive mob. The death toll is not known, and it was probably not as high as that of Stalin's purges in the 1920s and 1930s in Russia, but there is no doubt that several hundred thousand people, deprived of their rights by the government which should protect them, were accused, condemned, and murdered in legalized lynch courts during this first firm display of Communist power.

There was also a long term effect to immediate policy: the old gentry of the countryside, which for more than two thousand years had provided the social leadership of the nation and the natural source for administrators and advisers at every level of government, was destroyed and no man could now gain prosperity and influence by reason of his investment and ownership in land. The way to success and power in future lay with the administrative and ideological machinery of Communism, controlled from the Politburo in Beijing, and no social or political class remained to rival the new regime.

There were, however, less obvious and more resistant obstacles to the full effect of Communist authority. One of these was the tradition of *guanxi,* 'connections',[2] and the other was that of the family and clan.

In traditional China, beneath the formal structures of power, the important matter for any individual was the group to which he belonged, for that determined his potential to deal with authority and, most importantly, to preserve himself from the full effect of arbitrary power. The group could be regional, economic, or social, as with men from a particular province or smaller area, from a guild of merchants or manufacturers, or from a class background such as landowners, scholars, or officials (and often all three at once). It could also be personal, as with schoolmates, military comrades, or the clients of a powerful patron or, most essential of all, it could be based upon family and clan.

In public terms, the large associations, such as groups of merchants, or the city gangs, could be faced and brought under control by the Communist regime. More private connections, however, could also be useful and very influential in public life. Yuan Shikai had come to power with the aid of former subordinates in the Northern Army; Duan Qixiang, leader in the early republic, obtained much of his support from his fellow countrymen of Hefei in Anhui, and the pattern of alliance between Chiang Kaishek and his relatives of the Soong family, together with the sons of his former patron Chen Qimei, was no more than an example, at the highest level, of the basic personal association which could be replicated throughout China to the smallest village community.

It was, moreover, the family which could provide the best protection against the government, and the best means to exploit the system, if only because members of a family were naturally bound by indissoluble ties, not by individual agreement. Not all relationships were valuable, and an extended family or clan could well select those of its members who were useful and those which could be left to fend for themselves, but in the amorphous and uncertain society of China the family was one group which might be expected to hold together against external pressure; and it was, of course, strengthened by the morality of Confucian tradition.

Under the Communist regime, slogans and campaigns attacked private individual relationships as remnants of 'feudal' society, and people at every level were urged to direct all their energies to national service under leadership of the party. More directly, however, the government sought to break up the influence of the family upon its members, and the Marriage Law of 1950, one of the first to be proclaimed, not only gave effect to reforms long argued in the past, but also provided opportunity for direct involvement in the daily private lives of the people.

With such notable exceptions as the old empress-dowager, women in traditional China were largely deprived of a public role, and certainly of a political one. Such status as a woman did hold came as the mother, wife, or widow of a man, and authority was commonly exercized on her behalf by another man, a father, son, or brother. For the most part, the only circumstances in which a woman exercised real power was inside the household, supervizing, and often persecuting, other women.

The position of women in traditional China was based upon two considerations. Firstly, there was the masculine prejudice, common to most societies, which insists that women's place is in the home and their contribution is in all respects secondary to that of the male. Among the élite, it was an article of faith that women should be segregated from men, that they should submit to the authority of fathers and husbands, and that they should not receive education lest it give them inappropriate ideas. In a peasant household, the attitude was reinforced by the belief that male strength is more suited to hard labour in the field, even though women's contribution of planting and weeding is frequently of comparable value, and handicraft work such as spinning and weaving, generally reserved to women, may produce more value than the harvest.

The second factor, however, comes from the structure of a society which depends so much upon family and clan. In traditional China, a woman married away from home and took up residence in her husband's house, normally under the eye of her parents-in-law. Apart from social linkage, her own family obtained small value from the transaction — dowry was more important than bride-price — and the new wife was swiftly isolated in a strange community. The function of marriage, moreover, was to maintain the male lineage upon which the future depended, and a woman's status depended very considerably upon the sons she produced. Should she fail in this duty, a principal wife could find herself supplemented by a concubine, and there was no moral reason why a man might not take a younger favourite for pleasure as much as for progeny.

Though the idea of a harem tends to excite the imagination of monogamous Western males, the reality was far less attractive, for imprisonment, frustration, jealousy, and ambition could produce tension, cruelty, and sometimes death. Even in a poorer family, without the resources to maintain many women, several generations and siblings often shared one roof, and the oppression of the elders and the rivalry of other wives could have comparable effect. Sadly enough, it often appeared that suicide was the one way for a woman to obtain relief, and she could hope that such an action might bring shame to her persecutors and some enquiry from her male relatives.

Among revolutionaries influenced by liberal and foreign ideas, female emancipation had long been a popular cause. San Yatsen argued for the rights of women, and one of Mao Zedong's earliest published writings was a bitter criticism of the social system which had driven a young bride to kill herself. The new Republic did forbid foot-binding, and the example of the reformed élite had considerable effect in eliminating the custom among all classes of society, but female infanticide was less easy to discourage, and the famine and deprivation which afflicted the countryside in the 1920s and 1930s, followed by the turmoil of the 1940s, meant that many girls were sold by their families to become servants, concubines, or prostitutes. Traditional China was oppressive, and young females were at the bottom of the heap.

The Marriage Law of the People's Republic gave women equal rights of property, marriage, and divorce, abolished arranged marriages, established the age of consent at eighteen for women and twenty for men, and forbade the taking and keeping of concubines. Naturally enough, the immediate effect was varied and often limited: many unfortunate women who had been taken as secondary wives were driven from their homes to fend for themselves, often with limited success; many families continued to force their young people into marriages negotiated by their elders; and while prostitution was likewise outlawed, and the keeping of prostitutes is still often punished by death, it has certainly not been eliminated. Though divorce is legal, moreover, social pressure in most communities, encouraged by party policy, prevents the normal breakup of any marriage.

Just as the land reform programme, however, had destroyed the old pattern of power in the countryside and left the Communists with unrivalled authority, so too the Marriage Law, besides its formal emancipation of women, gave government and party the right to interfere at the basic level of private life. No longer was the family a closed entity, with the man holding all authority. On suitable occasions, women have been driven or inspired to criticize their husbands in public political terms, and children have been taught in their schools that they should be prepared to denounce their parents for errors and faults. And in exchange, where many women had been restricted to acting as guardians of conduct within their own homes, enforcing authority against junior members of the household, they were now encouraged to extend supervision against their neighbours; some of the most energetic agents for local Communist control have been the elderly ladies of the villages and city streets.

Despite such negative considerations, however, the vast majority of people found the unification and reconstruction of China an immense improvement on their past experience. The very establishment of peace brought instant gains in terms of security and prosperity, and the new regime offered both a coherent plan for development and also freedom from the foreigners. Regardless of the political tenets of her rulers, if China could maintain peace at home and security abroad her future seemed assured.

Soviet Ally

In the outside world, at first, the new regime in China was accepted and approved. When the People's Republic was proclaimed in 1949, the British government gave speedy recognition, partly to maintain the advantageous treaty agreements such as those which confirmed the status of Hong Kong, but most of all for the pragmatic reason that the Communist government clearly controlled the greater part of the country, and should be dealt with on a proper basis. In the first months of 1950, though the American government and some others, including the Australians, still delayed, there was no particular reason to expect that normal diplomatic contact would not be established.

The United States had not sent direct aid to the Nationalists against the Communists, and had specifically refused to get involved in the civil war, but there was some sentiment against abandoning a former ally. The China Lobby in Washington had not been able to commit the government to military support of Chiang Kaishek, but they could point to his refugee administration on Taiwan as still possessing Chinese soil, and describe it as a legitimate alternative to the Communists, like a 'government-in-exile' during World War II in Europe or the Nationalists at Chongqing during the years just past. Against these arguments, liberal advisers to the Democratic administration respected the ideals and achievements of the Chinese Communists, and they scored an early success in 1949 when the American ambassador, Leighton Stuart, was ordered to remain in Beijing after the Nationalists surrendered, to make contact with the new regime. This particular initiative, however, foundered on a misunderstanding, possibly deliberate on the Chinese side, and the ambassador was ordered to leave.

By 1949 and 1950, however, American distrust engendered by the Communist takeover in Czechoslovakia in 1948 and the sense of fear and betrayal which came with the Cold War in Europe extended very naturally to the Communist Chinese. Soon after the proclamation of his government, Mao Zedong claimed there were two sides in world affairs: the side of imperialism and the side of socialism; and the People's Republic of China would lean firmly to the side of socialism. As Western diplomats and officials, missionaries and merchants, suffered petty humiliations and harassment, Beijing signed a treaty of friendship with the Soviet Union in February 1950, welcomed Russian advisers and technicians, and negotiated agreements for the Soviet Union to assist in national development.

Even so, the obstacles to recognition at first reflected the internal situation of United States and other foreign countries rather than the actions of the Chinese government. In the middle of 1950, however, with the outbreak of the Korean War, the Communists found themselves threatened by invasion and diplomatically isolated.

In accordance with the agreement at Yalta, after Japan surrendered in 1945 the peninsula of Korea was divided between a Russian-dominated Communist North Korea and the Republic of Korea under United States auspices. On 25 June 1950 North Korea attacked the South with full military force.

It is possible the North believed they could win the war so quickly there would be no time for international action, and they may have been encouraged by a statement of President Truman that America would not fight the Chinese Communists. Their aggression, however, brought immediate reaction: the Americans put land and air forces into the battle, they obtained the endorsement of the United Nations, and within a few weeks the North Korean army was in full retreat. The Soviet Union, embarrassed and fearful for the fate of its protégé, was regularly outvoted in the General Assembly, and the Russians could hardly intervene to support North Korea against the troops of the United Nations to which they themselves belonged.[3]

In the view from China, the picture was extremely serious. As a corollary to intervention in Korea, President Truman revoked his earlier statement of unconcern and announced that the United States Seventh Fleet would patrol the straits of Taiwan, between the island and the mainland. Ostensibly, this was to force an end to the Chinese civil war by preventing further action from either side, but Beijing saw it as clear support for the defeated Chiang Kaishek, and was concerned at the threat now fast approaching the Yalu River, the border of Korea and Manchuria. Though the Russians had stripped the territory of almost all its machinery at the end of the Japanese war, Manchuria was still the predominant source of known minerals, including coal, oil shale, and iron ore, and its railway network equalled the total mileage of all the rest of China. That region was China's hope for future industrialization, and now, five years after its return to Chinese control, the United States, based upon Japanese soil and a declared ally of Communism, was moving an army northwards through Korea. In the last weeks of October 1950, even as MacArthur was promising that the troops would be home for Christmas, 'volunteers' from China poured south across the international boundary to oppose him.[4]

In that first onslaught the United Nations forces were driven back into South Korea, and in almost three years of bitter fighting the battle-line shifted across the centre of the peninsula. In July 1953, when the armistice was signed at Panmunjom, the border between the two Korean states was very close to the 38th parallel of north latitude which had originally separated them.[5]

The war itself, with all its destruction and misery, was inconclusive, and the Korean peninsula remained uneasily divided, much as it had been before the attack from the north. For China, America, and the rest of the world, however, the conflict determined a pattern of hostility for the next twenty years. Those countries which engaged in the war under the United Nations flag acquired bitter memories of courage and loss, and the reports of North Korean and Chinese treatment of prisoners, with physical maltreatment and mental torture designed to gain converts and turncoats by 'brainwashing' remained a black mark in the eyes of the West. For Americans in particular, the shock of comparative defeat in Korea at the hands of the nation they had supported against Japan such a short time before produced a bitter hatred and constant opposition to normal relationships with the Beijing government, whether in diplomacy between the two countries or in the United Nations as a whole. Inside the United States, under the first years of the Republican administration of President Eisenhower, the extremist Senator Joseph McCarthy and the Un-American Activities Committee investigated and persecuted any individual suspected of liberal or left-wing sympathies. Some of the most competent scholars of China, such as Professor Owen Lattimore, and well-informed State Department advisers, like John S. Service, were driven from their positions in this time of hysteria, and their moderate opinions were lost to the American people.

For China, the effect of the war was equally important. The government claimed that the troops fighting in Korea were volunteers, but propaganda made it very clear that this was a war to defend the nation against aggression, and victories against the West brought popularity at home and prestige abroad. For new nations like Indonesia and India, the spectacle of an Asian power fighting successfully against the old imperialists was a source of inspiration. In 1954, after the French defeat at Dien Bien Phu, the Beijing government was recognized as one of the major participants in the Geneva agreement which concluded that stage of the war in Indochina, and in 1955, at the Bandung Conference of Afro-Asian states in Indonesia, Premier Zhou Enlai confirmed the friendship of his people among the non-Western countries, and established China as a leader among these new emerging forces.[6]

In more general world diplomacy, however, China was isolated. The emotions aroused by the war in Korea made it impossible for the Beijing government to gain acceptance in the United Nations, and the permanent seat on the Security Council which had been allocated to China continued to be occupied by Nationalists from Taiwan. Under American pressure, moreover, recognition of the People's Republic by other countries slowed and stopped, and normal contacts with Communist China came to appear as a deliberate slight against the United States.[7]

Internationally, the situation was absurd, for Taiwan in the 1950s was economically and politically insignificant even in the limited region of East Asia. Both sides were fixed in their attitudes, however, and the United States fleet in the Taiwan straits prevented the

Communists from mounting an effective attack. In 1954 artillery on the mainland opened fire against those islands off the coast still held by the Nationalists, and in the following year the Communists occupied the Dazhen group near Zhejiang. In Fujian, however, Jinmen (Quemoy) near Xiamen and Mazu (Matsu) outside Fuzhou maintained stout resistance, and the chief result of the fighting was to confirm American support for the government in Taipei. For the People's Republic, separation of Taiwan from the motherland remained a source of bitterness and indignation. For the United States, given the hostility of Communist China, the Nationalist regime on Taiwan, dependent upon their support, provided 'an unsinkable aircraft-carrier', part of a ring of bases and allies from Pakistan and Thailand to Japan, by which they sought to monitor and hinder any expansion of Communist influence.

The influence of Communist China in the region was in any case limited by contradictions within its own policy. As an independent country which had freed itself after severe suffering from foreign aggression, China was a natural ally and even a leader among the new nations emerging from the colonial territories of South-east Asia, and Zhou Enlai demonstrated that policy well at Bandung. On the other hand, Beijing also criticized many of the new regimes as bourgeois, reactionary, and neo-colonial, and showed undiplomatic sympathy for local Communists, some of whom, as in the Philippines, were at open war with the government. Embarrassingly, moreover, though the Chinese paid lip-service to the cause of revolution, they could seldom give practical aid over any distance. During the Emergency in the Malaya of the early 1950s Beijing gave official support to the Communist rebels, many of them overseas Chinese, but they could do nothing to prevent the final success of British forces. And throughout South-east Asia, though the government of the People's Republic generally advised the overseas Chinese to avoid involvement in politics, there was always an uneasy feeling they might hold divided loyalties between their homeland and the country they lived in.[8]

Overall, therefore, though China was recognized as a leading nation in Asia, its Communist government was not entirely trusted, it was isolated from the general community of the United Nations, and its influence was limited by the entrenched opposition of the United States and its allies. This, in turn, rendered Beijing all the more dependent upon the Soviet Union. As the second Communist power, China was host to many delegations and conferences, and the Soviet Union returned territories and concessions, including the region of Dalian and Port Arthur, now the combined region of Lüda, and control of the Chinese Eastern Railway in northern Manchuria. In exchange, China's intervention in Korea had not only preserved the Communist regime in the north but also saved Stalin from a humiliating set-back. International prestige and local authority had been gained, however, at the cost of wider isolation, and for the next ten years, in general dealings with the outside world, China remained closely bound with the Soviet government.

Socialist Development

In June 1953 the first official census of China was taken since the last days of the empire. An incomplete survey from the end of the Qing dynasty, 1908–11, had shown a total of some 360 million, and from 1928 the Nationalists at Nanjing had produced occasional estimates, based in part on head-counts in regions under their control, but amended and adapted to fit in with official preconceptions. The figure published by the Ministry of the Interior for 1936 was just under 480 million, and in 1947, in the count associated with the elections held by the restored government at Nanjing, the figure was 460 million. None of these reflected a true census, and all figures for this century must be based on calculations from the Communist records of 1953. Technically, there are flaws in the methods used, but the results published in June 1954 revealed a population for mainland China of 582,603,417, one fifth to a quarter higher than any expectation.[9]

Official reaction to these figures has varied. Many leaders made speeches in favour of increasing population, somewhat in line with the Roman Catholic dictum that for every extra mouth there are two more hands to work, but there were other occasions on which birth control was advocated. No one programme, however, was followed consistently, and the number of young people reaching child-bearing age in each generation ensured that even with the most restrictive controls the increase will continue. In immediate terms, the apparent peace and prosperity of the 1950s brought a baby-boom comparable to that a few years earlier in the post-war West, and the effects were demonstrated fifteen to twenty years later.

Further on, after years of uncertainty, the 1970s saw the adoption of the present policy: that each family should have no more than one child. Predictably, in a marginal industrial economy, mechanical and medical techniques of contraception are not widespread, but there has been increasing discouragement of early marriage, and sterilization or abortion are approved and encouraged. Despite prohibitions and exhortations, however, the birth-control programme has been generally effective only among party members, officials, and the people of the cities, while the masses in the countryside have been consistently reluctant to accept restrictions on their self-bred work-force. In 1982, a second published census indicated that China contained 1,008,125,288 people, almost double the figure less than thirty years earlier.

The high and increasing population was thus a factor to be considered in all plans for the future. Given the restricted area of cultivable land, the First Five-Year Plan, from 1953 to 1957, aimed at development in two major fields: industrial production to provide a domestic supply of machinery and chemicals for agricultural use; and further expansion and improvement in the techniques and organization of agriculture, notably through the introduction of peasant co-operatives and collective farming.

In mining and industry, the gains were remarkable. The first factories relied heavily on Soviet capital and Soviet expertise, and much of the First Five-Year Plan could be traced to the Soviet favour of heavy industry, regardless of economic cost, but by 1957 the former production of Manchuria had been more than recovered, the established industrial centres of Wuhan and Shanghai had been expanded and developed, and new decentralized works had been set up in cities such as Luoyang in Henan and Baotou in Inner Mongolia. Major achievements of the period were the reconstruction of the Anshan Iron and Steel works in Manchuria south of Shenyang and the completion of two rail and road bridges across the Han River and the Yangzi at Wuhan linking north and south China for the first time without the use of a ferry. Geological surveys proved large reserves of oil in Xinjiang and elsewhere, and uranium and other minerals were found in significant quantity. Certainly there was still no meaningful comparison between China and the industrial West, but the progress outstripped most other developing countries, and many fields of industry and mining were able to satisfy a fair proportion of domestic needs.

Still more important for future planning, the inflation of the last years of Nationalist rule had been halted by the financial measures of the minister Chen Yun.

In December 1948, even before the final takeover, a People's Bank had been established in Communist territory, and the *renminbi* 'People's Currency' [RMB, or *yuan*] spread through the country as Communist control extended. Though the new currency was subject to the same inflationary pressures as had ruined the finances of the Nationalists, strict moves were taken to stabilize it through restrictions on the holding of cash by private individuals and companies, and by regulations compelling state undertakings to make all major payments through bank transfer and to deposit cash reserves with the central People's Bank. The Korean War brought danger of renewed inflation, but large bond subscriptions took further money out of circulation, while wages and savings were indexed to the cost of living. With full control over the country and with a proper administration of taxation and finance at the centre, the pre-Communist pattern of government deficit and heavy short-term loans was changed to a balanced budget. In 1955, with inflation fully under control, the old *yuan* were called in and exchanged at the rate of ten thousand old for one new *yuan*, and during the next twenty years, with firm government control and limited trade overseas, the exchange rate was held remarkably steady against foreign currencies.

For some time at least, in major cities such as Shanghai and Guangzhou, individual businesses were permitted to continue under state supervision, but through the early 1950s there was a steady transfer of plants and companies to joint private-state ownership, and by 1956 no large enterprise remained legally in private hands, though the former owner frequently continued as manager of the business. In that year the percentage dividend return on capital was fixed at an arbitrary 5 per cent, regardless of profits, and was later reduced to an effective 3.3 per cent. On the other hand, though the capitalists and businessmen of the cities were frequently intimidated and fined, the principle of income from dividends and private property was maintained into the 1960s, in striking contrast to the land reform programme and the later collective movement in the countryside.

At the time of the Korean War, the 'Five Antis' campaign against bribery, tax evasion, fraud, and the theft of property or economic secrets, brought heavy fines against many prosperous men, whether or not they were guilty of the charges made against them, and the currency exchange of 1955 was a means for the government to discover those who had hoarded illegal cash and to render their holdings valueless. Inevitably, there was black-market enterprise and similar small-scale illegality, but much of the supervision was effective, and the main development of Chinese industry and trade was held in the control of the state.

With strong emphasis on official inspiration, and a high degree of public response to slogans and personal exhortation, public hygiene was markedly improved. Even in the first days of conquest, the men of the People's Liberation Army had enforced sanitation in the areas they came to, and civilian cadres followed up their work. Despite a lack of qualified medical personnel, the Communists encouraged training at every level possible, and many people received basic instruction in elementary nursing and midwifery. The result was a tribute to a basic health programme carried through with minimal resources, for over wide areas diseases such as cholera and dysentery were almost eliminated and infant mortality showed a swift decline. On this basic level the energy of the government, and the pragmatic policy of making maximum use of the human and physical material at hand, was spectacularly successful.

In the field of education, policy concentrated on two points: improvement of literacy and the training of technicians. Throughout the country, there were evening classes of reading and writing, and school-children were urged to teach their elders. Of the three thousand characters which form the basic vocabulary of a newspaper, many were redesigned in simplified form, and a short, largely unsuccessful attempt was made to popularize an alphabetic script.[10] Under the Republic, despite official programmes, it is unlikely that the general literacy rate was higher than 25 per cent, and in many areas less than one per cent of women had received any education at all. Faced with such a backlog, and limited funds for schools, progress was comparatively slow, but it was claimed by the 1970s that three-quarters of the people could read and write.[11] For a government wishing to popularize its policies and spread its propaganda, the basic education of the people has high priority.

The majority of students selected for secondary schools and education were directed towards practical training in engineering and mechanics, but new and modern techniques were maintained and developed in more theoretical sciences, and also in archaeology and anthropology, while a concern for national culture brought new editions of the ancient classics together with impressive documentation of nineteenth and twentieth-century history. Major emphasis, however, was given to the training of men and women who could handle the technology of a new industrial society, and all work was judged by the needs and requirements of the state.

It was in agriculture, however, that the planners and politicians of the central government aroused the greatest controversy. At the end of 1953 it was announced that the traditional peasant life of the countryside, with each family responsible for the produce of its own fields, should be changed to a system of co-operatives, with all land held by the community and all labour given to common service.

At first, the principle of collective farming was brought in gradually, chiefly through the extension of 'mutual-aid teams' which had been established in the north during the early years of Communist settlement. In 1952, in north and north-east China, it was estimated that some 20 per cent of peasants belonged to mutual-aid teams which operated all the year round, and that another 50 per cent took part in such co-operative ventures at the vital times of seeding and harvest. With propaganda and encouragement from local cadres, the system of mutual aid was extended to the rest of the country.

For poorer peasant families, with limited capital to buy tools and insufficient funds to hire additional labour when it was needed, the programme offered substantial advantages. There were, however, problems: animals and property held in common were less well cared for than those of private individuals; and the rudimentary division of labour, with no classification between skilled and unskilled work, caused disagreement between energetic and competent farmers and those who appeared lazy and inefficient. At first, when the system was voluntary, many prosperous peasants preferred to rely on their own resources, and some who joined a mutual-aid team withdrew from it later. In a way of life as traditional and as individualistic as farming, it is not surprising that an organized general system should have shown considerable weakness.

In theory, there was to be a steady progression from the mutual-aid teams towards larger village units, agricultural producers' co-operatives, but it soon became clear there was substantial resistance in the country, and the pace of advance was slowing down. In July 1955 Mao Zedong urged greater speed towards full collectivization, and these words of authority put pressure on local cadres and recalcitrant peasants. The principle of voluntary co-operation was not quite abandoned, but political and social forces were now used to the full. In March 1956 it was claimed that half of the rural population was enrolled in co-operatives, and, in July 1957, 90 per cent of all peasants were said to have joined the new system. Collectivization was largely complete, and in 1958, on the eve of the Great Leap Forward, there began the further organization of the commune system, designed to unite as many as a hundred villages in one economic unit.

For a Communist government, there are three great advantages to rural co-operatives and communes. Firstly, on strictly economic grounds, the pooling of resources among low-capital peasants and rural workers gives opportunity for shared development and the purchase of machinery which would otherwise be too expensive for any individual or small family group, while communes with a working population of several thousand people could generate the capital resources of a small city. A second point is the control and prevention of private wealth and excess profit: where all capital is owned by the community, and a unit of labour is the base for sharing proceeds, no member of the community can establish economic predominance over his fellows; and with financial resources held in common there is small opportunity for the money-lender and usurer. In this way, the new system confirmed the effects of land reform, and so long as it was maintained there were no means by which one man could become rich through capital investment in another's land and labour.

Finally, for central government, perhaps the most important advantage of the commune system was the opportunity it gave for direct control through party members and cadres. Where individual farmers had formerly decided for themselves such matters as seedtime and the crops to be sown, or the maintenance of fields and earthworks, these questions were now decided at general meetings with strong guidance from the party leadership. By their control of the co-operative and commune system, planners could determine national policy in agriculture, and by grants of capital, or supplies such as tractors and fertilizer, they could control the development of each region. Furthermore, the organization of collective units, each with its allocation of tax payment and its quota of production for compulsory purchase, made the collection of revenue more reliable and straightforward than it had been under previous governments.

The disadvantages, of course, for any plan to reorganize agriculture into a vast industrial mechanism lie in the vagaries of nature, in individual resentment and distrust, and in the inevitable inefficiency and lack of realism of some officials. It was often found that figures had been exaggerated and falsified, and, although production certainly increased, the levels of improvement were not always up to those expected and required. On many occasions, when policy called for particular crops and a certain programme for their treatment, the poor results justified individual doubts over the wisdom of the changed techniques. Most importantly for members of the farming community, government requirements always received top priority. Each household had poultry and pigs, the traditional and basic source of meat and protein, and a small strip of land for raising vegetables, but the demands of work in the public fields often made it difficult for individuals to find time for their own affairs.

In the earlier years of the century, rural families had supplemented their meagre income by domestic manufacture, handicrafts, carving and weaving, and by specialities such as silk-farming or tea-growing. In themselves, these hardly appeared significant, and they were now discouraged by the local cadres as they appeared to compete with work on crops for the whole community, but as much as 50 per cent of production in many regions came from these sources, and official disapproval of such private enterprise brought a serious drop in farm income and a considerable loss to the whole economy. In one locality of Hubei province for which figures are available, it appears that the income of 368 farming families declined from RMB 28,000 in 1951 to RMB 14,000 in 1955, when they were all organized into a co-operative, and national production figures confirm the tendency : in 1957, China produced 95,000 tonnes of silk compared to 200,000 tonnes in 1932, and tea production was 150,000 tonnes in 1957 against 200,000 tonnes in 1932.[12]

Overall, during the First Five-Year Plan the economy of China probably achieved a growth rate of some 6 per cent from the low base at which it started after the disruption of international and civil war, and there is no question that the new prosperity was shared throughout the whole society more effectively than ever in the past. This progress, however, had been obtained at the cost of considerable cruelty to those regarded as enemies of the people, and there were many occasions when arbitrary political decisions prevented the best use being made of available resources and opportunities. It was nevertheless general policy that people should be persuaded rather than ordered into action, and every device of propaganda and social pressure was used to secure acceptance for each new project. In this rather one-sided fashion, the government was concerned to maintain popular support, but its chief interest lay with the programme which its theorists and planners determined, and by the end of the First Five-Year Plan in 1957 there was resentment among the people and some uncertainty among the rulers concerning the speed and the manner of the country's development.

Government from Beijing

By the defeat of Japan and their victory over the Nationalists, the Communists had secured control of Manchuria, and both Moscow and Beijing agreed that Xinjiang remained Chinese. However, all along the great frontier with the Soviet Union there are areas of disagreement, sometimes because the terms of ancient treaties are difficult to establish on the ground, often because the treaties themselves may be called into question by the Chinese as agreements forced upon their predecessors by Russian imperialism and aggression. The most recent of these treaties is that which settled the independence of Outer Mongolia in February 1950, and neither the Communists in Beijing nor their Nationalist rivals on Taiwan have quite accepted the loss of that territory.

During the 1930s official policy of the Chinese Communist Party recognized the essential independence of the non-Chinese minority peoples, the Mongols, the Uighurs and other groups in Xinjiang, and the Tibetans. In theory, the outer territories of China, which had never been closely linked to the central government of China Proper, would join at most in a loose federation, and might even establish themselves as independent. The practice of power, however, brought a very great change, and in the early 1950s the international disfavour which China had gained through her intervention in the Korean War was enhanced and confirmed by the story of Tibet.

There is good argument that Tibet had never been under the effective control of a Chinese government. There were periods when imperial officials at the court of the Dalai Lama, traditional priest-ruler of the territory, wielded great influence, but Tibet itself was not incorporated into the Chinese empire by direct administration, and the old vassal relationship was not markedly closer or more consistent than that of other countries such as Korea and Thailand, now fully independent. And although the Chinese and Tibetan languages are distantly related, their traditions and literature, their culture and religious beliefs, are markedly different.

On the other hand, the loose theocratic government of Buddhist monks in Tibet was in no position to establish a national state with real independence, and the high cold plateau, with its scattered settlements in deep river valleys, could never be anything more than a buffer or zone of conflict between the powerful states—India, Russia, and China—which surrounded it. To forestall any chance that a foreign power might stir up trouble there against their new government, the Communists proclaimed their intention to 'liberate' Tibet, and in October 1950, after a short attempt at negotiations between Chinese and Tibetan authorities, the People's Liberation Army was ordered into the country 'to free its people from imperialism and aggression'.

There was, at first, little opposition. The people of Tibet were in no state to resist, and the Chinese promised that they would continue to enjoy their traditional autonomy and freedom of religion. Very soon, however, the value of these assurances became doubtful, for the Beijing concept of 'autonomy' entailed direct supervision by the central government, and though the Communists claimed to offer freedom of worship they could not tolerate the dominance of Tibetan government and society by the feudalistic monasteries, which held the land in superstition and serfdom. The Chinese introduced modern techniques of medicine and other aids which had long been forbidden or remained the prerogative of priests; but they also forced a new and alien way of life upon a traditionalist and conservative society. There was bitter resistance and guerilla warfare, culminating in a major uprising in 1959. As that revolt was finally crushed, the Dalai Lama fled to India and the Chinese imposed their own administration maintained by the soldiers of the People's Liberation Army. Since that time, the Beijing government has held Tibet with soldiers and garrisons, it has encouraged immigration by Han Chinese, and it has put down local resistance with steady brutality.

In other regions of 'national minorities' resistance proved less bitter, but the pressure against these peoples and their separate traditions has been equally firm. Except during the Cultural Revolution, the Buddhism of Mongolia and the Islam of Xinjiang have been officially tolerated by the Chinese rulers, and the non-Chinese peoples of the south are protected against discrimination and encouraged to progress. In areas where there is a significant number of non-Chinese the government is maintained by a province-level 'autonomous region' or through a smaller unit controlled directly from Beijing. At the same time, there is no question that government shall remain in Chinese hands, and any talk of independence and federalism has long disappeared. In Mongolia and in Xinjiang, mining and manufacturing centres have been developed at Baotou and Urumqi, and the government has continued to encourage migration of Chinese into the new areas of settlement. With these new industries, and the expansion of irrigation agriculture, the Mongols and the Uighurs found their grazing grounds occupied and their traditional economy rivalled and threatened, and in Inner Mongolia Chinese settlers now outnumber the Mongols by more than eight to one. In south China, the tribes-people of Guangxi, Guizhou, Yunnan, and other isolated regions were also subject to Chinese colonization, and in almost every autonomous territory, supposedly ruled for the benefit of the non-Chinese national minorities, more than half the population is Han Chinese. With their drive for new resources and new areas of development, the Chinese Communists' control of their non-Chinese subjects has surpassed that of the old imperial governments.

According to the provisional Constitution first promulgated in 1949 as the 'Common Programme', and extensively revised in 1954, the government of the People's Republic was based on a pyramid system of People's Congresses, with franchise extending to every man and woman over the age of eighteen, excluding only convicted criminals and declared enemies of the state. Local congresses elected representatives upwards to the provincial level,[13] and the process continued to the National Peoples' Congress at Beijing, with additional representation from national minorities and units of the armed forces. This first constitution remained in force, with some amendments, until the period of the Cultural Revolution, when it was replaced in 1965 and again in 1978. In 1982, a fourth constitution largely restored the provisions of 1954.

The First National People's Congress met in 1954. In theory the Congress holds office for four years, with meetings of two weeks each year. For the rest of the time, the 1,200 delegates of the National People's Congress transfer their power to a permanent Standing Committee which acts as a secretariat, prepares and conducts elections, supervises the executive government, and appoints judges to the Supreme People's Court. The actual business of the state is carried on by the State Council, whose members include the premier, deputy premiers, ministers, vice-ministers, and chairmen of commissions, and the State Council itself is subject to general supervision from the official head of state, the chairman, or president, of the People's Republic of China. Mao Zedong was elected to this post in September 1954.

Beside this government structure the Communist Party maintains a parallel arrangement of local representation and national authority, and members of the party control the functions of government at every level. By the party constitution established in 1956, the chief body is the Central Committee, with members and alternates amounting to a hundred or more, meeting twice a year during its term of five years. The Central Committee when out of session delegates power to the Politburo of some twenty members, and the Politburo in turn entrusts immediate control to the Standing Committee of the Chinese Communist Party, the small group of fewer than ten men who hold the highest executive position in the People's Republic.[14] In 1954, besides the post of national Chairman and Head of State, Mao Zedong acted also as Chairman of the Central Committee, of the Politburo, and of the secretariat of the Chinese Communist Party,[15] and the linkage between party hierarchy and national government has been maintained in this fashion at each level of administration since the establishment of the People's Republic.

It must be recognized, moreover, that the organization of government was in no way designed to reflect the wishes of the people. Representation from local to national level is not intended to encourage popular debate which might be heeded by the rulers but is, on the contrary, a structure for the dissemination of command. Though the regime does encourage people to participate in politics, participation does not involve issuing requests or demands to the government, but the energetic, often compulsory, support and performance of policies determined and sent down from above.

In this respect, once policy has been decided by the leaders of the state, the people are instructed by campaigns, orchestrated from the centre, as to the attitudes they must adopt and the actions they should take. Campaigns may take various forms: they can be as general as the party 'Rectification Campaign' at Yan'an in the 1940s, the 'Cultural Revolution' of the 1960s, or the 'Four Modernizations' of the 1980s, but more limited ones which are particularly common are those of the simple numerical slogan, such as the 'Five-Antis' campaign of 1952, ostensibly against profiteers from the Korean War, but actually planned to bring all private entrepreneurs under control, or that in public health against the 'Four Pests'—rats, sparrows, flies, and mosquitoes. Similarly, as a device for education and inspiration among the masses, the government made frequent use of models, either by competition among real units of the army or sections of a factory, or by more general examplars, such as the ordinary soldier Lei Feng, who did exist, but who was exaggerated by propaganda into a humble, selfless hero, devoted to the thought of Mao Zedong.[16]

It should also be observed that no Communist regime ever recognizes any limits to the authority of those who hold political power. Just as the citizen is involved in politics only through direction from above, and there is no concept of the rule of law which might protect individual privacy against the state, so too the formal structures of the state and the party, expressed in their 'constitutions', are always subject to revision or may, in appropriate circumstances, be ignored. Indeed, revision or re-drafting of the constitution has been a feature of several party congresses, and there have been many occasions when provisions of the national constitution have been ignored. Unlike the American Constitution, for example, which can only be amended by complicated political processes, and requires detailed legal interpretation, a Chinese constitution may best be regarded as the current arrangement, to be followed just so long as it is convenient, while the judicial system, like everything else, is ultimately influenced by the politics of the day. As in all Chinese tradition, the government of the People's Republic is an apparatus of powerful men and weaker institutions.

For the first years of government, however, the unity among Communist rulers, which had played so great a part in the days at Yan'an and in the campaigns of civil war, was maintained. Whatever disagreements there may have been in the 1930s and 1940s, political leaders such as Zhou Enlai and military commanders like Zhu De and Peng Dehuai had shown their ability and loyalty and appeared content to serve under Chairman Mao. Just one incident disturbed the pattern: in 1953 Gao Gang, who had held almost independent power in the north-west before the Long March of 1935, and who had been appointed chief of the Communist Party in Manchuria after the Communist victory in 1949, was accused of plotting a separatist movement to gain personal control in the north-east. Gao Gang committed suicide and a number of his supporters in Manchuria and eastern China were imprisoned. In contrast to Stalinist Russia, however, Gao Gang's disgrace was not followed by any widespread purge, and political murder was not an accepted technique of the new regime.

In fact, the great achievement of the Communists was the stability they gave to the people whom they controlled. During the 1950s the population of China reached 600 million, the highest of any nation on earth, while the area of the country placed it second or third in the world. Government and administration of such great numbers and vast expanse would tax the abilities of any regime, but through the enforcement of land reform and their social discipline the leadership achieved a unity which had defied the governments of the old republic, and this in itself at once raised China to a reasonable prosperity at home and a considerable position of influence in the outside world.

By the middle of the 1950s, however, the Communists in China had taken advantage of immediate opportunities and were approaching the limit of swift development. Internationally, China's power was recognized but restricted, while progress at home depended now on a slower process. Among the people of the cities and the peasants of the countryside, the government had established socialist control of society and the economy, and even at the lowest levels, in the smallest villages, people generally accepted the advice of their cadres. Scarce resources were being distributed as widely as possible, and for the first time in living memory the fear of starvation had been lifted from the masses of the people. On the other hand, though the first years of reconstruction and reform had brought well-deserved success, tensions between those who wished to progress faster and those who felt the speed was too ambitious and too harsh were gradually coming to a head.

The Hundred Flowers

The policies of the Communist Party had encountered no substantial opposition during its first seven years of power in China, but some people were dissatisfied with the unified system of government and the Communist economic programme, and there was increasing dissent in the wider Communist world. At the Twentieth Congress of the Soviet Communist Party in 1956 the Russian leader Nikita Khruschev shocked his audience, and observers everywhere, by a public denunciation of the errors and cruelties of Stalin, who had died only three years earlier. Khruschev also attacked his 'personality cult', and a new principle of collective leadership, with no single predominant figure of power and ideology, was announced as the model for Communist government.

In Eastern Europe, the apparent loosening in the rigid structure of Communism produced general unrest among peoples who had suffered control from Moscow since the end of World War II. In Hungary, discontent culminated in riots and rebellion which was firmly put down by Russian tanks. Fairly obviously, the repressive features of the Communist system in Europe had aroused individual and intellectual discontent which waited only for an opportunity to break out, and any Communist regime, east or west, would have to deal with this or come to terms with it.

These policies and events were of considerable importance to the Chinese government. In September 1956, at the first session of the Eighth National Congress of the Party,[17] the new party constitution reduced emphasis on Mao Zedong as a political leader and as an intellectual guide. In contrast to the Seventh Congress in 1945, Mao's thought was no longer described as the doctrine of the Communist movement, and his authority was considerably weakened from that implied by the National Constitution of 1954. Though he was still

Chairman of the People's Republic and Chairman of the Chinese Communist Party, he was no longer chairman *ex officio* of the Central Committee, the Politburo, or the Secretariat of the Party. As First Vice-Chairman, Liu Shaoqi was entitled to perform many executive functions of the Chairman of the Central Committee and the other subsidiaries; and Deng Xiaoping, the new General Secretary of the Party, was now also the Chairman of the Secretariat, the centre of administration. Furthermore, a six-member Standing Committee of the Politburo was established, sharing the authority which had formerly been vested primarily in Mao Zedong.

Though the new arrangements did remove the absolute supremacy from Mao Zedong's position in the Communist Party and the state, he was still the chief force in practical policy and possessed a decisive voice in theory and philosophy. In May 1956, however, in line with the new policies of relaxation and collective leadership, the Propaganda Committee of the Party Central Committee reported a speech by Chairman Mao urging that among artists and writers 'a hundred flowers should blossom', and among men of science 'let all the schools of thought contend'. Despite this encouragement to criticism and debate, most intellectuals had become cautious from their experience of Communist rule, and they showed no early response to the invitation. Even Guo Moruo, the recognized Marxist writer and scholar of the 1920s and 1930s who was now President of the Academy of Science, warned that criticism and proposals for change must harmonize with the policy of the country and the party.

Late in 1956, however, it began to appear that the government was indeed considering a policy of relaxation, and from February 1957, reports on a speech by Mao 'On the Correct Handling of Contradictions Among the People' appeared to confirm the hopes of those who looked for some liberalization. The speech appeared to invite criticism of the party and the government, and it was implied that discussion could be tolerated and even welcomed so long as it was not purely hostile and destructive. Gradually, but with increasing enthusiasm, university professors and students joined in expressing their objections to the rigidity of the party apparatus, to the excessive reliance upon political dogma at the expense of open debate, and to the inefficiency, arrogance, and selfishness of Communist cadres. By the middle of May denunciations of the government were published in official newspapers, and universities were covered with posters and filled with mimeographed pamphlets and meetings around soap-box orators leading criticism of almost every aspect of the current regime. On 31 May 1957 the People's Daily reported a statement by a university professor that:

You [the Communist Party] should not be arrogant and conceited, and you should not distrust the intellectuals. If you do good work, that is all very well; if you do not, then the masses can strike you down, can kill Communist Party members, and can destroy you. And no one would say they were unpatriotic, for the Communists no longer serve the people. The end of the Communist Party would not be the end of China, and even if the people no longer accepted the leadership of the Communist Party, that would not be a betrayal of the nation.[18]

The rulers of China had probably never expected such response to their loosening of the reins, and the situation was evidently getting out of hand. For a few weeks the party had given no guidance to the debate, but at the end of May official posters stated clearly that no support could be given to words or actions against the policies of socialism. In the first weeks of June supporters of the main Communist line came forward again to criticize those who had taken the lead in finding fault with the government. By the end of July the former critics were under personal attack as rightist elements, they were harangued and humiliated in public until they confessed to their faults and their mistaken ideology, and students, professors, and some non-Communist leaders paid public penance for their errors. The distinguished sociologist Fei Xiaotong, who had originally expressed the fear that the Hundred Flowers might suffer from a political freeze, now thanked the party for 'beating me awake before it proved too late'; and the former Secretary-General of the Peasants' and Workers' Party, one of the non-Communist groups which retained a token existence in the People's Republic, made a weeping, broken self-confession before the National People's Congress of June 1957.[19]

For non-Communist intellectuals, writers, thinkers, professors, and students of the universities, the destruction of the Hundred Flowers marked the end of an era and the finish of a dream. In the first years of Communist power, despite government control over much that was said and written, there had appeared some room for independence, and there was always the justification that the country could ill afford open debate and argument when unity was still fragile, economic progress was slight, and enemies stood at the borders. In these circumstances, the All-China Federation of Literary and Art Workers, which controlled literature in the same fashion as the Union of Writers of the USSR, could justify its insistence on the propaganda value of literary works, and in the field of liberal arts, though scholarly themes and theories had to be fitted to Marxist views of history and politics, there was valuable work to be done. This period, however, came to an end in 1957, when the government and the party realised how little sympathy they had earned among intellectuals, and when the intellectuals recognized the way they had been tempted and then betrayed by their rulers.

There is nothing very surprising about this: no government like the Communist regime in China can tolerate open and vocal opposition, and the enforcement of censorship is limited only by its power. In Chinese tradition, the scholars and writers of the imperial dynasties had long accepted the official interpretation of Confucianism, and maintained their role as generally unquestioning supporters of established order. For many observers, however, and for the people involved, the real tragedy was the end of that one generation, which began in the early republic and flourished in the late 1920s and 1930s, when scholars like Hu Shi and writers like Lu Xun had broken from the restrictions of tradition and had spoken and written with a depth of feeling and a scholarly understanding which gave China a respected place in the world community. It was clear that a time of hope had ended, and the intellectuals could never trust the Communists again, nor look for freedom under their rule.

In the realm of ideas, even as Communists praised Lu Xun for his love of the people and his biting satire, they interpreted his words for their own doctrines and they criticized and persecuted his former colleagues. Of the essayists and novelists who opposed the Nationalists from Shanghai in the 1930s, many of them Communists or members of the League of Left-Wing Writers, few published significant work under the Communists, and the end of the Hundred Flowers was marked by an attack on the woman writer Ding Ling, a Communist supporter from the days of Yan'an but too independent a personality to accept the strict line of propaganda: she and a number of associates were purged from the party. Ba Jin published his last major work of fiction in 1947; Qian Zhongshu, author of a brilliant satirical novel *The Besieged City*, also published in 1947, brought out his first work under the Communists only in 1958, and that was a collection of poems from the ancient Song dynasty and a preface with extensive quotations from Chairman Mao. Lao She now produced short plays of social realism which were praised by party authorities. Of them all, the two most successful were Mao Dun, who became Minister of Culture but wrote very little after he received that appointment, and Guo Moruo, long-time member of the party, who kept his position through all political changes and, as a 'grand old man', became the dominant literary figure and recognized interpreter for scholarship in China. His calligraphy was admired second only to that of Mao Zedong.

In the words of Lu Xun himself, 'John Stuart Mill declared that tyranny makes men cynical; he did not know that a republic makes them silent.'[20]

Notes

1. Yang, *Chinese Village*, pp. 138–9.

2. An excellent chapter on *guanxi* is given by Jacobs, *Local Politics*, pp. 40–60.

3. At the beginning of United Nations intervention, the Soviet Union could have vetoed the proposal in the Security Council, but its delegation at that time was boycotting meetings in protest at the exclusion of the new Chinese Communist government. Once the intervention had been approved, the United States was able to override later vetoes by reference to the non-Communist majority in the General Assembly.

Considering the international isolation which China suffered from its intervention in Korea, some analysts have argued there was a plot by Stalin to force exactly that situation: once China had gone to war against the United Nations, the regime would be all the more dependent upon the goodwill of the Soviet Union, and could offer no rivalry to its leadership of the world Communist movement. This was indeed the result, but it may be debated whether Stalin was so prescient, or whether he really wanted to take such a risk for what appears to be only a marginal gain.

4. Many of the troops used by China in the Korean War had been Nationalists, captured in Manchuria during the civil war. Their courage and devotion casts a critical light on the quality of their former leadership — and the war in Korea gave the Communist government a good occasion to use up their strength.

5. From the American point of view, a major decision of the Korean War was the unexpected dismissal of General MacArthur by President Truman in April 1951: an act which confirmed the authority of the civil government over an extremely popular military leader who was the central authority of the Occupation regime in Japan; and which also ensured, equally important, that American power would not extend the war across the border into Manchuria. Though the affair was of great concern to the Americans and had considerable effect on Japanese understanding of democratic authority, it is doubtful if the Chinese appreciated its importance to themselves.

6. At the Plenary Session of the Asian-African Solidarity Conference at Bandung on 19 April 1955 Zhou Enlai publicized the 'Five Principles of Peaceful Coexistence' by which the People's Republic of China would judge friendly or hostile countries. The principles had already been enunciated in a communiqué on trade and intercourse with India twelve months earlier, but they were now spelt out as the central theme of China's foreign policy, and they have been referred to on many occasions since. The five principles were: mutual respect for each other's territorial integrity and sovereignty; mutual non-aggression; mutual non-interference in each other's internal affairs; equality and mutual benefit; and peaceful coexistence. From the point of view of Beijing, of course, acceptance of the first and third of these principles would basically entail an enforcement of the Chinese position in Tibet and of the sovereignty of the Communists over Taiwan.

7. Though the question is of more interest to students of Australian than Chinese history, it seems reasonable to give brief consideration to Australia's recognition of Beijing.

In the late 1940s, when the Communists took over the mainland, the Australian

Labor government was under attack by the Liberal and Country Party opposition for its sympathy with Communism, and in the federal elections of December 1949 the Labor Party was defeated and Mr Menzies, later Sir Robert Menzies, became Prime Minister. The Labor government had not been anxious to add material for the right-wing attack by a speedy recognition of the new Communist regime in China, and the Liberal Party under Menzies had made the outlawing of the Communist Party one of the chief planks in its electoral platform. In such a mood of Cold War excitement no Australian government was in a position to take quick action, and by the time a decent interval had passed the Korean War had begun, with an Australian contingent in the United Nations forces, and recognition was out of the question. On the other hand, for the next sixteen years, though Australia received an embassy from the government in Taiwan there was no Australian ambassador accredited to Taipei. In 1966, again largely as a result of internal domestic political pressure on the eve of an election, the Liberal government under Prime Minister Holt established an embassy to the Nationalist regime. In 1972, immediately after the victory of the Australian Labor Party under Mr Whitlam, diplomatic relations with Taiwan were broken off and ambassadors were exchanged with the People's Republic of China.

In this pattern, with the question of recognition for Beijing being determined internally by party politics and externally by official attitudes towards the United States, Australia is typical of many small countries. Significantly, though the Liberal governments remained faithful to the American connection and did not recognize Beijing nor supply the Communists with specifically strategic goods, they showed no hesitation in selling large quantities of the Australian wheat crop during the l960s, and mainland China became for a time one of Australia's chief trading partners.

8. Traditionally, a person of Chinese descent could still be regarded as a Chinese subject or citizen, even though his family might have emigrated long ago and settled for generations overseas. At times, both in the 1930s and since the Communist victory on the mainland, this policy has brought embarrassment to the Nationalists and caused many Chinese in South-east Asia, and the governments of the countries where they live, to be wary and resentful of policies and propaganda from Taipei. The Communists in Beijing have generally shown good faith in their claims to relinquish ties with overseas Chinese but they have not always been able to avoid suspicion and they have occasionally, in moments of stress, appeared to court it. Besides the Malayan anxiety of the 1950s, Beijing also expressed concern at treatment of the Chinese minority during the anti-Communist military coup in Indonesia in 1965, and at the time of the Cultural Revolution in 1966–7 there was considerable propaganda directed against the involvement of the overseas Chinese in South-east Asia. There have been several occasions when the rhetoric from Beijing has exposed the overseas Chinese to great difficulty in the countries where they live, but there is no cause to believe that the Communists have ever attempted to organize the overseas Chinese communities as a Fifth Column. For the overseas Chinese themselves, there is often a strong pull where relatives are still living in China, and also from the concept of China as the ultimate homeland: for both the Nationalists and the Communists, remittances from overseas Chinese have provided a significant quantity of foreign exchange. It is a sad fact, however, that although the overseas Chinese have often been embarrassed by the policies of Beijing and Taipei, they have never received effective protection or support from either government.

9. On the population statistics of twentieth century China, see Ho, *The Population of China*, particularly pp. 73–97.

10. It is, of course, possible to express the sounds of Chinese in alphabetic form, but the variety of tonal values, the high number of homonyms, and the great difference between dialects have hindered the use of an alphabetic script. The northern Mandarin dialect, described now as *putonghua*, 'common speech', has obtained wide use as a second language in schools and official communication, even in the southern provinces of China, so the dialect problem is being gradually overcome. On the other hand, the

programme for literacy in characters achieved such good early results that romanization has not been pressed.

11. As an example of the literacy programme in action at the grassroots level, I may cite the elderly woman I met at a commune outside Beijing in 1973. Then in her sixties, she had learnt to read and write only since the Communist takeover, but she claimed to know some two thousand characters, enough to follow a newspaper. She had recently begun a literacy class of her own, to teach other old people.

12. From Yang, *Chinese Village*, pp. 233–5, quoting official Communist publications.

13. Some municipalities such as Beijing and Shanghai, and some autonomous regions such as Guangxi, Xinjiang, and Tibet, hold provincial status.

14. On the establishment of the Standing Committee in 1956, see also p. 214 below.

15. It is somewhat confusing that the Chinese term *zhuxi*, 'chairman', is used as the title both for the head of state as well as for the leading position in Government committees. In English, the title 'Chairman of the People's Republic of China' would be better understood as 'President', but such a translation is seldom used. It may be observed, also, that Mao Zedong can always be described as Chairman Mao: though he ceased to be Chairman of the People's Republic after 1958 (p. 230 below), he continued as Chairman of the Chinese Communist Party.

As an exercise in historical etymology, the phrase *zhuxi* literally means 'chief of the mats'. Chairs were not commonly used in China until a thousand years ago, and before that time people knelt or sat on mats on the floors, as in Japanese tradition today.

16. Lei Feng, an ordinary soldier who died in an accident, was first celebrated in a People's Liberation Army campaign of 1963, but his example has been used in many different guises, and has lately been revived as an ideal of the lowly, loyal worker for the Communist cause.

17. The second session of the Eighth National Congress of the Chinese Communist Party was held in May 1958: see p. 224.

18. Quoted in Chen, *Thought Reform of the Chinese Intellectuals,* pp. 146–8, and Mehnert, *Peking and Moscow*, p. 187. For fuller English translation, see *Survey of China Mainland Press (SCMP)* 1553, pp. 19–20.

19. See, for example, Mehnert, *Peking and Moscow*, pp. 192–3, also pp. 184 ff, and Schurmann and Schell, *Communist China*, pp. 157–62 for general descriptions of the Hundred Flowers movement in the universities, and *SCMP*, e.g. nos. 1558 and 1560 for the debate and rectification of the China Democratic League and the Peasants' and Workers' Democratic Party.

20. Lu Xun, 'Odd Fancies' in 'And That's That', from the *Selected Works.*

The attitude of Chinese Communists towards Lu Xun is ambivalent. They praise his criticism of the Nationalists, and place great store on his occasional notes and telegrams of encouragement to the Communists, but they glide over the fact that Lu Xun never joined the Party. There are museums and memorials to him in Beijing, Shanghai, and Guangzhou, all associating his literary career with the struggle for revolution. Politically, Lu Xun is seen as a spokesman of the anti-establishment forces in China, and his criticisms were used during the Cultural Revolution.

The sentence I have quoted, however, can hardly be taken as approval for thought control. The Communists explain that Lu Xun was objecting only to the Nationalist republic of the 1930s. Observers outside China are free to make their own judgements.

Further Reading

Major English-language sources for the history of Communist China are the official publications such as *Peking Review* (renamed *Beijing Review* in 1979), *China Reconstructs*, and the news items and bulletins of New China News

Agency, *Xinhua* (formerly transcribed as *Hsinhua*). The United States Consulate-General in Hong Kong has published a *Survey of Chinese Mainland Press (SCMP)*, renamed *Survey of People's Republic of China Press* in 1974, which translates articles from the *People's Daily (Renmin ribao/Jen-min jih-pao)* and other newspapers and also a parallel series on magazines such as the Communist Party journal Red Flag (*Hongqi/Hung ch'i*), all subject to regular indexing.

Also in Hong Kong, the Union Research Institute has published several books and research papers, including *Who's Who in Communist China,* and the weekly *Far Eastern Economic Review* is an authoritative source for information and analysis. The Institute of International Relations in Taiwan has an English-language journal *Issues and Studies,* and the Contemporary China Centre of the Australian National University, Canberra, publishes *The Australian Journal of Chinese Studies.*

Among many other journals dealing with the People's Republic, the most important is *China Quarterly,* published by the Contemporary China Institute of the School of Oriental and African Studies at London University: each issue contains a Quarterly Chronicle and Documentation.

On economics:

Donnithorne, Audrey, *China's Economic System,* London, Allen and Unwin, 1967.
Howe, Christopher, *Employment and Economic Growth in Urban China 1949-57,* Cambridge, Cambridge University Press, 1971.
Schurmann, Franz, and Schell, Orville, *Communist China: revolutionary reconstruction and international confrontation, 1949 to the present,* Penguin, 1968 [third in the series of *China Readings,* preceded by *Imperial China* and *Republican China*].
Wu, Yüan-li (ed.), *China: a handbook,* Newton Abbot, David and Charles, 1973.

On society:

Jacobs, Bruce E., *Local Politics in a Rural Chinese Cultural Setting: a field study of Mazu township, Taiwan,* Canberra, Australian National University, 1980.
Meijer, M.J., *Marriage Law and Policy in the People's Republic of China,* Hong Kong, Hong Kong University Press, 1971.
Myrdal, Jan, *Report from a Chinese Village* [first published in Sweden 1963], Penguin, 1973.
White, Lynn T. III, *Careers in Shanghai: the social guidance of personal energies in a developing Chinese city, 1949-1966,* Los Angeles, California University Press, 1978.
Yang, C.K., *Chinese Communist Society, the Family and the Village,* Boston, MIT Press, 1965 [contains two works on the first effects of the Communist

regime on Chinese society, *The Chinese Family in the Communist Revolution,* and *A Chinese Village in Early Communist Transition*, both published first in 1959].

On intellectuals:

Chen, Theodore H.E., *Thought Reform of the Chinese Intellectuals,* Hong Kong, Hong Kong University Press, 1960.

Hsia, C.A., *Gate of Darkness,* Seattle, University of Washington Press, 1968.

MacFarquhar, Roderick, *The Hundred Flowers Campaign and the Chinese Intellectuals,* New York, Praeger, 1960.

On foreign policy:

Fitzgerald, Stephen, *China and the Overseas Chinese: a study of Peking's changing policy, 1949-70,* Cambridge, Cambridge University Press, 1972.

Mehnert, Klaus, *Peking and Moscow,* London, Weidenfeld and Nicholson, 1963.

Whiting, Allen S., *China Crosses the Yalu: the decision to enter the Korean War* (second edition), Stanford, Stanford University Press, 1968.

9

Red and Expert, 1958–1965

FOR the intellectuals of China, the Hundred Flowers incident and its repressive end brought humiliation and the loss of hope, but for the people as a whole the misfortunes of university professors were of little concern. Observers in the West and other outsiders regarded the persecution of scholars and writers with contempt, but the masses of Chinese, so lately recovered from decades of misery and war, gave priority to strong government and economic development.

In 1959, however, the political force of the party, and the authoritarian nature of government, brought economic disaster. Led by Mao Zedong, but with minimal opposition and general acquiescence at the highest level, the Communists embarked upon an accelerated programme of development, largely disregarding real restraints, and determined to break the bonds of economics. One can see, in the Great Leap Forward, an attempt to recreate the spirit of Yan'an on a national scale: to reject models of development imposed from outside and, to mobilize once more the masses of peasantry, encouraging local self-reliance and spreading basic technology among all the people. In practice, however, the gains proved illusory, and failure to obtain proper information about the true situation in the country, combined with an insensitive determination to maintain a particular political line, brought disruption, misery, and starvation to the people. As the immovable facts of economics met the irresistible will of the party, China was crushed between the two brutal forces.

In the aftermath of this débâcle, as millions of Chinese starved and died, Mao Zedong and his political associates were compelled to give way, while more practical leaders such as Liu Shaoqi and Zhou Enlai sought to repair the damage and restore the economy. Internationally, however, the slogans of self-reliance and independence had already broken the alliance with the Soviet Union; partly through disagreement about the exaggerated ambitions of the Great Leap, but also through Russian reluctance to grant Mao the prestige he expected, or to support his military adventurism against Taiwan.

In the early 1960s, however, even as the country struggled from its economic misfortunes, the government demonstrated its independence. Besides the quarrel with the Soviet Union, which raised a number of disputes along the borders, there was a short victorious war against India, which destroyed the prestige of its leader Nehru and ensured there would be no further support from that quarter for rebels seeking freedom in Tibet.

Further afield, China looked for a position of leadership among the Afro-Asian community, and spoke with President Sukarno of Indonesia about the New Emerging Forces which would rival the neo-colonialist powers of the United Nations. In 1965, however, when Sukarno was overthrown by a military anti-Communist coup, the Chinese lost their chief ally, while closer to home they could offer only limited support to their Communist neighbours in North Vietnam as the United States increased its military commitment to aid the South. For the time being, the foreign policy of self-reliance and the spread of revolution had brought isolation, encirclement, mistrust, and frustration.

The Great Leap Forward

By the end of the First Five-Year Plan in 1957 the Communists had achieved real success. At constant 1952 prices, the gross output value of agriculture rose from RMB32,590 million in 1949 to RMB48,390 million in 1952 and to RMB60,530 million in 1957, while the output value of industry rose from RMB14,020 million in 1949 to RMB34,330 million in 1952 and RMB78,390 million in 1957. Moreover, though much of the improvement could be attributed simply to peacetime recovery after years of war, the factory sector had been greatly enlarged: between 1952 and 1957 handicraft production increased less than 15 per cent, from RMB4,720 million value added to RMB 5,380 million, but factory output almost trebled, from RMB6,450 million to RMB 17,260 million.[1]

Organizing this development, China had accepted advisers and loans from the Soviet Union, who emphasized capital investment rather than current consumption. Twenty per cent of resources went to capital, twice that of India and almost as much as the forced industrialization of Stalinist Russia in the late 1920s, and ninety per cent of these funds went to heavy industry, notably to large enterprises such as the Anshan Iron and Steel Works in the north-east and similar complexes at Shanghai, Wuhan, Baotou, and other major cities. Agricultural production and light industry were left to local initiative, and the pattern of the economy followed the Soviet style of centralized, strong bureaucracy.

Despite the success of the planners in establishing an industrial base, there were questions of overall strategy and the relevance of the Soviet example. Unlike those in the Soviet Union, the Communists in China obtained chief support from the peasantry of the countryside: they had no strong base in the cities, nor was there any great reservoir of skilled labour. The new government had eliminated its debt to the West, but shortage of capital meant that investment funds must be sought abroad. The Soviet Union, however, gave no grants, only loans, with interest repayable in large and swift instalments, and frequently restricted to projects under direct Soviet control or guidance. Though the government needed the money, Chinese gratitude was tempered by the fact that the assistance they received compared unfavourably with that given to satellites in Eastern Europe, and there was a growing feeling that what was good for the Soviet Union was not necessarily good for China.[2]

During 1956 and 1957, the last years of the First Five-Year Plan, there was discussion on future development, and at the Third Plenum of the Central Committee of the Communist Party in October 1957, Mao Zedong forced through his policy of mass mobilization, to engage the whole of China in a struggle for progress. Late in that year a number of measures urged decentralization and the spread of medium-size, small, and light industries, and in February 1958, when the National People's Congress issued its call for a 'Great Leap Forward' at the start of the Second Five-Year Plan, it was emphasized that the whole population should work to accelerate production beyond any previous achievement. In May 1958, at the second session of the Eighth National Congress of the Chinese Communist Party, the line was far more radical than at the first session in 1956, and though Mao's constitutional position was unchanged, his leadership was central to the new policy.[3]

The concept of the Great Leap Forward was a complete contrast to the strategy of the First Five-Year Plan, but there were reasons for the new approach. In many respects, the country remained too backward, with too great a population of poor peasants, to benefit fully from a programme based on heavy industry. It was hoped, moreover, that if the people of the countryside were mobilized there was a chance the nation might overcome those restrictions, lack of capital and lack of skilled labour, which bound her in a traditional economy. The comparative success of collectivization in the countryside, and general overfulfilment of industrial targets in the First Five-Year Plan, gave many leaders a heady sense of self-confidence, and they were convinced they could build upon recent achievements much faster than their foreign and technical advisers would allow.

At the centre of the new strategy lay the system of people's communes, developed from co-operatives and collective farms already established, and now proclaimed as a vital unit of the economy. In the first months of 1958 communes were formed by the amalgamation of co-operatives in Hebei, Henan, and southern Manchuria, and by summer the principle had received official endorsement from the Central Committee. With speed reminiscent of the collectivization movement two years before, communes spread across the country, and at the end of November it was announced that over 99 per cent of households were members of a commune. There were 26,000 communes in the countryside, and more in the cities, often centred around a major factory unit, with facilities such as canteens and kindergartens.

The communes brought official authority and propaganda closer to the citizen than ever before, setting up special committees, with heavy party representation, to deal with finance, security, health and welfare, education, and entertainment. In theory, the commune was the essential unit for all activity: it had directive and planning control over lower levels such as production brigades and production teams, but it operated closer to the masses of the people, with more influence than the counties and provinces at higher level. Most importantly, the rural communes, with an average population of some fifty thousand people, were perceived as the smallest unit which could accumulate capital for investment. Where the co-operative movement had sought only to reorganize agriculture on more efficient socialist lines, the commune was seen as the key by which industry in the countryside could supplement that of the cities.

Many aspects of the new policy were valuable. Emphasis on local initiative and mobilization of labour brought improvement for agriculture, particularly in irrigation, drainage, and power schemes, while small factories producing farming equipment gave local people confidence in their abilities and freed them from ignorance and awe of modern technology. In almost every district there were local plants to repair or build machinery, and the Great Leap was a take-off point for the spread of mechanization.

At the same time, however, there were dangers in the new pattern of development, and excess enthusiasm was the most serious. The Great Leap Forward was presented by propaganda as the will of the

people mobilized to overthrow the negative, limited ideas of impractical theorists. Communist inspiration, the quality of being 'red', was
lauded at the expense of the unimaginative 'expert'. A sense of struggle towards the common goal produced prodigies of effort, but failure to consider the warnings of experts rendered much of the labour
pointless or counter-productive. Two examples were often quoted:
on the North China Plain, a vast program of well-drilling to raise water
for the crops brought brackish water to the surface, increased the salin-
ity of the land and reduced fertility over wide areas. And the programme for 'backyard blast furnaces' collapsed when it was discov-
ered that small-scale plants, with erratic temperature and quality control, could never produce reliable pig-iron and were useless for the
production of steel: in practice, these amateur efforts turned good ore
into trash.

More important than these short-term failures, the Great Leap
Forward, by the speed with which it expanded the enterprise and by
its encouragement of local authority, disrupted all administration.
The complexity of the economy had already put pressure on ministries
of the central government: state and joint-state enterprises, for example, had increased from 3,000 to more than 60,000 between 1949 and
1958, and it was extremely difficult for the bureaucracy, with a limited number of competent cadres, to supervise so many varied units.
With decentralization, moreover, came increasing fragmentation and
confusion of reports and policy-making. Of all the symptoms for
potential disorder, the most dramatic was the collapse of Chinese
statistics.

Up to 1957 the figures produced by the Chinese statistical bureau
had been generally reliable, showing a remarkable but accountable
rise in prosperity from one year to the next. In 1958, however, observers
of the Chinese scene were amazed, and supporters and spokesmen for
the Beijing government were elated, by reports of immense gains in
production. The output of steel in 1957 was 5.35 million tonnes, but
in 1958 the target was set at 11 million tonnes, and 30 million were
planned for 1959. In 1957 the output of coal had been declared at 130
million tonnes; in 1958, it was doubled to 270 million tonnes. Overall
industrial output, RMB 78,390 million in 1957, jumped to RMB
130,279 million in 1958, and production of food grains rose from 195
million to 375 million tonnes in a single year. Some experts doubted
these figures, and their caveats were confirmed in August 1959, when
the government announced that grain production in 1958 had been
not 375 but only 250 million tonnes, while 3 of the 11 million tonnes

of steel had come from backyard furnaces and was below industrial quality. Both the planning and the boom in the economy were clearly out of hand, and the information received at the capital was quite inaccurate. From the beginning of 1960 statistical publication from China ceased almost completely, and was not restored until the early 1980s.

Fig.2 Grain Production in the Great Leap Forward

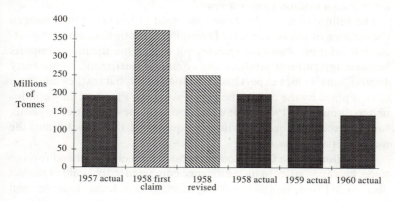

Of all the factors which destroyed the momentum of the Great Leap and brought ruin to the Chinese economy, poor administration — impetuous, arrogant, and ill-informed — was at the centre of the problem. Under the new arrangement, cadres and party members in the provinces, communes, towns, and villages had major responsibility for developing production and reporting results to the capital, and many of them failed on two counts: firstly, because they refused to adapt their orders and expectations to reasonable advice; and secondly, through reluctance to make proper reports of the real situation. With touching but groundless enthusiasm, each unit vied with the next to announce higher levels of production, and it took some time before planners at the centre could recognize and admit to the full deception which was being practised.

The terrible thing was that administrators and cadres at every level believed their own reports, and they based ambitious plans on their own false optimism. The one good harvest of 1958 actually yielded 200 million tonnes, but even the revised figures of government claimed 250 million tonnes: as a result, great areas of agricultural land were turned over to cash crops such as cotton, and China embarked on major exports of grain. Partly through less favourable conditions, but also because less acreage was sown, production in 1959 declined to 170 million tonnes; but net exports of grain rose from 2.7 million tonnes in 1958 to 4.2 million tonnes in 1959. In 1960, production was down to 143.5 million tonnes, lower than even the desperate years of the 1940s, and people were starving to death, yet the government still exported 2.6 million tonnes. At last, in 1961, food was purchased overseas, and the country became a net importer, at an average of more than 4 million tonnes a year.

The failure of agriculture, and the need to import grain, changed the balance of overseas trade. Through the 1950s China had imported industrial plant and equipment, but from 1960 the major imports became agricultural produce and chemical fertilizers. In the early 1950s China's chief export had been foodstuffs, but textiles had risen to first place by the end of the First Five-Year Plan and they continued so through the 1960s. But where formerly they were used to obtain foreign exchange for new industrial equipment, they now became the means to pay for food.

The failure of the Great Leap in terms of development and foreign trade, however, pales to insignificance compared to the scale of human disaster. At first, refugees to Hong Kong told of local shortages and frequent hunger, and economic observers used various snippets of information to judge the overall problem, but until the 1980s, when Chinese statisticians revealed the catastrophic collapse of grain production from 1959 into the 1960s, there were no official figures to confirm or deny their guesswork. Now, assessing the effect of the crisis, estimates indicate that between 16 to 27 million people died as a direct result of famine, and national population fell by 10 million in 1960. This was the heaviest mortality of the century in China, including the miseries of the 1920s, and it was proportionately worse than the disasters of collectivization in the Soviet Union under Stalin.[4]

Perhaps the most frightening aspect of the story is the way in which political control, so well established by the Communists for implementing its programmes, could ignore and defy reality, even as millions of people hungered and died. It was, indeed, the political strength of the regime that proved a decisive factor in bringing economic disaster, for the techniques which had forced development forward during the first years of power were now used in arrogant error to make the situation worse.[5]

Apologists have argued correctly that, despite the failures of the Great Leap Forward, China had taken a large step to reduce the centralization of the economy, to encourage local initiative, and to remove dependence on the Soviet Union. The changes were achieved, however, only at enormous cost in human suffering, and one lesson of the misfortune was that political power and authority should be used with restraint in dealing with a nation and an economy as vast and complex as China. Sadly, Mao and his associates refused to learn that lesson.

The Struggle for Power

The Great Leap Forward brought two major changes in government: an immediate effect was considerable alteration in the hierarchy of power; and in the longer term, the authority of the centre was weakened against that of the provinces.

Throughout the century there have been instances of provincial separatism, for great distances and poor communications make things difficult for any central government. After their victory in civil war the Communists had exercised power more effectively than their predecessors, but there was no reason this situation should continue indefinitely. In particular, though communications were improving, the pressures on the bureaucracy and the difficulties of recruitment were affecting administration, while the policy of the Great Leap was designed to discourage centralism. Between 1957 and 1959, provincial authorities were granted control of many light and medium industries, particularly those for consumer goods, hitherto administered from Beijing, and these smaller plants were generally more profitable. In financial terms, there was a new relationship between the provinces and the capital: while the provinces obtained local control of their budgets, the only revenues collected directly by the central government were customs dues and the profits of industry under its own control, mostly low-yield heavy industry. Furthermore, whereas credits received from Soviet loans in the early 1950s, channelled through the national government, had given power to the capital in the allocation of resources and the control of development, new policy called for all borrowing to cease and existing loans to be repaid: so the central government had to seek funds from the provinces. As a result, where the centre had held initiative in planning and policy, local cadres could now seek a degree of autonomy, and they had increasing power to determine what revenues would be transmitted to the capital.

The failure of the Great Leap, and the disastrous famine which accompanied it, placed further strain on the authority of central government, and put the peasant communities and agricultural provinces of China into a stronger bargaining position. As the situation gradually improved, many regions preserved a high degree of autonomy, and by the mid-1960s the Chinese were speaking of local 'kingdoms' and 'duchies' in the provinces. No one failed to pay lip-service to the government in Beijing and the authority of the Chinese Communist Party, and there was no move for full independence, but the tendency to separatism was strong, and a constant problem for the national leadership.

At the centre of power, there was resentment and criticism of Mao Zedong. Mao's authority had been somewhat reduced by the reorganization of 1956, sharing power among the members of the Standing Committee, and in December 1958 the Sixth Plenum of the Party Central Committee agreed to his 'request' not to take another term as Chairman of the People's Republic and Head of State. Liu Shaoqi was designated for that office, but Mao continued as Chairman of the party, and the Great Leap Forward was based upon his philosophy: the 'Maoist' line was characterized by propaganda and slogans encouraging intensive effort towards a series of individual goals and production at very high levels. At the same time, moreover, Mao's policies emphasized the initiative of the party at the expense of regular government structure, and many officials, notably Premier Zhou Enlai and the veteran economist Chen Yun, objected to what they saw as a hasty and distorted programme, preferring a more even development, in line with previous years, and less hostility towards the Soviet model. By 1959, as evidence gathered that the Great Leap was in difficulties, the division of opinion became increasingly serious.

In August 1959, at the Plenary session of the Central Committee of the Chinese Communist Party held at the mountain resort of Lushan, matters came to a head. Peng Dehuai, Minister for Defence and one of the early commanders of the Red Army, argued for continuing close relations with the Soviet Union, criticized the whole concept of the Great Leap, and spoke of leftist mistakes brought about through 'petty-bourgeois fanaticism'.[6]

Peng Dehuai's major criticisms were expressed in a letter to Mao, which Mao then published to the committee as a catalyst to compel debate; and at the same time, with a firmness and hostility hitherto unknown in the senior ranks of government, he forced a choice between himself and his critic. Faced with an ultimatum, the leadership was compelled to support the Chairman. Though it was already clear the Great Leap Forward was going wrong, men such as Liu Shaoqi, still

awaiting confirmation as head of state, and Deng Xiaoping, appointed Secretary-General of the Party in 1956 with Mao's support, had
earlier endorsed the new programme, and they were reluctant to risk
sudden political change. Whatever his faults, Mao held immense prestige as leader of the revolution, and open rejection would add to the
strains the country was already suffering. So Mao was able to isolate
Peng Dehuai, arranged his dismissal as Minister for Defence, and
replaced him by his own nominee, Lin Biao.

Peng, however, had been defeated by political tactics, not because
his arguments were wrong, and as more practical counsels prevailed
Mao and his supporters in the party were removed to the 'second line'.
Liu Shaoqi was formally endorsed as head of state by the National
People's Congress of 1959, and by 1960 Zhou Enlai and Chen Yun,
with a Small Group for Finance and Economics, were supervizing the
salvage work of reconstruction. Many ideas of the Great Leap, such
as backyard blast furnaces and universal community kitchens, had
already disappeared through inefficiency or unpopularity, but the
commune was retained as a unit for the organization of society and
the economy. At the same time, however, the regime accepted private plots for peasants and a small free market as natural and necessary, while the allocation of 'work-points', by which rural communes
assessed each person's contribution as the basis for distributing the
annual profits of production, was confirmed as a system akin to regular wages. Workers in factories received bonuses and incentives for
technical skill and high production, management ability were emphasized at the expense of political enthusiasm, and great care was taken
in accounting, to encourage individual initiative in a general socialist framework.

Under this guidance, and with improvement in the seasons, farming was on the road to recovery in 1962, and by the mid-1960s the
economic situation was largely restored. In 1964 agricultural production regained the average level of the 1950s, the harvest of 1965
equalled that of 1957, before the disaster of the Great Leap, and results
thereafter showed a slow but fairly steady increase, roughly keeping
pace with the natural increase of population.

Despite the failures of the Great Leap, industry had been less dras-
tically affected than agriculture, and recovered more swiftly. By the
middle of the decade, industrial output had doubled the figures of
1957, not only in heavy industry such as coal, steel, cement, and factory equipment, but also in medium and light industry and consumer
goods. In particular, by 1963 the exploration of oil fields in central

Asia and the north-east, and energetic development of refining, had made China basically self-sufficient in petroleum, and brought a parallel growth in related industries, including chemical fertilizers, pesticides, pharmaceuticals, and synthetic fibres.

Though the situatuion had thus improved by the mid-1960s, the speed of economic development had been seriously reduced. Average annual growth through the early 1960s, little more than 3 per cent, was less than that of many Asian countries, and compared sadly with almost 9 per cent under the First Five-Year Plan. The balance of foreign trade had shifted from import of capital goods to the import of food, and volume had declined considerably: trade with the Soviet Union and other members of the European Communist bloc went from 74 per cent in 1955 to 36 per cent in 1964, and the change reflected both an absolute reduction in dealings with the Communist countries and also the increase of grain and other imports from such non-Communist countries as Australia, Canada, France, and Mexico. Overall, from a high point of some US$4.3 billion in 1959, China's foreign trade dropped to no more than $2.7 billion in 1962, and it was not until the late 1960s that the levels of ten years earlier were regained.[7] On the other hand, foreign trade had held only marginal influence on the internal economy of the People's Republic, and it was proportionately less important than in non-Communist Asian countries such as India and Pakistan. At a low level of development, without extensive foreign credits, trade was limited to those goods seen as essential to the country and to stockpiles against future food shortages, and it was basic policy that China should live off her own resources.

Despite the slogans of self-reliance, however, much of the recovery after the Great Leap had been achieved on foundations laid by the First Five-Year Plan. The end of Soviet co-operation had interrupted many works, but after some delay the Chinese were able to complete their major plants, and it was the earlier development which allowed priority areas to be maintained and made new ventures possible.

In contrast to mining and secondary industry, agriculture was harder and longer affected. Apart from the absolute death toll of the famine, and the social disruption which accompanied it, there was continuing shortage of consumption goods, not only cereals but also vegetable oils and cotton cloth. Moreover, even as industrial centres returned to their former position, and there was increasing pressure for renewed industrial development on the lines of the First Five-Year Plan, the countryside continued to suffer, and by the middle 1960s the state was reducing its prices for compulsory purchase of farm products. Rural consumption per head was less than that in the cities, and remained well below the levels of 1957, while investment in the agricultural sector dropped from over 20 per cent in 1962 to less than 15 per cent in 1965, and continued to fall in following years.

So although one object of the Great Leap had been to spread development throughout the countryside, the failure of the movement had fallen most heavily upon the peasants, and the reconstruction which followed had offered them only marginal benefits. By and large, where the mistaken policies of the Great Leap had set back the pattern of development established in the 1950s, reconstruction continued as it had before once those political errors were removed. As a programme for recovery from induced disaster, the policies of the government were pragmatic and reasonable, but they produced disproportionate industrial growth at the expense of the vast majority of the people—the peasants engaged in agriculture—and details of the approach, notably the encouragement of private plots and personal incentives, fell far short of the Communist tradition inspired by Mao Zedong.

Soviet Enemy

In the early days of power in China, faced with hostility from the United States and other Western countries and influenced by the prestige of the Soviet Union under Stalin, it seemed only natural that the government in Beijing should play its international role as a member of the Communist bloc. In the early 1950s the conflict in Korea confirmed China's isolation from the rest of the world, and the Soviet Union became the only important source for development finance. As late as the 1960s some observers in the West still insisted on the essential unity of the Communist world, and argued that signs of disagreement were no more than ripples on the surface of a deep understanding.

In fact, however, the causes of conflict between the Communist powers were real and important to both China and the Soviets. There was always a difference between the Beijing government and the satellites of Eastern Europe: in Europe, the troops which defeated Hitler's Germany in World War II acted as sponsors and enforcers of Communism in such embattled countries as Poland, Hungary, and Romania; in China, the Communists had maintained their own fight against the Japanese, and only Manchuria had been occupied by Soviet armies. In the civil war which followed the defeat of Japan, Soviet aid was of marginal value, and the victorious Communists owed no real debt nor depended in any vital fashion on the approval of their colleagues in Moscow,

The People's Republic had reason to be grateful for diplomatic and economic support from the Soviet Union, but there were grounds for discontent. Despite the treaty of 1950 which confirmed the separate independence of the Mongolian People's Republic, many Chinese still regard Outer Mongolia as a natural part of the territory to be ruled from Beijing, in much the same category as Manchuria, Xinjiang, and Tibet, and tsarist expansion in the nineteenth century had left a legacy of 'unequal treaties'. Including the Mongolian sector, China's northern frontier stretches 7,250 kilometres, from the Pamir and Tian Shan mountains to the Ussuri River and Vladivostok, and almost every part of the border has been determined by Russian power and reluctant Chinese acquiescence. Between the two countries stretches the longest land frontier in the world, and this, regardless of politics and philosophy, would be sufficient to cause some tension.

Until the death of Stalin in 1953 the potential disagreement remained largely concealed, for the apparent advantages of alliance outweighed feelings of national resentment. By 1954, however, there were already references to Chinese dissatisfaction, and approving recollections of the promise by Lenin in the early 1920s, that the Soviet Union would renegotiate all the old border agreements. The leaders of China, and Mao Zedong in particular, regarded themselves as equal in authority to Stalin's successors, and they were prepared to court some hostility in order to advance their own position.

In 1956, however, at the Twentieth Congress of the Communist Party of the Soviet Union, when Secretary Khruschev published his denunciation of Stalin, the new divisions of Communist policy and philosophy were made very clear. The Chinese, as we have seen, echoed some opposition to the Stalinist personality cult by their rearrangement of party organization at the Eighth Congress in September of that year, and they supported the moves towards independence among the small states of eastern Europe. By their attack on Stalin's dominance of the Communist world, Khruschev and his colleagues had produced an argument that could readily be turned against themselves, and the Chinese were quite prepared to see Russian policy embarrassed when it enhanced their own position. Besides the dispute over boundaries on the map, there was new reason to criticize the rulers in Moscow, both for betrayal of the Communist tradition maintained by Stalin and for the harshness towards their dependencies in Europe. From the point of view of Mao Zedong and his partisans, the mantle of philosophical authority should now be transferred from Stalin in Moscow to Mao in Beijing, but it was also essential the Soviets should be restrained from enforcing their power upon Communist governments which disagreed with them.

In October and November 1956, for the few short weeks of rebellion in Hungary, it appeared the delicate game might have got out of balance, and that a crisis had appeared which would affect the very survival of the Communist world. The Chinese criticized the Soviets for curbing the independence movement in Poland, but they were compelled to support military intervention in Hungary. Without this endorsement, the rulers in Moscow would have faced heavy pressure against such interference, but Chinese support, whole-hearted against the Hungarian renegades, gave the Soviet Union valuable prestige in eastern Europe and ensured that other Communist states would neither protest nor dissent. With Britain and France uselessly involved in Suez and quarrelling with America about the invasion of Egypt, the Warsaw Pact countries faced down the rest of Europe and snatched back power for Communism in Hungary.

For a time, therefore, relations between the two great Communist powers were almost restored, and through 1957, as Khruschev confirmed his authority and the Soviet Union established a lead over the United States with man-made satellites and the missiles that launched them, the friendship was maintained. In November 1957 Mao Zedong visited Moscow,[8] and at an international gathering of Communist leaders he proclaimed the favourable situation where 'the East Wind prevails over the West Wind', and acknowledged the leadership of the Soviet Union in the socialist camp.

Despite apparent goodwill, however, there was still uncertainty between the two governments, and increasing concern in Moscow at the thought of atomic weapons controlled by Beijing. In the period of new-found friendship, the Soviets showed some willingness to share information on nuclear arms, and Chinese reports claimed that an agreement in October 1957, just before Mao Zedong's visit to Moscow, confirmed this understanding. During the November meeting in Moscow, however, Mao Zedong discussed atomic war in political, almost optimistic, terms: if the worst came to the worst and half of mankind dies, half would still remain, while imperialism would be razed to the ground and the whole world become socialist.[9]

By 1958, moreover, it seemed Mao might be preparing the holo-
caust which he claimed to fear so little, for in August, with massive
bombardment of Jinmen and Mazu, the islands off the coast of Fujian,
the Communists prepared an attack on the advance positions held by
the Nationalist forces on Taiwan. On the first day, 23 August, the
Nationalists claimed 41,000 shells had been fired at their defences,
and fighting continued over the following months with aerial dog-
fights and small-scale naval combat. Since the United States fleet was
on station in the area of Taiwan, the Communist Chinese could make
no serious attack unless the Soviet Union offered sufficient threat to
ensure America's neutrality.

It is not likely that any official statement was made, but as days
went by it became clear the Soviets were not prepared to use their
current military lead to risk a war on Beijing's behalf. The bombardment continued with occasional flurries of activity, and in January
1959 the Nationalists reported another day of 33,000 shells, but the
meaning of the incident was gone, and the attack died down to a token
shelling on alternate days. In June 1959 the Soviet Union cancelled
all agreements concerning the sharing of nuclear weapons with China,
and in September, when Khruschev met President Eisenhower at Camp
David, it was clear that the Soviets were prepared to deal with America
regardless of their ally's interests and expectations.

In documents published to justify Beijing's version of the disagreement, the meeting at Camp David appears as a turning point,
and indeed it marked the beginning of a process of *détente* which
came to concrete form in the Nuclear Test Ban Treaty of 1963. From
America, Khruschev flew to Beijing, where he added fuel to the flames
of Chinese resentment by a hint they should consider a 'two-China'
policy with Taiwan. The Chinese were now in open disagreement
with Moscow. Internally, the dismissal of Peng Dehuai was a further
blow at Soviet prestige, for Peng had been recently in Russia and was
known as a old comrade. Internationally, China expressed support for
North Vietnam's demand for unity with the South, in defiance of
Soviet caution on the issue, and in a series of meetings early in 1960,
notably at Bucharest, they criticized the Soviet regime for its betrayal of Marxist ideals.

Through 1958 and 1959, moreover, the Great Leap Forward had
asserted popular will against the theories of economics, and rejected
the Soviet model of centralized planning. As the Chinese sought to
eliminate dependence on outside aid, they became increasingly suspicious of Soviet policy, while the Russians were embarrassed by the

implied scorn of their expertise, and were also indignant at the pride with which the Chinese described their commune system as ideal Communism, claiming that in this ideological aspect they had surpassed the Soviet Union. In July 1960, seeking to demonstrate the importance of his support, Khruschev gave orders for all technical assistants to leave China and return home.

Though Soviet technicians had been criticized for their sense of superiority and their conservative attitudes, their work had been valuable even within the patterns of the Great Leap. The unexpected decision for withdrawal was bitterly resented by the Chinese, and made a final break in the ties between the former allies. The loss of expert advice and the drain on central finances through the withdrawal of Soviet loans did weaken the economy, and had particular effect, as we have seen, on major projects for industrialization. It was three or four years before the Chinese had overcome their technical problems and gained sufficient experience to restore production, while the loss of Soviet support added to the misfortunes of the early 1960s.

From this time on, almost unbelieving, the rest of the world watched a conflict of arguments between the two Communist powers, spirited and indignant on the Chinese side, more defensive on the Soviet, but certainly deep-felt and sincere. Occasionally, as in the developing conflict in Vietnam, the two governments co-operated to provide supplies, but they were united only by opposition to a common enemy and such a limited agreement gave no sign of further settlement. At intervals, there were incidents along the frontier, and the war of propaganda, seeking support both within the Communist bloc and outside, had its effect on every Communist party in the world. By the middle 1960s, after unsuccessful attempts had been made either to restore a measure of unity or to force the Chinese from the Communist movement as heretics, the schism was confirmed. Moscow and her allies held the majority, but the Chinese line was accepted in Albania, and the revolutionary doctrines of Chairman Mao gained enthusiastic interest and support in under-developed countries and among all the restless idealists of the world.

The New Emerging Forces

In 1959, as the Chinese government entered the final stages of its break with the Soviet alliance, and began to perceive the first signs of economic crisis, it was faced also with rebellion inside its borders. In March there was a widespread rising in Tibet, with rioting in the cities and guerilla resistance and banditry in the mountains and valleys. The People's Liberation Army quickly crushed organized resistance, and by the beginning of April the Dalai Lama had fled to India for refuge, but the sparks of revolt continued for some time in distant parts of Tibet, and the Chinese found it necessary to increase local garrisons and put guards along the frontier as a means of restricting the movement of supplies and reinforcements to the rebels.

Reports of repression and violence aroused world-wide concern and sympathy for the Tibetans, and India showed little hesitation in granting asylum to refugees. For the past ten years Indian policy had offered understanding and support for China's international position, a logical extension of her traditional non-alignment. Historically, India had accepted the Chinese claim to Tibetan territory, and protests at the time of the invasion in 1950 had been directed rather at the use of force than at the principle of occupation. Since that time, Nehru had frequently acted as patron to the Chinese and interpreter of their interests among Asian countries and in the United Nations, and the good, though somewhat distant, relations between the two governments had been celebrated in the streets of Indian cities by the cheerful slogan *Hindi-Chini bhai-bhai*, 'Indians and Chinese are brothers'.

Behind this pleasant facade, however, there were difficulties. China was naturally suspicious of the tolerance which India showed to refugees from Tibet, while in the general field of international affairs they resented Nehru's patronage and were jealous of his prestige in Asia and the rest of the world. India, moreover, though regularly preaching non-violence, non-intervention, and neutrality, had established a remarkable record of warfare in her own sub-continent: murderous riots during the partition from Pakistan in 1947, the invasion and destruction of Hyderabad and other princely states, and the forceful occupation of disputed territory in Kashmir. In the early part of 1962, when the Indian army seized the small Portuguese colony of Goa, the triumphant approval of the people left no doubt of their support for military adventure in any cause of national ambition. From this pattern, when natural disagreement on the border between India and Tibet came to a head in the early 1960s, there was little goodwill, considerable misunderstanding, and small attempt to negotiate.

China and India have two comparatively short stretches of common frontier. Only in the desolate triangle of the Aksai Chin in the north-west, between India, Kashmir, and Tibet, and in India's Northeast Frontier Agency, where the present border is based on the so-called McMahon Line, is there opportunity for direct disagreement.

The McMahon Line was established at the Simla Conference of 1913 and 1914, when Tibet accepted British Indian maps despite protests from a weak republican government in Beijing: where the McMahon Line runs along the crest of the ranges the Chinese would put the border at the bottom of the ridge on the Indian side, in the lower valley of the Brahmaputra. Both Nationalists and Communists had maintained the argument, though in the 1950s Beijing had offered to take the McMahon Line as a basis for negotiations.

At first the Indian government had little interest in serious discussion of any Chinese proposals. In 1957, however, it was revealed by maps in Chinese newspapers that China was constructing a road through the alpine wastes of the Aksai Chin to improve communications between Tibet and western Xinjiang. The news was not published in Delhi until 1959, but the Indians then made strong protest at this incursion into territory they regarded as their own; though many observers remarked that their hold on the ground must be very loose if they had not noticed the construction work much earlier. The matter remained unsettled in both areas of debated ground, and from 1959, as the Chinese stationed troops along the border for control of infiltration and aid to the rebels, there were increasing disputes between police posts, frontier pickets, and patrols. Still refusing any real negotiations, Nehru kept 'nibbling' (as the Chinese described it) along both sectors of the frontier, and in the summer of 1962, following their success against Portugal in Goa, the Indian army was ordered forward to occupy the whole area which they claimed.

In October the Chinese struck back, and with swift success they demolished India's positions and destroyed the pride of her army. Ill-equipped for mountain fighting, and demoralized by the enemy's blows, the Indian troops were driven helter-skelter from the high ground and cut off from supplies in the valleys. At one point, it appeared that the Chinese in the north-east could cut communications along the Brahmaputra valley and encircle the whole Indian army in that region, but then, having established their superiority on the battlefield, the Chinese abandoned all they had gained and withdrew once more to the McMahon Line.

Militarily speaking, the Chinese had no advantage in holding the territory they had conquered, and it was possible their communications problems in the difficult country might have left them exposed to an Indian counter-attack. In fact, they had gained all they wanted: no Indian troops would venture again to contest the Chinese border and, probably of more immediate importance, any thought of intervention in the affairs of Tibet was clearly gone for ever. On a wider scale, moreover, China had pricked the bubble of Nehru's pretensions. In rage and humiliation the Indians accused the Chinese of aggression; but it was clear that the policy of non-alignment and the prestige of India as arbiter of international justice had received a crippling blow. Fewer countries now looked to India as leader of Asia and the Third World.

On the other hand, the Chinese victory had gained them as much fear as respect. Several governments in Asia found Chinese military action a cause of anxiety, and regarded with distrust the increasing development of her diplomacy and interests abroad. The hostility of India encouraged Pakistan to seek alliance with China, and that former member of the American-inspired South-East Asia Treaty Organization now looked for support to the leading Communist power in Asia. Indonesia, in a turmoil of optimistic nationalism, sought and found Chinese approval first for the claim to western New Guinea, West Irian, and then in aggressive 'confrontation' with the new federation of Malaysia. For a short time, in sympathy with the government in Beijing, President Sukarno withdrew his country from membership of the United Nations, and spoke instead of an association of the 'new emerging forces' which would provide a defiant alternative to the neo-colonialist powers of Europe and America.

Further afield, less intensively but with considerable skill, the Chinese established a measure of support among the new nations of Africa, and their policy gained them acceptance among discontented rulers and rebellious peoples as the one important nation which was concerned with the have-nots of the world.

Most remarkable of all, on 16 October 1964, near Lop Nor in Xinjiang, the Chinese exploded a nuclear device. The development of the bomb was a major scientific achievement, for the Soviets, though they provided an atomic reactor for Beijing in 1958, had refused assistance with military technology. Admittedly, many of the senior Chinese atomic scientists had been trained in Europe, America, or Russia, but there was no question that possession of the bomb raised China at once to a sure position among the great powers.

Though China had nuclear weapons, however, she had yet no great number and no certain way to deliver them, while it is very possible that the possession of such force gave her rulers a better understanding of the dangers of atomic war than they had shown in the past. From the middle 1960s, though the words of Beijing lost nothing of their rhetoric, the deeds were careful and limited. Above all, closest to home, the government displayed both caution and a sense of reality in their policy towards war in Vietnam, where the growing strength of Vietcong insurgents in South Vietnam had by 1965 stimulated massive American technical and military aid and intervention. Despite occasional violations of her air space by American planes, which may have been unintentional but which were certainly as serious as similar incidents at the time of the Korean War, the Chinese made no move to intervene with force. They continued to provide transport facilities for supplies from Russia, with some equipment and anti-aircraft units of their own, and they offered verbal encouragement to North Vietnam against American attacks, but they made no move to halt the build-up of American arms in the South. Their decision was wise, and the advantage of hindsight reveals the disaster of America's failure against the guerillas and soldiers of the Vietcong and their allies in Hanoi, but in 1965 it was remarkable to observe how ineffectual China appeared while open anti-Communist war was developed near her borders.

In the second half of 1965, moreover, despite growing international prestige and evident military power, China's position in South-east Asia suffered one major set-back. On 30 September there was open fighting between the Indonesian Communist Party and the army. Though a number of the leading military men were killed, the soldiers crushed the Communists in a bloody counter-action. At least a hundred thousand people died in weeks of killing throughout the country, and rioting against the Communists, clearly linked with the pro-China policy of Sukarno, extended to Indonesians of Chinese race. As the new military government ignored Chinese protests and continued its massacre of enemies, President Sukarno, suspected with some justice of involvement in the attempted coup, was removed from executive power, and in 1966 the period of 'confrontation' was formally ended by an agreement with Malaysia and Singapore.

With the fall of Sukarno, finally forced from office in 1967, and the establishment of a bitterly anti-Communist government in Jakarta, the Chinese had lost a major ally and, at the same time, with the increasing American presence in Vietnam and the uncertain menace of Soviet hostility along the northern frontier, Beijing had cause to feel encircled. The recognition of the People's Republic in January 1964 by the French President de Gaulle, the first such courtesy from a Western country since the Korean War, was significant mainly as a gesture of opposition to the United States, and the Chinese accepted it with reserve, while the association with Pakistan did little more than confirm the enmity of India and saddle China with a comparatively weak ally whose religious doctrine and military government held small sympathy or natural appeal. For the time being at least, though China had established her independence and her position as a voice of the developing countries, she had many open enemies and few reliable friends.

Notes

1. Figures from *Ten Great Years*, Foreign Languages Press, Beijing 1960, pp. 87, 118–19.
2. More than half the assistance in the early 1950s was for the purchase of Russian-made weapons for use in Korea. The Chinese were not grateful that they were obliged to borrow from their ally to fight a war in the common cause.
3. See p. 213–4 above.
4. See, for example, *Cambridge China* 14, pp. 370–2 [Lardy, 'Economy under Stress'], also citing Coale, *Rapid Population Change in China*, p. 70.
5. At the end of 1958 there was debate whether grain production had reached 500 million tonnes, but no one had confidence in the reports, and it was Mao who chose the arbitrary figure of 375 million: *Cambridge China* 14, p. 379 [Lardy, 'Economy under Stress'], citing Lardy and Liebenthal, *Chen Yun's Strategy for China's Development*.
6. Documentation of disagreements at this time was not published until the Cultural Revolution in the second half of the 1960s, when the letter of Peng Dehuai, for example, was used not only against Peng himself, but also against Liu Shaoqi and Deng Xiaoping. By that time, any taint of criticism of Mao was damning.
There are a number of accounts of the debate and of Peng's letter, but the authorized one is in *Peng Dehuai zishu*, translated as *Memoirs of a Chinese Marshal*.
7. From US Congress Joint Economic Committee, *People's Republic of China: An Economic Assessment*, p. 343; also Donnithorne, *China's Economic System*, p. 318–19.
8. This was the second and last occasion that Mao Zedong travelled outside China: the first trip, also to Moscow, was from December 1949 to March 1950.
9. Quoted in *Peking Review*, No. 36, 6 September 1963, p. 10, cited by Schram, *Mao Tse-tung*, p. 291.

Further Reading

On economics:

Ten Great Years, Foreign Languages Press, Beijing, 1960.

Coale, Ansley, *Rapid Population Change in China 1952-1982,* Washington DC, National Academy of Sciences Press, 1984.

Lardy, Nicholas R., 'The Chinese economy under Stress 1958-1965' in *Cambridge China* 14.

Lardy, Nicholas R., and Liebenthal, Kenneth, *Chen Yun's Strategy for China's Development: a non-Maoist alternative,* Armonk NY, Sharpe, 1983.

Oksenberg, Michel (ed.), *China's Developmental Experience,* New York Academy of Political Science, 1973.

US Congress Joint Economic Committee, *People's Republic of China: An Economic Assessment,* Washington, 1972.

On politics:

Hinton, William, *Fanshen,* New York, Monthly Review Press, 1966.
MacFarquhar, Roderick, *The Origins of the Cultural Revolution: Volume 1: Contradictions among the People 1956–1957*, New York, Oxford University Press, 1974. *Volume 2: The Great Leap Forward 1958–1960,* New York, Oxford University Press, 1984.
Peng, Dehuai, *Memoirs of a Chinese Marshal*, Beijing, Foreign Languages Press, 1984
Schram, Stuart, *The Political Thought of Mao Tse-tung* [second edition], New York, Praeger,1969.

On foreign policy:

Gittings, John, *Survey of the Sino-Soviet Dispute,* Oxford, Oxford University Press 1968.
Maxwell, Neville, *India's China War,* London, Cape, 1970.
Wilson, Ian (ed.), *China and the World Community*, Sydney, Angus and Robertson, 1973.

10

Cultural Revolution, 1965–1975

THERE is an old Chinese saying, that you can conquer the empire on horseback, but you cannot rule it that way. By the early 1960s, Mao Zedong and his colleagues were faced with a modern version of that truth: the success of the Communists in rebellion and civil war had been based upon an ideal of nationalism and a dynamic of revolution, and Mao's own leadership provided the inspiration for military and political triumph. Yet with the establishment of formal government, some of the zeal for revolution had been lost, and economic and social progress, though swift by normal measures, had not fulfilled the dreams of the revolution. On the other hand, when Mao sought to impart new energy to the process, to break through the barriers marked out by cautious advisers and planners, his Great Leap Forward foundered in disaster. And as the engineers and bureaucrats worked to repair the damage of that misplaced enthusiasm, Mao himself was compelled to withdraw from any active role.

As the former system was restored, however, the danger increased that it could lose its reason for existence. When Mao fought for the revolution, he had not intended that it should provide just another government for China, and the example of the 'revisionist' Soviet Union was simply discouraging. If a revolutionary party lost its true ambition to change the world, then the heart had gone out of the movement and it was no more than a façade for self-serving men to maintain themselves in power and comfort. So Mao saw his dreams fading, and in 1965 he used his personal prestige and his political authority in one more attempt to reverse the process.

If that was the ambition of Mao Zedong, however, the people from whom he sought support had problems and ambitions of their own. At the highest level, the disorder which was created as the Chairman sought to criticize and shake out his own party and government presented opportunities for personal advancement, and the grand, simple design was readily distorted by private and factional ambition.

Far below, moreover, the doctrine of individual expression in the revolutionary cause, and defiance of repressive authority, was only too eagerly received among the masses of the people, suffering already from the frustrations of recession, with its slow reconstruction, and the loss of faith after the Great Leap Forward. In particular, young people at every level had found themselves shut out from work and opportunity, struggling against one another for position, while their elders were maintained with an 'iron rice-bowl'. There was tension in the schools and universities, in the streets and the factories, and the ideals of Maoism came like a spark thrown into tinder.

Neither Mao nor his junior followers could have succeeded against the entrenched power of the civil state if they had not been able to obtain the tacit support of the army, and in this process Mao's protégé Lin Biao, appointed Defence Minister after the dismissal of Peng Dehuai, played a critical role. In Beijing, Shanghai, and elsewhere, when the Red Guards and revolutionary rebels took to the streets, the army did nothing to stop them. So civil authorities were overwhelmed by the mobs, and within a few months the government and party had been devastated.

After the first shock, however, the excesses of the revolutionaries brought counter-attack from people who preferred a steady life to incoherent and destructive enthusiasm. So the army was ordered to support the Maoists, and when the soldiers showed reluctance they too were criticized and attacked. In July 1967, however, in the incident at Wuhan, the local garrison refused instructions and turned against the left. Though the mutiny was put down, the army had shown its strength, and as the Cultural Revolution was reorganized under military control the radicals were driven from the streets and young people were sent to exile in the countryside.

In the government, Lin Biao was the first beneficiary of this development, but Mao Zedong was increasingly uncertain of his devotion to the cause, and by 1971 Lin Biao had been weakened and destroyed. There followed an uneasy period of contradiction, as Mao sought to establish a revolutionary succession, as rhetoric and propaganda from the left dominated cultural debate, while the country at large yet sought a measure of order. At the same time, moreover, diplomatic initiatives from the United States gave an opening to the outside world: on the one hand a splendid success in international politics and an opportunity for prosperity and advancement through trade; on the other, a break in the closed political community which had allowed such free play to the radical ideals of revolution.

For people outside, particularly the intellectual radicals of the West in the late 1960s, Maoism was a form of inspired anarchy which might oppose the stultifying, often brutal forces of capitalism, and the ideals of puritan simplicity and selflessness for the cause provided a frequent debating point against the prosperous West, uncertain of its own morality. Those judgements, however, were made against a background of liberal democracy: inside China, there was no such relief. In open debate, radical revolution may hold attractions, but it is just as oppressive as any other doctrine when imposed by force from above.

So in the end, the dreams were fading but the damage remained. Throughout the country, the betrayal of the young people, the persecution of intellectuals and officials, and the despair which was inflicted upon their families brought a loss of knowledge and energy, of initiative and morale, which meant not only that the ten years from the mid-1960s to the 1970s had been wasted, but that the skills and potential of a whole generation had been lost.

During the Vietnam War, an American officer explained that 'We had to destroy the village in order to save it.' In the Cultural Revolution, Mao sought to save the revolution by destroying the government and the party. Tragically, the destruction was successful, but salvation did not follow.

'Bombard the Headquarters!'

By the mid-1960s the economy of China had largely recovered from the Great Leap Forward and the famine which followed. Production and trade had risen to new levels, while both agriculture and industry were improving in mechanization and output. Administratively, the State Council of ministers under Premier Zhou Enlai controlled the central government, but basic policy was dominated by the influence of Liu Shaoqi, Chairman of the People's Republic. Mao Zedong, still Chairman of the Communist Party, held a supervizing role, but published policy was controlled by Deng Xiaoping as General Secretary, and he did not always consult the Chairman.

Despite these political changes, and the comparative public eclipse of Chairman Mao, commentators remarked that since the establishment of Communist government in China the men at the top had remained very much the same. The disgrace of Gao Gang in 1954 and the dismissal of Peng Dehuai in 1959 had not involved many of their colleagues, and the arrangements by which Liu and Deng took power entailed little more than a reshuffle of offices. There had been no public display of ill-will within the leadership and in contrast to many other countries, notably the Soviet Union after the

death of Stalin, the People's Republic of China had shown remarkable political stability.

Within the Communist movement, however, and particularly among
people close to Chairman Mao, there were many who disapproved of
the cautious stability of official policy and the tendency towards
middle-class values. To Mao and his supporters, such concentration
on economic progress and political stability was a retreat from the
ideals of socialism, while the emphasis on workpoints in the communes and bonuses in the factories, and a decline in the fervour of
revolutionary art and propaganda, were warnings that the nation was
in danger of abandoning its heritage from the Communist victory.
When Mao Zedong proclaimed the foundation of the People's Republic
in 1949, he made it clear that that achievement was only a first step
in the long road to true communism, and he still felt it his duty to
keep China on that course.

Through the early 1960s, though Liu Shaoqi and Deng Xiaoping
held political power, there was considerable debate in literature, newspapers, and journals. In September 1962 Mao Zedong established a
socialist education movement, urging class struggle as a theme for
art and literature, and in following years he repeatedly criticized writers and scholars who ignored the need for socialist propaganda. Some
intellectuals fought back, notably three members of the Beijing
Municipal Committee, Wu Han, Deng Tuo, and Liao Mosha. Deng
Tuo published poetry and articles under the general title 'Conversations
in the Evening at Yanshan', satirizing Mao as a man who forgot his
promises, and as a child writing doggerel verse about the East Wind
and the Great Red Sun, while all three writers showed sympathy for
the former Defence Minister, Peng Dehuai, and his opposition to
Mao's policy of the Great Leap Forward.

In 1959 Wu Han published a short essay on the theme of Hai Rui,
an official of the Ming dynasty expelled from court for criticizing the
foolish policies of the emperor, and for supporting the peasants against
aggressive and greedy landlords. *Hai Rui Reprimands the Emperor*
was an early attack on the policies of the Great Leap Forward and
appeared even before the dismissal of Peng Dehuai. In 1961, moreover, Wu Han published a play, *Hai Rui Dismissed from Office*, and
the analogy was very clear.

Since Wu Han was Vice-Mayor and Deng Tuo Secretary of the
local government, it was assumed these literary attacks had approval
from the Mayor, Peng Zhen, and some sympathy from even higher
ranks in the government; for Peng Zhen was an old supporter of Liu

Shaoqi.[1] Peng Zhen's control in Beijing, however, was so firm that Chairman Mao is said to have exclaimed he had no room 'even to put in a needle'.

Madame Mao, Jiang Qing, however, had influence and old associates in Shanghai, and in the summer of 1965 Mao went to urge the local branch of the party against 'bourgeois reactionaries'. On 10 November, in the Shanghai journal *Wenhui Bao*, the local editor of the *Liberation Army Daily*, Yao Wenyuan, published an article attacking Wu Han's play. Still only in his thirties, Yao was noted mainly for his minimal literary taste and his enthusiasm for Communist jargon, but it was generally believed that Mao Zedong had inspired the attack, and Jiang Qing had supervised its composition.

Throughout the campaign which followed, one theme remained constant: by its very nature the Communist Party and the government of China was dedicated to the cause of revolution; and this was sufficient to ensure a propaganda victory for Mao Zedong and his supporters. As Chairman of the Party, regardless of his weakness in direct administration, Mao Zedong was the conscience and the prophet of Communist China, and from the last months of 1965 he used his personal prestige in a ruthless drive for power and a relentless demand that his ideal programme for continual revolution towards ultimate communism should be carried forward in defiance of all other considerations of government, society, and individual preference.

Central to the full 'Maoist' doctrine is the concept of the class struggle, the idea that any person, no matter how fine his Communist service, is always tempted towards middle-class security and élitism, and that this tendency must be constantly opposed, criticised, and overthrown by the true proletarians of the masses. When Yao Wenyuan attacked Wu Han, he claimed that the play about Hai Rui was lacking in class-consciousness, for it was manifestly absurd that a feudal minister of the Ming dynasty could ever have spoken in the real interests of the peasant classes: by their very use of historical analogy for comment on modern politics, Wu Han and his colleagues failed to show proper understanding of Communist principles, and the fact that Hai Rui had spoken for the rights of peasants to their land showed only that they wished to restore the reactionary system of private tenure from the days before the communes. Wu Han, Deng Tuo, and Liao Mosha, now described as the 'Black Gang', were denounced in one article after another, and on 30 December 1965 Wu Han confessed his failure to recognize class struggle. With this break among their opponents, the critics pressed harder. By April 1966 Wu Han,

Deng Tuo, and Liao Mosha had been arrested and made published confessions, and at this time the *Liberation Army Daily* published an editorial on the 'Great Proletarian Cultural Revolution'.

The lines of conflict were now roughly established: on the one side, Chairman Mao and Premier Zhou Enlai, with the support of the army under Defence Minister Lin Biao; on the other the official head of state, Liu Shaoqi, with his associate Peng Zhen and the cadres and officials who had held control of policy since the early 1960s. The attack was maintained on the theme of true revolutionary theory, with Yao Wenyuan as spokesman in Shanghai and Mao's personal secretary Chen Boda as editor of the official journal *Red Flag*. Following the disgrace of Wu Han and Deng Tuo, an article in *Red Flag* on 16 May 1966 accused Peng Zhen and his senior colleagues of supporting the counter-revolutionary intellectuals, and of planning to return Peng Dehuai and his revisionist line to power and authority. At the beginning of August, at the Eleventh Plenum of the Central Committee of the Communist Party, Mao Zedong appointed Lin Biao as his first vice-chairman, while Liu Shaoqi was demoted to eighth place in the party hierarchy. A few days later, Mao and Lin stood side by side to review the first contingents of Red Guards, and Mao with his own hand wrote out the wall poster, 'Bombard the Headquarters!' against Liu Shaoqi and Deng Xiaoping.[2]

The attack, however, had already moved from literary and political debate to practical physical force. Early in June the garrison commanders around Beijing were persuaded to support the new movement, and under their protection the cultural revolutionaries took control of government offices and the national newspaper *People's Daily*.

At this stage, the army was showing no more than benevolent neutrality for the Maoists, but the effect was none the less devastating. As the mobs came against their offices, the administrators were manhandled and humiliated, and they could look for no assistance in maintaining law and order, nor seek redress from the government which they served. In one department after another, helpless officials were either driven from their work or compelled to accept the dictates of self-appointed leaders, guardians of the new revolution. Whatever the ideals of the thought of Mao Zedong, the critical success of the Cultural Revolution was not based upon the quality of debate but on arrogant, uncontrolled force in the streets.

So the headquarters of government had been bombarded and taken by the Maoists, and Liu Shaoqi no longer held personal power. The Maoists and their allies controlled Beijing, and the pattern of the new order was established in November 1966 with the formation of the Cultural Revolution Group of the Party Central Committee, dominated by Mao's personal supporters, allied to the army through Lin Biao, and to the formal administration through Zhou Enlai. Among supporters of the new regime who came to prominence or gained in power were the propagandist Yao Wenyuan, the Minister of Security Xie Fuzhi, and another veteran secret-service official, Kang Sheng, an old friend of Jiang Qing, who had earlier defended Peng Zhen but now changed sides and took a senior position in the new government. Among those who fell from power with Liu Shaoqi were Luo Ruiqing, also a senior secret-service man and then chief of staff of the army, who was arrested on a vague charge of plotting at the beginning of 1966, attempted to commit suicide soon afterwards, but survived to be abused and humiliated by the Red Guards at the end of the year. And the provincial chieftain Tao Zhu, for many years head of the party and government in Guangzhou, made the mistake of coming to Beijing to share in the spoils of the coup but was himself eliminated from power by the Maoists.

Though the position in Beijing was essentially secure, it was still based on the personal following of Mao Zedong, and it could not be certain of support from the provinces. Even within the Cultural Revolution Group, Mao's need for loyalty was shown by the appointment of his secretary, Chen Boda, as chairman, and of his wife, Jiang Qing, as first vice-chairman and leading member for cultural affairs. At no time in the past had Jiang Qing held prominence in the party or the government, but from this time on the arts, literature, and above all the theatre were dominated by the taste of that former film actress from Shanghai.

Still more important was the role of Zhou Enlai. As Chairman of the State Council, he was head of the formal administration, yet he approved the attack on Liu Shaoqi, and when radicals attacked the bureaucracy, he sought only to play a moderating role and made speeches to support the new revolution. His motives have been the subject of complex debate: did he plan to succeed Liu Shaoqi and, later, the ageing Mao Zedong? Was he still inspired by the revolutionary ideals of earlier days, or did he simply fear that he might be the next one to be attacked? People compared him to the branch of a willow, which bends before greater force but retains its strength for the future. In any event, his support was critical at the beginning of the revolution, his presence gave an acceptable face to much of the turmoil which followed, and at his death, before his Chairman, he became a hero of the masses: that, perhaps, was what he always wanted.

The question remained, however, whether officials in the country as a whole, who had gained and held power under the government of Liu Shaoqi and the established practices of the Party, were prepared to accept the full terms of the new, radical, vision for revolution; and if they did not do so what force could be mustered to persuade them. For the next two years and more, the whole of China was riven with the struggle between Communist government and Communist ideology.

The Red Guards and the Army

For Mao Zedong and his supporters in the Cultural Revolution, the victory in Beijing and the access to power in the centre which went with it was not enough to ensure success. Mao Zedong regarded himself, rightly, as a leader of revolution and an important theorist of Marxism. To a man of his ideals, political success, whether it came at the end of civil war or as the result of a *coup d'état,* was only preliminary to a complete reform of thought and behaviour among the people. The 'Great Proletarian Cultural Revolution' had just that intention: to change ways of thinking and to alter the established patterns of culture so that people should achieve a true proletarian outlook and a genuine commitment to ultimate communism. In these terms, the class struggle can never end, for the term 'class' ceases to refer to any particular social group: it describes the tendency for each individual to become unduly proud of status or ability, or which encourages the search for personal, selfish security, without concern for the general well-being.

Mao Zedong, therefore, seized authority in the government not just to satisfy his own drive for power but also because he believed the Chinese state was losing its sense of purpose and was turning towards complacent conservatism, giving more attention to worldly production than to spiritual values, and heading steadily towards the bureaucratic arrogance of the Soviet Union. If this were true, then revolution must be raised not only against individuals who held power but also against the whole apparatus of the state.

Chief of Mao's weapons in the struggle for new revolution in China was the organization of Red Guards. The term had been known since the early days of guerilla fighting in Jiangxi, but in August 1966 the young people of China were urged to join in bands of Red Guards, taking the thought of Chairman Mao as their guide and touchstone of revolution, and opposing by mob demonstration all those in authority who lacked revolutionary fervour and so 'followed the capitalist road'. Such an accusation could be levied against almost every Communist cadre and government official, for anyone with responsibility was naturally inclined to seek order and stability, and any official in charge of finances was cautious in his use of government funds. It was only a few years earlier that idealist fervour had brought disaster from the Great Leap Forward, and the men who held power in the middle 1960s were not anxious to return to that period of misfortune.

On the other hand, regardless of political principles, the people, particularly the young, had cause for resentment and suspicion of their leaders. Like any non-elective government, the rulers of Communist China were divorced in sympathy from the people under their charge, and accusations of élitism and bureaucracy were often well founded. At the lowest levels of administration, on the factory floor and along the docks, there were petty regulations and frequent fines; in the communes there were many occasions when party directives about sowing and harvest, work-points and allotments, were administered without good sense; and in schools and universities discipline was firm, teaching methods strict, and staff maintained power without comradeship against their students.

It has been well observed that education in modern China inherited three unfortunate traditions: the rote-learning of ancient scholarship, the dogma of Christian missions, and the dictatorship of Marxist Russia.[3] By the middle 1960s, moreover, the effects of the past fifteen years were beginning to affect the outlook and ambitions of young people. The opportunities of expansion in the 1950s had largely been taken by their elders, and the depression which followed the Great Leap Forward meant that chances for employment in the future were heavily restricted. Education, with its opportunity to go further, was critical, and there was desperate jealousy in the classroom as competition for places at university, and the good career that might lead to, became more intense. There was division between the children of cadres, who might expect advancement because of their excellent background, and those who were compelled to rely merely upon their own abilities; and there was fierce competition to demonstrate revolutionary fervour, as a means to gain approval among superiors.[4] Not surprisingly, therefore, the Red Guard movement found its most active support among the students of secondary schools and in the universities. For these young people, the revolutionary past was a childhood story, but the new movement offered a chance to relive that excitement and to rebel with success against the restrictions of life under arbitrary rule. Indeed, by curious but natural contradiction, some of the most energetic of the radical revolutionaries were able young people of mediocre or bad class background who had been excluded from advancement by the prejudice which favoured the established party élite and their children.

At first, in the closing months of 1966 and the beginning of 1967, the Red Guards were most active in Beijing and Shanghai, and they concentrated largely on abusing Liu Shaoqi, Deng Xiaoping, and their known supporters. Gradually the attacks shifted and broadened in scope, and in every province and city the party and government came under attack.

The struggle for power, however, was neither straightforward nor simple. In Shanghai, for example, though the city had been the launching place for the initial campaign against Wu Han, the Red Guards and 'revolutionary rebels', recruited from junior workers in the factories, received little popular support. The authorities gathered loyalist volunteers who fought pitched battles with the Maoists in the streets, and disrupted the city by massive strikes and power cuts. In every work unit and district the contest was maintained with wall newspapers and public debates. Both sides claimed to support the ideals of Chairman Mao, and officials were able to bolster their position by distributing funds from the public treasury to their supporters in the streets and factories. In November 1966, however, two members of the Shanghai party committee, Yao Wenyuan and Zhang Chunqiao, proclaimed themselves as leaders of the rebellion, and they received the support of Mao Zedong and tacit approval from the army. On 5 January the rebels seized the offices of the important journal *Liberation,* and their victory was hailed by the press in Beijing as the 'January Revolution'. On 5 February the Maoists at last succeeded in overthrowing the former administration and proclaimed the establishment of a new Shanghai Commune: two days later, orders came from the capital that the term 'Commune' was premature, and on 24 February a Revolutionary Committee was announced instead, with a three-part union of Maoist revolutionaries, rehabilitated cadres from the old regime, and representatives from the People's Liberation Army.

Events in Shanghai showed the pattern of the Cultural Revolution: until the last stages of the struggle for power, the revolutionaries were in the minority and their opponents, supporters of the local regime, had wide approval. Regardless of whether they were misled or mistaken, the masses for whom the Maoists claimed to be fighting were reluctant to join their cause, and ordinary people sought stability rather than revolution. In Mao Zedong's own message to the army in Shanghai, published in the *Liberation Army Daily* on 25 January 1967, he urged the troops to 'support the left' even if leftists should be in a minority, so it was acknowledged the revolution might need to be carried through even against the short-term wishes of the people themselves.

The problem, of course, came with the role of the army. Though any minority force as youthful, energetic, and inspired as the Red Guards could always cause confusion, and would certainly gain a modicum of popular support, any efficient local organization could hold them at bay and bring other forces against them, and the situation in Shanghai had been solved only by intervention from the capital with support from the army. No one could be sure that the authority of Beijing and the loyalty of the troops would be so certain in other centres, and the situation became more tense in every region as local rulers realized the terms and the scale on which they must face attack.

Through the first half of 1967 the struggle raged in almost every part of the country, with varying and fluctuating success and with frequent reports of bloodshed. Revolutionary committees, the outward sign of the Cultural Revolution, were set up in Shanghai and Beijing, in Shandong and Shanxi, in Heilongjiang in the far north-east, and in Guizhou in the south-west. In Yunnan, however, the party leader maintained vigorous opposition, arresting Red Guards and sending them to labour camps, until he was defeated and committed suicide in July. In Henan, there was widespread rioting in Zhengzhou, the provincial capital, and traffic on the north–south railway came to a standstill. In Xinjiang in the north-west, the military ruler Wang Enmao was able to force the suspension of the Cultural Revolution in his region, and he and his colleague from Tibet, Zhang Guohua, both made their peace later with the central government. In Guangzhou, the fall of Tao Zhu broke the control of local authority, with shooting in the streets and the disgrace of his lieutenant Zhao Ziyang, while from Sichuan there came reports of fighting at Chengdu, rebellion in Chongqing, and guerilla warfare in the countryside.

In every district where Maoists and other groups contended for power, the central government had ordered that the army should take responsibility for local government, and military control commissions were set up in all provinces and cities where no revolutionary committee had yet been established. On the one hand, however, the army was supposed to stay neutral while the political struggle was fought out by civilians; but at the same time the troops were not to hinder the revolutionaries in their battle for the thought of Mao Zedong. Ultimately the two requirements were self-contradictory, and there were constant reports of local army units which took law and order as their prime concern, and which were quite prepared to put down opposition by force, regardless of whether the disorderly mob was a Red Guard unit or not. In January 1967 the old fighting general of the civil war, He Long, was denounced and purged by the Red Guards as a 'revisionist counter-revolutionary'. The incident shook the army's non-aligned position, for Marshal He was widely respected and admired as one of the heroes of the past, and his disgrace did little to endear the revolutionaries among the soldiers.

At Wuhan, in July 1967, tensions came to head. There had already been rioting and disturbance in Hubei province, and the railway crossing of the Yangzi had been closed twice in June. As in other industrial centres, factory workers had no enthusiasm for the Maoist programme, which expressed itself in immediate terms by abolishing bonuses and other incentives for production. At Wuhan the major conservative force was known as the 'One Million Warriors', many of them former soldiers, and though its members may not have equalled that number they had majority support in the metropolis of two million people. On 14 July, however, Zhou Enlai was in Wuhan on a tour of investigation, and he gave orders that the army should support the Red Guards against the One Million Warriors. Xie Fuzhi and Wang Li, two senior members of the government in Beijing, remained in Wuhan to supervise the carrying out of these instructions.

On 20 July, however, Chen Zaidao, chief of the Wuhan Military District, gave his support to the One Million Warriors. Xie Fuzhi and Wang Li were arrested, and although they obtained their release within twenty-four hours the provincial forces had shown their independence. For the next two weeks there were tense manoeuvres around Wuhan as units of the army, navy, and air force under central command moved against the locals, both the One Million Warriors and the ten thousand soldiers of Chen Zaidao's command. The defenders were doomed to defeat, but they maintained their positions until August, and reports in the Beijing press could not conceal they had wide support among the citizens of Wuhan.

Chen Zaidao had flown to Beijing to put his case at the capital, complaining that the political situation was so confused it was impossible to follow central government instructions, and that there should be some concern for good order. He was disgraced and dismissed, and the mutiny at Wuhan brought a short flurry of denunciations and purges among officers of the army suspected of similar mistaken disloyalty to the Cultural Revolution. For a few weeks in August, Jiang Qing and the ultra-leftists sought to purge the army, but the incident was decisive in changing the course of politics to the other direction: no matter how eager the Maoists might be to establish loyalty on their own terms, they could not destroy the one organization which preserved the unity of the country from the anarchy of the Red Guards. To maintain the movement the rebels must have tolerance from the army, and the army had now shown a little of its strength and was demanding some price for its acquiescence.

So there was now a counter-reaction, and its success was confirmed by an order of 5 September, encouraging local military units to play a greater role in the reform and reconstitution of provincial, municipal, and lower-level government. Regional commanders such as Huang Yongsheng in Guangzhou and Xu Shiyou in Nanjing appointed themselves chairmen of their provincial revolutionary committees, and new committees were formed with increasing speed throughout the country, but the mixture of revolutionary rebels, rehabilitated cadres and representatives from the army swung steadily to the favour of military men. As demonstration of this new political power, in October 1967, less than two months after his triumphant return from Wuhan, the Maoist emissary Wang Li was disgraced and purged, on the very grounds, formerly acceptable, that he had accused the army of harbouring revisionist and capitalist elements.

Comrades in Arms

The Wuhan Incident, and the show of strength by the army, was the beginning of the end for the rebels and the Red Guards, but it was many months before the Cultural Revolution was brought under control. For the young people and fanatics of Maoism, self-righteousness and anarchy brought heady delights, with posters, banners and slogans, with the thoughts of Chairman Mao in a little red book, with the power to humiliate ordinary citizens for petty or non-existent errors in political thinking and behaviour, the power to dress political cadres and university professors in dunce's hats and parade them through the streets, the power to attack foreigners, imprison them and beat them, all with the support of government authority. In this regime of misrule, private dwellings were ransacked and looted, temples, gardens and objects of art were defaced and destroyed, and innocent men and women were exposed without protection to the screaming, mindless fury of an hysterical mob.

Overseas, the reports of the Red Guards and the realities of official Chinese behaviour lost friendship and contacts everywhere. In one African country after another, as the government found that the local Chinese embassy was encouraging sabotage and subversion of their authority, the diplomats were sent home. In British Hong Kong, an obvious target for nationalist feeling, there was a wave of strikes, riots, and bombings, and in London, outside the offices of the Chinese missions, clerks and guards came into the street to fight pitched battles with policemen. In August 1967 in Beijing, Red Guards attacked and burnt the British Chancellery and occupied the Chinese Foreign Ministry by force.

Gradually, however, but with increasing strength and firmness, the army established control, and the government which had supported the young rebels now turned away from them. Most universities remained closed, but schools were reopened, and workers in factories, who had earlier continued to receive wages while they were acting as revolutionaries, were now paid only when they were engaged in regular employment. By their hundreds and thousands, young Red Guards were sent out from the cities. Some were persuaded they should emulate the Long March of the 1930s, carrying the thoughts of Chairman Mao among the people, but all were eventually dispatched to work with the peasants and to display their enthusiasm in physical labour on the land.

It was hardly surprising that the young people who had joined the call for total change and new development should feel suspicious and betrayed, and through 1968, as the changes spread through every level from factories and communes to cities and provinces, their fears were shown to be well justified. The army held control at each level, and the new revolutionary committees, quite unlike those of earlier days, contained many of the same party members and cadres that the Cultural Revolution had originally sought to remove from power. In September it was proudly proclaimed that every province had appointed its own revolutionary committee, and the Cultural Revolution had thus achieved total success; but the reality was very different, for the revolutionaries had been pushed into the background and in several places the commander of the local military was also chairman of the revolutionary committee and chief of the party.

Many Red Guards had been settled in the countryside, but others turned in protest, and through much of 1968 there was open fighting between soldiers, revolutionaries, and rival factions still striving for power. Refugees to Hong Kong brought stories of disorder, and bound corpses of young dissidents were washed ashore. By the end of 1968, however, the period of popular action had passed away, and propaganda throughout the country was developed to encourage popular hatred against the disgraced Liu Shaoqi and admiration for the army.

In higher levels of government the situation was still fluid. On the one hand, the leaders of the army had made conditions for their support of the government, and the first targets of attack were some of Mao Zedong's closest associates. Not only Wang Li, but several other members of the Cultural Revolution Group were purged from power and public life, and of the major civilian figures who had risen to prominence there remained only Chen Boda, Yao Wenyuan, Xie Fuzhi, and Zhang Chunqiao, together with Madame Mao, Jiang Qing.

The authority of Mao Zedong, however, as theorist and unquestioned leader of the Communist movement in China, still managed to prevent a full attack against the Cultural Revolution he had sponsored. Equally important in practical terms, Lin Biao had gained time to enlarge his support in the army. As Minister for Defence and ally of the revolutionaries in the first coup of 1966, Lin Biao had already acquired a considerable following, but from 1968 he was arranging appointments in local military regions so that his personal allies would hold high position. On the civilian side, Premier Zhou Enlai maintained cautious support for Mao Zedong and his associates, but he was under constant pressure from radicals criticizing the central government. Zhou Enlai had generally defended his people against such attacks, but he achieved only mixed success. The Red Guard takeover of the Foreign Office was a notable defeat, many ministers were forced from office, and in the State Council and the bureaucracy, by abolitions and amalgamations, the number of ministries was reduced from forty-nine to twenty-two. On the other hand, though the civil service was demoralized and decimated by the purges, and Zhou Enlai was often in danger of direct personal attack, he maintained his position as indispensable head of the functioning government. All authority was weakened by the Cultural Revolution, but Zhou Enlai could be said to have gained from the crisis, partly through the fall of his rivals but also from the renewed loyalty of those subordinates whom he was able to save from their enemies.

In April 1969, at the Ninth Congress of the Chinese Communist Party, the country, the government, and 1,512 delegates had a chance to take stock.[5] Predictably, the chief function of the meeting was to confirm the disgrace of Liu Shaoqi, who had been finally and officially dismissed as Head of State in the preceding October. The complex apparatus of party secretariat and administration was eliminated, and the simplified structure was dominated by the figure of Chairman Mao . The balance of power in the new Central Committee, however, was now held by the military, for out of 170 members and 109 alternates, 110 positions were held by serving officers.

Lin Biao achieved a central position as link-man between the civilian government of the Cultural Revolution and the power of the armed forces which he should control, and the party constitution named him formally as official successor to Chairman Mao. At the same time, however, chieftains of the great military regions also gained influence at the capital: among them Huang Yongsheng of Guangzhou, Xu Shiyou of Nanjing, and Chen Xilian of Shenyang, all members of the Politburo elected from the Central Committee. In the Politburo, of twenty-one members with four alternates, there was a majority of military men, but only three were clear supporters of Lin Biao. The five-man Standing Committee of the Politburo comprised Mao Zedong, Lin Biao, Chen Boda, Zhou Enlai, and Kang Sheng, all leaders of the Cultural Revolution, but the composition of the larger body showed that the situation was still unstable and undecided.

Ping-pong Diplomacy

A first result of the Cultural Revolution was to intensify the international isolation of Beijing, and to halt all attempts by diplomats on either side to establish normal relations. Among obvious indicators of China's support in the outside world was the annual vote on membership of the People's Republic in the United Nations to replace the government in Taipei. In 1965 the numbers in favour of Beijing equalled those against, 47 each side with 20 abstentions. In 1966, however, with six new member states in the Assembly, the vote for Beijing was 46, with 57 opposed and 17 abstentions, and there was no improvement in the situation for several years to come.

For some countries, like the France of President de Gaulle and the small east-European state of Albania, approval of the Beijing regime was as much a matter of defiance, either of the Americans or of the Russians, as a real indication of sympathy for Mao Zedong and the Chinese Communists. In the Third World, however, and notably in neighbouring countries of Asia, the question seemed more practical and serious. In September 1965, Lin Biao published *Long Live the*

Victory of the People's War, widely praised in China for applying Mao Zedong's guerilla theory of revolution to foretell the destruction of the Western capitalist 'cities' by the Third World 'countryside'. Though the essay could be interpreted as a statement of inevitable historical process rather than incitement to aggression and revolution, the concept was hardly popular with non-Communist governments, while loud Chinese support and actual military aid sent and transmitted to the Vietcong and the North Vietnamese in their attack on South Vietnam showed active assistance in the expansion of revolution. The Chinese, indeed, always claimed they did not interfere in other sovereign states; but it was equally true that Beijing maintained public support for 'progressive forces' throughout the world, whether or not this might appear to contradict the principles of peaceful co-existence.

Throughout South-east Asia, moreover, there was a perceived threat from overseas Chinese, many of whom were disfranchised and disadvantaged by national governments. Local Chinese were killed in Burma, Cambodia, Malaysia, and Indonesia, and although Beijing, unlike the Nationalists, disclaimed responsibility for citizens of other countries, the Foreign Ministry was always prepared to protest against such anti-Chinese riots, and the Communist government had always shown sympathy for left-wing movements. By the time of the Cultural Revolution, rulers of these countries were concerned at the support which Mao Zedong Thought might attract among local Chinese and indigenous radicals. In December 1966, for example, *Peking Review* published a letter from some overseas Chinese held in an Indonesian military prison after the 1965 coup against Sukarno:

Let the enemy beat and injure us and bloody our bodies all over, but they cannot shake our faith and our warm love for our motherland and Chairman Mao. Our motherland is our strongest support and Chairman Mao is the reddest sun in our hearts... Without fail we shall emulate the proletarian revolutionary rebel spirit of the Red Guards — the vanguards of China's great proletarian cultural revolution — their daring to think, to speak out, to act and to break through.[6]

Regardless of official policy, many young people saw Chairman Mao as the answer to their personal and racial sense of oppression, and they presented a threat to the security of other governments.

With the Red Guards' seizure of the Foreign Ministry in 1967 the situation became still more confused, but by 1968, as control was established by the army, there were signs of return to normal procedures. Instead of international revolution, and general opposition to reactionaries and revisionists, there was increasing attention to the Soviet Union. For two weeks in February 1967 the Soviet Embassy in Beijing was besieged by Red Guards, and there were constant reports of harassment and maltreatment of Soviet citizens and of Chinese suspected of being friendly to them. Against the heresies of the 'new Tsars', the thought of Mao Zedong was seen as the true line of Marxism–Leninism, and Communist China was now armed for nuclear war: since the first atomic blast at Lop Nor in 1964, the Chinese had tested missiles in 1966 and a hydrogen bomb in 1967.

The Chinese leaders, however, had cause for concern. They knew how vulnerable they were to a limited strategic blow with either nuclear or conventional forces, and when Soviet tanks occupied Czechoslovakia in August 1968, the rulers in Moscow showed that they too had opinions about unity and orthodoxy and that they were prepared to take military action, even against other Communist countries, to secure obedience. Since 1966 the Russians had been moving troops against the northern borders of China, and by the winter of 1969 open war seemed possible.

The immediate cause of the conflict was a small island known as Zhenbao to the Chinese (Damansky to the Russians) which lies in the Ussuri River on the borders of Heilongjiang and the Soviet Maritime Province. The region is very sparsely settled, but there had been occasional attacks on fishing boats and huntsmen who crossed the disputed borders, and in March 1969, when the river was frozen, Russian and Chinese soldiers fought small-scale engagements. Both sides claimed victory, but the skirmish was not allowed to develop into anything more dangerous.

Nevertheless, there were rumours that the Soviet military had urged a pre-emptive strike, firstly to eliminate nuclear installations, and secondly to humiliate China, with or without the loss of territory, in the same way China had burst the bubble of Nehru's pretensions by the war of 1962. No such campaign took place, but the national and ideological conflict remained unsettled, and the military concern combined with the quietening of the Cultural Revolution to produce a general easing of tensions with other countries and some search for friends and allies against Moscow. Through 1970, the Chinese government showed signs of seeking contact with the West, and in April 1971 the American team visiting Japan for the World Table Tennis Championships, in which China is an undisputed leader, was invited to visit the People's Republic. This was the first semi-official visit between the two countries since the early 1950s, and the ping-pong players were followed rapidly by small groups of scientists, scholars, and journalists, many, but not all, chosen for their left-wing sympathies. In July, after a secret preliminary visit by his adviser Dr Kissinger, President Nixon of the United States announced that he himself would visit China early in 1972.

Even the announcement of such a trip made an immediate difference to the whole pattern of international politics, and the growing *rapprochement* between China and America gave support to Beijing against the threat from the Soviet Union. Still more directly, the United States' playing of the 'China card' meant a drastic realignment of its own policies. Immediately, the essential rationale of the war in Vietnam was gone: how could the loss of more American lives be justified by the need to combat Asian Communism, when the President was visiting the headquarters of that very movement? By the end of 1972 United States forces had withdrawn from Vietnam and in early 1975 the Communists captured Saigon and united the country.

The visit of President Nixon took place in February 1972, though the United States did not yet give full recognition, and retained its concern for Taiwan.[7] The Beijing regime, however, had already gained swift rewards of international goodwill and security. Inevitably, the old opposition of the United States no longer carried weight in the United Nations, and in October 1971 the People's Republic was accepted as the representative of China, and acceded to the permanent seat on the Security Council. In the following September, after Nixon's visit, Prime Minister Tanaka of Japan also went to Beijing, and Japan switched formal recognition, though it still retained close commercial contact with Taiwan.

It was now easy for other United States allies such as Australia and West Germany to make formal contact with Beijing, and countries such as Britain, France, and Canada, which had recognized Beijing earlier under different circumstances, developed relations further. Closer to home, in South-east Asia, the Philippines and Singapore were variously cordial and always cautious, Indonesia under Suharto was little interested, and only Malaysia proceeded to full recognition in 1974. India, still bitter from the war, remained hostile and indeed signed a pact with the Soviet Union, while the Chinese position on the subcontinent was weakened by the dismemberment of Pakistan and the foundation of Bangladesh in December 1971. These, however, were matters which could now be dealt with in regular terms, for the enterprise of Nixon and Kissinger had removed the People's Republic from twenty years of isolation and made it possible for China to play a proper role in the world at large.

The Fall of Lin Biao and the Progress of Jiang Qing

In the excitement and euphoria which accompanied Chinese development on the international scene, however, it was not always appreciated that the regime was still unsettled and insecure, and had barely grappled with the self-inflicted problems of the Cultural Revolution.

Indeed, despite the new openness to contacts abroad, it was a feature of 'China-watching' that economic and other statistics remained unavailable, while political conflict took place behind closed doors. Interpretation was constantly bedevilled, both for Chinese and for outside observers, by long delays between formal events, real changes in power, and official statements of justification or denial. At every level, China remained a closed society, with published information designed not so much to give the facts, but to tell the people what they should think, and with gaps of knowledge filled by rumour and speculation.

In August 1970, Chen Boda, leader of the Cultural Revolution and personal secretary of Mao Zedong, disappeared from public view, and in May 1971 he was attacked as a charlatan and a false Marxist, of the same nature as Liu Shaoqi. Far more seriously, in September 1971 it was announced against all precedent that the National Day parade in Beijing on 1 October would not be held, and on 30 September a newspaper in Outer Mongolia revealed that a Chinese military aircraft had crashed in that territory shortly after midnight on 13 September. From that time on, Marshal Lin Biao was seen no more in public, but almost a year later the Chinese press announced the former leader had made an unsuccessful attempt at a *coup d'état*, and he and his party had been destroyed as they fled.

The true series of events is still unclear, but it appears Mao had begun to distrust Lin Biao's ambitions, and the disgrace of Chen Boda, who had given Lin some political support, was a preliminary warning move. Some of Lin Biao's associates may have sought to arrange a military coup, but the position was already too weak, and Mao's supporters had adequate opportunity to move against them.[8] Other officers, notably Huang Yongsheng, chief of staff for the armed forces and head of the Guangzhou military region, also disappeared, returning to public life at the end of 1980, to be tried and convicted for conspiracy. Lin Biao's chief allies, however, had been members of the central military staff, and commanders of the other great military regions had been persuaded not to interfere in such active politics. Despite his active military service in the past, and the fact that, as Minister for Defence, he had controlled senior service appointments for more than ten years, Lin Biao had not been able to establish a position in the state which was fully independent of Chairman Mao. His fall reflected both the problems of a political general and those of an ambitious protégé who loses the trust of his patron.

By the end of 1971, Mao Zedong and Zhou Enlai were the only two members remaining of the five-man Standing Committee of the Politburo established by the Ninth Party Congress in April 1969. Lin Biao and Chen Boda had disappeared, and Kang Sheng was mortally ill. Despite his defeat of the conspirators, Mao's direct authority was naturally weakened, for they were his decisions which had given the traitors such power. For the time being, formal administration was in the hands of the Premier Zhou Enlai, with cautious respect given to the interests of military leaders in the provinces. On Army Day, 1 August 1972, the rebel commander of the Wuhan Incident in 1967, Chen Zaidao, evidently returned to favour, took part in the official banquet in Beijing, and in the following year the former villain Deng Xiaoping returned to a modest place among the leaders at various ceremonies.

For Mao Zedong, eighty years old in 1973, and already suffering from Parkinson's disease, problems of the future were only intensified by the destruction of Lin Biao. Whom should he now choose as his successor, to prevent the Chinese people returning to the sloth and self-satisfaction of the past? For the time being, though the revolutionary committees which now governed the country were little more than the old regime with a military slant, the rhetoric remained, and the Cultural Revolution Group dominated by Jiang Qing remained the symbol and driving force to permanent revolution guided by Mao's ideals.

Jiang Qing and her two Shanghai associates, Zhang Chunqiao and Yao Wenyuan, had general control of the machinery of propaganda, but they were all on the margin of formal government. In August 1973, however, when the Tenth Congress was held to reconstruct the party after the destruction of Lin Biao,[9] these three leaders of the Cultural Revolution obtained Mao's approval to seek position for themselves. All were elected to the Politburo, and Zhang Chunqiao became a member of the Standing Committee. A protégé of Jiang Qing, moreover, a former worker in a Shanghai cotton mill who had taken a lead in the earlier days of the Cultural Revolution at street and factory level, was made second vice-chairman of the party, next only to Mao and Zhou Enlai. Wang Hongwen was at this time thirty-seven years old, and the new triumvirate of leadership was acclaimed as an alliance of the old, the middle-aged and the young: a model somewhat confused by the fact that Zhou Enlai was only five years younger than the octogenarian Mao.

Despite the claimed success of the Tenth Party Congress and an apparent restoration of order in the government, provincial authorities, and notably the great military regions, still held considerable autonomy. The People's Liberation Army had somewhat withdrawn from its dominance over the civilian administration, and government and party organisation in the provinces appeared largely restored to civilian hands, but regional commanders had acquired great authority in their districts. In January 1974, however, in an unexpected major reshuffle of military appointments, the Chief Political Commissar Li Desheng was transferred from Beijing to replace Chen Xilian in the north-east, Chen Xilian took over in Beijing, and Xu Shiyou exchanged places with his opposite number, Ding Sheng, in Guangzhou. None of the generals was demoted, but the transfers broke their connections with established bases (Xu Shiyou had held his post at Nanjing for sixteen years), and they were seen as an attempt to weaken the local connections of military power. It could be observed, moreover, that while Jiang Qing and her left-wing colleagues had never found sympathy among professional military men, they had good contacts with militia groups in Shanghai and the other major cities.

To some extent, it was in related political terms that one could interpret the campaign of propaganda which dominated discussion in late 1973: the attack and denunciation of the ancient sage Confucius, and praise for the unifying First Emperor of Qin. For the most part, the First Emperor of Qin, who ruled China at the end of the third century BC, was lauded for his opposition to Confucianism, for his burning of non-government books, and for burying alive some scholars who opposed him, but it was possible to see an indirect warning to 'feudal lords' in the provinces that they should not press independence too far. More directly, however, in the tradition of Maoist ideals, the anti-Confucian movement was one more attempt to overthrow the traditional conservatism of ordinary Chinese, and to convince them that the future lay rather with the dynamics of Communism than with reactionary teachings of the past. The propaganda was supported by some scholarship and quality of debate, seeking to show that Confucius' doctrines were based upon intellectual élitism, on social injustice to the masses and to women, and on the general pattern of a slave-owning society. The programme, however, was not helped by associating anti-Confucianism with criticism of the traitor Lin Biao, for the affair was displayed too clearly as just one more campaign, and the force of the argument was soon lost. Family grave-mounds could still be seen across the open fields of the communes, and the traditions of the past again proved stronger than the slogans of the present.

But all these matters were short-term and ephemeral. Despite, and indeed because of, the disorder he had brought upon the government and the party, Mao Zedong was again at the centre of politics. So long as he lived his opinions could determine the fortunes of his colleagues in the leadership and of the people as a whole.[10] Now, however, he was old and ill, and the rival factions jockeyed for position — cautiously, so as not to disturb him — while they waited for the Chairman to die, and wondered when that crisis would come.

Revolutionary Society

As ten years of turmoil approached their end, the effects of the political struggles could still be seen in the community at large. The Great Proletarian Cultural Revolution had been well-named, for the effects were widespread, and they continued to influence both the daily lives of the people and the ideas which they were urged to adopt.

Central to the Maoist ideal is the concept of egalitarian service under a single line of thought. Moreover, no one except the leader could always be correct, and everyone must therefore expect to suffer criticism and correction. At first the attacks had been made against the managers of the state. Deserted by the government and party which they served, they were criticized, bullied, and humiliated, they were often physically beaten, many of them were sent to prison, and some of them died there. At the highest level, Liu Shaoqi had been arrested, tortured, placed in solitary confinement, and there he died of illness and neglect, surrounded by his own dirt. Other, lesser figures, suffered comparable misery, as a whole generation of cadres who had been trusted in the past was betrayed, and their families shared their fate.

On 7 May 1968, after the attacks had reached their peak, Chairman Mao issued a directive calling for special camps to be set up where unemployed and disgraced officials could be sent for rehabilitation. At first, these 'May Seventh Schools' were little more than concentration camps, but they were later established on a regular basis, with a six-month course of practical farm-work, generally in poor pioneering country, and an intensive study of the Thought of Mao Zedong. Very much like spiritual retreats in the Christian movement, this training in ideology came to serve not only for reform but also for advanced indoctrination, and after a time it became mandatory that any person, factory manager, team leader or teacher, who had or hoped to obtain a position of responsibility in the Chinese state, should take part in such exercises.

Soon, just as cruelly, the young people who had brought the disorder were forced into line. Following the crisis at Wuhan in 1967, as the army took control, many of the radicals were killed, and other young people were sent away from the cities to work also in the harsh conditions of the country. Some appreciated this exile for what it was, others accepted the move with ideal enthusiasm, and sought indeed to serve the common people in the fields. Sooner or later, however, all were disillusioned, for there was a limited role that outsiders could play in the closed communities of peasant society, they were often despised for their lack of practical knowledge and their irrelevant ideals, and the work they were set was drudgery, without any hope of long-term achievement.

Some young people found service in their new community which helped them return to their schooling. In particular, basic city education often allowed them to act as accountants of work-points or as 'bare-foot doctors': practical nurses with six months training in a commune hospital, who then supervised public health and provided first aid. Others, without such expectation or opportunities, drifted back to the cities. With no right to rations or even a place to sleep, the wanderers led a marginal existence outside the law as an urban underclass. The majority, however, remained where they had been sent, and as policy was changed in the later 1970s, they were able to seek to restore the wreckage of their lives.[11]

By this time, however, the damage had been done. Most universities began classes again in 1970, but students were fewer than in the past, courses and faculties were more restrictive, and great attention was paid to the need for work experience both before and during university. Apart from politics—studying the thought of Chairman Mao, which was compulsory—science departments such as physics, electronics, and engineering were far more popular and better staffed than the humanities, and the constant emphasis on matters of immediate practical use, combined with renewed discipline and still punctuated by mass meetings, produced little but the most limited training.

The staff of the universities, and the intellectual world of China as a whole, suffered the same abuse as the managers and men of affairs. Distinguished senior scholars were persecuted by their students and junior colleagues, and though some areas of scientific research received special protection, few of the leading writers and thinkers of China escaped imprisonment, often with solitary confinement. In such an environment, under continuing official pressure and mutual surveillance, the universities were slow to recover, and it was a long time before there was the slightest sense of self-confidence.

Besides the attack on intellectuals, the Cultural Revolution saw a deliberate attempt to force propaganda upon all the people of the country. In the second half of the 1960s, *Quotations from Chairman Mao Zedong*, the 'little red book' of extracts from his writings, overthrew all previous publishing records: 700–800 million copies were made, one for every man, woman or child in the country; study groups were held in the fields and factories, in schools and offices, and even on trains, aircraft, and buses, to read and discuss and praise the teachings of the leader.

The intensity of propaganda extended to every form of art and literature. Sculpture and painting, already affected by the 'socialist realism' of Russia, now developed in poster and Christmas-card style, with a few faint echoes of traditional Chinese design overwhelmed in Communist symbolism. Though generally banal, some posters achieved a measure of artistic success, and there were a few pleasing primitives produced by common people inspired with daily toil. The major work of the time, however, reprinted in every pictorial magazine, was a painting of 'The Youthful Mao Zedong on the Road to Anyuan' (to lead the Hunan miners' strike in 1922), which appeared in 1968 as the co-operative production of students in an art school. It has the predictable merit of any work of art composed by a committee and in fact the chief organizer of the strike had been Liu Shaoqi.

Traditional Chinese opera, which had been maintained and encouraged by the regime of the 1950s, was also eclipsed, and ancient themes were swept away by the reviving doctrine of Yan'an that all literature and art must serve the people and their revolution. So there was now a strict diet of Communist propaganda, recalling the days of struggle against the Nationalists and the Japanese, or presenting examples of heroic dock-workers and dam-building in the communes, while as a symbol of the influence of Madame Mao each new drama included a leading female character, frequently named either Jiang or Qing. Though echoes of traditional forms could still be found, each piece was presented for the purpose of propaganda, all characters were shown as either Communist perfection or unrelieved villainy and cowardice, and splendid costuming, brilliant acrobatics, and skilful special effects were used to hammer a lesson home with every scene.

Some of these lessons were obvious: one theme of drama might show the strength and popularity of the People's Liberation Army; another would tell of heroic self-sacrifice by simple workers or peasants in the communal cause. Behind such simple motifs, however, there were inherent and serious contradictions in the philosophical structure of the Cultural Revolution, and in the direction which it gave to the life of the people. On the one hand, the people were inspired to defy authority — but they were obliged to seek their inspiration from the dictates of a single leader and one pattern of thought. On the other, the people were urged to express themselves — but every work-place and living unit was controlled by a committee, with a network of informers and no set limits to the level of interference and intervention which the community could enforce against people's private lives.[12]

Apart from the ideals of rebellion, the Cultural Revolution was designed to enhance the power of the young, the weak, and those disadvantaged by the formal structures of the state. It emphasized, moreover, an ideal peasant ethic — hard work with the hands, social community, and economic co-operation — as opposed to the artificial and potentially corrupt world of the cities. In this respect, Mao and his followers were seeking to restore the magical days of simplicity and struggle at Yan'an, and indeed to 'rectify' the whole of China in the same fashion as the formative campaigns of the early 1940s. So the Cultural Revolution, with emphasis in the communes, sought to develop the agricultural sector. Young people sent to the countryside gave additional hands to work in the fields, and many districts were encouraged with capital and other assistance to improve their land and increase local mechanization. At the same time, learning from the tragedy of the early 1960s, the ideologues paid some attention to the essential importance of agriculture. When the Cultural Revolution came to the country, it did so with less intensity than in the cities, and mass meetings and demonstrations were seldom allowed to interfere with the essential business of farming. Overall, production continued to rise by some 3 to 4 per cent a year; not a high rate, but more than the increase of population.

Like the Great Leap Forward, moreover, and in the same spirit as Yan'an, people were encouraged to self-reliance and local initiative. One celebrated model was the Red Flag Canal, constructed through tunnels, aqueducts, and a channel in the cliffs to water the farmlands of fifteen communes in Linxian, among the eastern ridges of the Taihang Shan; but it was the small commune of Dazhai in Shanxi, transforming barren hills to a fertile, terraced and irrigated land, and the great Daqing oil-field in the far north-east that were most widely praised as models of self-sufficiency: 'For industry, learn from Daqing; in agriculture, learn from Dazhai'.

Though it was later revealed that the 'self-sufficient' commune of Dazhai had been favoured with official credits and supplies, and Daqing was found to have been extravagant and inefficient, the propaganda of selflessness and mutual responsibility produced some effect. Throughout China, and particularly in the north where the danger of erosion and flooding is greatest, tree-planting programmes gained a new lease of life, and there were many small-scale dams and conservancy projects, with dredging and land reclamation. There was emphasis on the provision of basic schooling and on medical care: besides the 'bare-foot doctors', each commune was expected to have a hospital and each brigade an outpatients clinic, and a medical insurance scheme was established in 1970.

At the same time, just as in the days of the Great Leap Forward, there was a general lack of planning, much wasted effort, and ideology itself produced diseconomies. In rural communes, labour in the public fields was urged as a moral good, and individuals were discouraged from skilled handicrafts, even though such work had always been valuable to the farm economy. Similarly, in the cities, many women were compelled, as a matter of principle, to spend their time in 'street factories' on sweated labour producing low-value, handmade items of marginal value to the community. Individual initiative was discouraged, and there was constant argument about what time might be allowed to the people for work on their own tiny plots of ground and to care for their limited number of privately-owned animals. Sadly but inevitably, people everywhere proved more interested and energetic in working on their own account than for the amorphous concept of the common good, and no arrangements for communal wages and shared prosperity could match the incentive, now disapproved, of individual achievement and personal profit.

On a larger scale, official policies of equalization meant that more advanced centres of production such as Shanghai and the north-east were stripped of their profits to subsidize more backward regions, and were given no incentive or funds for investment. Shanghai in particular, though it had been the place of origin of the Cultural Revolution movement, was suspect to the radicals on account of its advanced economy, and was restricted and hampered in its development. In many respects, the city remained much the same as it had been in the 1940s, and some of the cruellest incidents of the period took place when the local red gangs turned on the sophisticated people of that once great metropolis.

As the country returned to some form of order, however, and the *rapprochement* with the United States took effect, the opening to the outside world was reflected in foreign contact and trade. At the height of the Cultural Revolution, when faction fighting in streets and factories delayed and halted production, China's foreign trade had dropped from US$4,245 million in 1966 to $3,895 million in 1967 and $3,860 million in 1968. By 1969 the position was recovering and in 1970 it returned to the level of 1966. Then came a marked increase. From a high point of US$4,680 million in 1971, a level never reached before, trade in 1972 jumped fifteen per cent to US$5,715 million, and even allowing for international devaluation of the US currency the rise was substantial. Thereafter, however, development came on a totally different scale: to $15 billion in 1975 and $21 billion in 1978.

From the beginning of 1972, moreover, a steady stream of business and trade delegations, particularly from Japan, visited China, and their talks concerned not only the exchange of goods on traditional lines but also the establishment of new plants with foreign capital and the training of Chinese in new techniques at home and abroad. As foreigners came to seek trade, and the government sought opportunity for development, the years of isolation drew to a close, and new ideas broke once more through the barriers which had sealed China apart for so long. Indeed, the critical point for the Cultural Revolution came not with the incident at Wuhan, nor with the fall of Lin Biao, nor with the rise of Jiang Qing, but with the return of foreigners bringing goods and ideas from outside.

Overall, indeed, though one may seek to discern some historical logic in the Cultural Revolution, it is as well to recognize that the apparent pattern of events emerged from a very human chaos within the country. Mao Zedong is said to have observed that if he had known what would happen as a consequence of his first big-character poster in 1965, he might not have written it, and the confusion which followed was surely beyond any one person to comprehend. There was rivalry in the leadership, tensions among young people, and frustration in society at large. As the formal government lost its power, this medley of individual fears and ambitions brought a level of confusion and excitable violence that was quite unintentional, and certainly not planned. Behind all the explanations, the propaganda and the rationalizations, the basic fact of the Cultural Revolution was that Communist China went out of control.

Notes

1. Though their surnames are the same, Peng Zhen and Peng Dehuai were not related. Peng Dehuai came from Hunan, from the same district as Mao Zedong, Peng Zhen from Shanxi.

2. The Chinese phrase *dazi bao,* literally 'big-character poster' can refer, as here, to a placard with a single slogan. The term, however, also describes a detailed denunciation put up on a wall or notice-board, together with a similar reply, and the meaning is extended to wall-newspapers and other forms of information and propaganda.

3. This pedagogical style could still be observed at a tertiary teacher's training college in 1973: students aged nineteen or twenty stood to attention before their teacher, raised hands rigidly to offer answers to questions, and chorused their work together. The style would have appeared repressive in a Western primary school of the 1940s.

4. Refugees from the competitive classrooms told later how moral lessons encouraged young people to emulate the selflessness of Lei Feng, who would wash his comrade soldiers' dirty linen: as a result, to gain points of morality, students would snatch others' clothing and bring it back washed, while hiding their own so no one could steal such a march on them.

5. According to rules approved in 1956 the Congress should have met every four years, but no attempt was made to call the meeting in 1960 or 1964, and though it had been announced that the Congress would be held in 1968 it was again delayed until the following year.

6. *Peking Review,* 16 December 1966. The coup in Indonesia had taken place on 30 September 1965, too early to be influenced by the Cultural Revolution. In so far as Beijing encouraged it, this was 'normal' support for left-wing activities abroad. The full Maoist spirit came later, and did nothing to help already damaged diplomatic relations.

7. In the Shanghai communique of 28 February 1972 the United States acknowledged that Chinese on both sides of the Taiwan Strait maintain there is one China, and that Taiwan is part of China. The Americans, however, affirmed their interest in a peaceful solution to the problem of division, and undertook to withdraw their forces and installations 'as tension in the area diminishes'. The Chinese, for their part, did not insist that the United States abandon Taiwan.

8. In July 1972 a number of documents published in Taiwan purported to prove the complicity of the Lin Biao group in a plan for armed uprising against the government. The most important item, 'Summary of the 571 Engineering Project', calls for a coup in Beijing against Mao Zedong. Though the validity of the material was at first uncertain, the Communist press later acknowledged it and made frequent use of it in propaganda against Lin Biao. (The phrase '571' is a pun: the three digits are expressed in Chinese by the sounds *wu qi yi,* and the same three sounds, written with different characters, have the meaning 'plan for a military uprising'.

9. Unlike the Ninth Congress, which had been prepared with announcements, slogans, and editorials, the Tenth Congress began in secrecy during the middle of August. Later, from 24 to 28 August, formal public sessions were held, and for months thereafter the people were urged to celebrate the meeting with parades, songs, and dances on every possible occasion.

10. It was not inappropriate that document 571, discussed in note 8 above, described Mao by the code-name 'B-52'. At that time in Vietnam, the American B-52 bomber was noted for its ability to fly far above the clouds, out of range from the ground, dropping devastation at random among the unfortunate people underneath.

11. As one example, I cite my local escort in Henan in 1978. In 1965 he had been chosen for language training in Beijing as a cadet to the diplomatic service. After the short period of political excitement he was sent down to a commune, where he worked as a labourer for several years. Because he had some English, he was later employed in a local school, and then, as tourism expanded, he found appointment as guide in a provincial city; a long way from the career he had expected thirteen years before.

12. An example may be taken from the birth-control programme. In each community, a list of the names of every married woman was kept, and often displayed, at the local medical centre, with an indication of what form of contraception she was using. If a woman fell pregnant, and was not entitled to a child at this time, she would be 'counselled' by a member of the local committee. Such a counsellor would call on the family, and could discuss, hour after hour, night after night, the need for restraint in child-bearing. In almost all cases, such argument was successful, and the woman would be persuaded to have an abortion. The central point in this procedure, and the critical reason for its success, is that Chinese society had no acceptance of the idea that the pregnant women could ever claim personal privacy, tell the unwanted intruder to go away, or silence the barrage of propaganda in her own home.

Further Reading

On the early development of the Cultural Revolution

Asia Research Center (compiler), *The Great Cultural Revolution in China,* Tokyo, Tuttle, 1968.

Barnett, A. Doak, with Vogel, Ezra, *Cadres, Bureaucracy and Political Power in Communist China,* New York, Columbia University Press, 1967.

Dittmer, Lowell, *Liu Shao-ch'i and the Chinese Cultural Revolution: the politics of mass criticism,* Los Angeles, California University Press, 1975.

Hunter, Neale, *Shanghai Journal,* New York, Praeger, 1969.

Pye, Lucian W., *The Spirit of Chinese Politics: a psychocultural study of the crisis in political development,* Boston, MIT Press, 1968.

Robinson, Thomas W. (ed.), *The Cultural Revolution in China,* Los Angeles, California University Press, 1971.

On young people and the Red Guards:

Bernstein, Richard, *Up to the Mountains and Down to the Villages: the transfer of youth from urban to rural China,* New Haven, Yale University Press, 1977.

Chan, Anita, *Children of Mao: personality development and political activism in the Red Guard generation,* Seattle, University of Washington, 1985.

Dai Hsiao-ai [pseudonym], *Red Guard,* New York, Doubleday, 1970 [compiled from interviews in Hong Kong by Gordon Bennett and Ronald N. Montaperto].

Rosen, Stanley, *Red Guard Factionalism and the Cultural Revolution in Guangzhou,* Boulder Co., Westview, 1981.

On later political development:

'20 Years On: four views on the Cultural Revolution' in *China Quarterly* 108, 1986.

Bridgham, Philip, 'The Fall of Lin Piao' in *China Quarterly* 55 (1973).

Brugger, Bill (ed.), *China: the impact of the Cultural Revolution,* Canberra, Australian National University Press, 1978.

Domes, Jürgen, *China after the Cultural Revolution: politics between two party congresses,* Los Angeles, California University Press, 1977.

Joffe, Ellis, 'The Chinese Army after the Cultural Revolution: the effects of intervention' in *China Quarterly* 55 (1973).

Leys, Simon [pseudonym for Pierre Ryckmans], *The Chairman's New Clothes: Mao and the Cultural Revolution* [first published in French, 1972], New York, Viking, 1979.

Rice, Edward, *Mao's Way*, Los Angeles, California University Press, 1972.

Robinson, Thomas W., 'The Wuhan Incident: local strife and provincial rebellion during the Cultural Revolution' in *China Quarterly* 47 (1971).

Wich, Richard, 'The Tenth Party Congress: the power structure and the succession question' in *China Quarterly* 58 (1974).

On the livelihood of the people:

Tachai – standard bearer in China's agriculture, Beijing, Foreign Languages Press, 1972.

Taching – red banner on China's industrial front, Beijing, Foreign Languages Press, 1972.

Chan, Anita, Madsen, Richard and Unger, Jonathan, *Chen Village: the recent history of a peasant community in Mao's China,* Los Angeles, California University Press, 1984.

Hinton, William, *Shenfan,* New York, Random House, 1983.

Howe, Christopher (ed.), *Shanghai: revolution and development in an Asian metropolis,* Cambridge, Cambridge University Press, 1981.

Leys, Simon, *Chinese Shadows*, New York, Viking, 1977.

Orleans, Leo E., *Every Fifth Child: the population of China,* Stanford, Stanford University Press, 1972.

Parish, William L., and Whyte, Martin King, *Village and Family in Contemporary China,* Chicago, Chicago University Press, 1978.

On foreign affairs:

Selected Documents No. 9, US Policy toward China, July 15, 1971–January 15, 1979, Washington DC, The Department of State, 1979.

Hsiao, Gene T., 'The Sino-Japanese Rapprochement: a relationship of ambivalence' in *China Quarterly* 57 (1974).

Macfarquhar, Roderick, *Sino-American Relations, 1949-1971,* New York, Praeger, 1971.

Van Ness, Peter, *Revolution and Chinese Foreign Policy: Peking's support for wars of national liberation,* Los Angeles, California University Press, 1970.

11

Economics in Command, 1976–1989

BY the middle 1970s, the energies of the Cultural Revolution were exhausted, and the examples of foreign achievement brought by trade and political exchange had made many people dissatisfied with the policies of the past. The country, however, could not absorb the full implications, nor take advantage of the opportunities, so long as the Maoist regime remained. In a transient, uncertain time people wondered when the old man would die, and what would follow when he did.

Demonstrations after the death of Zhou Enlai early in 1976 showed wide popular resentment, and strengthened the hands of the 'pragmatic' faction represented most obviously by Deng Xiaoping. When Mao died in September, his radical protégés, notably his wife Jiang Qing, were left isolated and unprotected, and a few weeks later they were arrested and disgraced.

Gradually, Deng Xiaoping and his allies increased their influence, and by 1980 the Cultural Revolution had been formally ended, the Thought of Mao Zedong had been brought under critical control, and Mao's compromise successor, Hua Guofeng, was removed from power. The country was embarked upon a programme of development which relied on a reduction of central and communal control, and the encouragement of individual incentives and enterprise. Both in the cities and the countryside, there were new opportunities for personal prosperity, and real penalties for failure.

The task for the government, and for China as a whole, however, was made all the harder by the legacy of waste and disillusionment from the years of the Cultural Revolution, and the forced transition from a planned economy, no matter what its shortcomings, into a liberal proto-capitalist model presented massive difficulties. One was the bureaucratic structure of the government itself, which was often inadequate to deal with private enterprise, whether at home or in negotiations with foreign companies. Furthermore, the new freedoms granted to individual and local enterprise meant that the state lost access to the funds it needed to develop and balance the national economy. So there were problems with the infrastructure required for modernization, such matters as transport, communications, and energy supply; and at the same time there was growing resentment among the people at the inequalities brought forth by the new system and the inadequacy of government either to control them or alleviate distress.

There was, moreover, some contradiction in the establishment of an open economy, with weakening of central control, while the political structure of the state continued under the dominance of a single party and a closed leadership. There is no essential reason why a comparatively free market society cannot be governed by an authoritarian regime, and the opposite example of the Soviet Union and eastern Europe, where political liberalism brought instability and economic collapse, is not encouraging, but many people, and particularly intellectuals, sought a greater freedom of expression and argued forcefully that real development, to match the West, required the openness of democracy.

By the latter part of the 1980s, as the difficulties of development on a low industrial base with a high population intensified economic and social tension, a new generation of students began to agitate for greater freedom. In May 1989 the pressure came to a head with massive demonstrations and the occupation of Tiananmen Square in Beijing.[1] The demands and the conduct of the protesters, however, were quite unacceptable to their rulers, and as both sides remained intransigent, the leaders of the government called in the troops. In one of the great human and political tragedies of modern China, the old men of power used the tanks of the people's army to crush the young idealists in the heart of the people's capital.

The Struggle for Succession

The death of Zhou Enlai, on 8 January 1976, marked the beginning of active contest for control of the future. Almost 78 years old, Zhou had been suffering for some time from cancer, and Deng Xiaoping was in line to succeed him. Deng had been the number-two public enemy at the beginning of the Cultural Revolution, but he was a well-qualified and experienced administrator, and although his return to favour and influence as vice-premier in 1973 had been observed with jealousy by Jiang Qing and her left-wing allies, there was no one in the senior leadership to match his ability, and he had his own allies within the party. As Zhou Enlai became physically more frail Deng acted as effective head of government.

The loss of his patron, however, exposed Deng Xiaoping to immediate threat. Within a few weeks of Zhou's death, he disappeared once more from public view, and there were new attacks against unrepentant officials who yet maintained the capitalist road. On the other hand, despite this success, no one in the radical faction had any wide acceptance: Zhang Chunqiao, second vice-premier after Deng, was their leading candidate, but he was unpopular in the army and among senior officials, and it was a sign of this weakness that Hua Guofeng was named acting premier on 6 February.

Hua Guofeng had been First Secretary of Mao's home province, Hunan, and he became Minister of Public Security in the aftermath of the Lin Biao affair. He was, however, only sixth vice-premier, and although he held Mao's favour[2] he was a compromise candidate, not widely known, so the supporters both of Deng Xiaoping and of the radicals might have hoped to influence and eventually control him.

A few weeks later, at the end of March and the beginning of April, during the period of the autumn festival when the Chinese traditionally pay tribute to their ancestors, crowds of people went to the Monument to the People's Heroes in Tiananmen Square, at the centre of Beijing, to lay wreaths in memory of Zhou Enlai. The movement appears to have been largely spontaneous, and as the process continued the authorities, increasingly uneasy, ordered that the tributes be removed. On 5 April a crowd of a hundred thousand gathered in furious demonstration, chanting praises for Zhou Enlai. As excitement mounted, rioters attacked police and the militia which sought to restore order, while slogans abused the government and even attacked Chairman Mao.

This Tiananmen Incident was a new and unwelcome development, the first time for many years that people had gathered of their own accord and shown their true feelings. In first reaction, on 7 April, Deng Xiaoping was once more disgraced, and the following day Hua Guofeng was confirmed as Premier, with demonstrations organized in his favour. Despite such manoeuvres, however, the show of public support for Zhou Enlai's model of leadership had circumscribed Hua's role, and the immediate loss for Deng Xiaoping gave him strength in the longer term.

Towards the end of April, Mao formally designated Hua as his successor, with written instructions presented to the Politburo, and Hua sought to develop his position with propaganda emphasizing his personal humility, and his resolve to maintain Mao's programme cautiously and sensibly. In July, when the old general Zhu De died, there was another period of mourning and honour for the heroes of the past, and at the end of that month, after an earthquake devastated much of the region north-east of Tianjin, Hua Guofeng showed sympathy and energy among the people of the afflicted region.

The earthquake at Tangshan was a sad natural disaster, with enormous property damage and half a million people dead or injured, and it had political implications beside the tragic cost in lives. Chinese tradition has always held earthquakes and such phenomena as portents of disorder, but this occasion was particularly embarrassing. Maoists had claimed that Chinese popular wisdom was superior to much that the West had to offer. Partly through economy, for example, partly through belief, herbal remedies and acupuncture had been utilized side by side with Western medicine throughout the countryside. Moreover, some months earlier there had been an earthquake near Shenyang, and the national press was full of stories of how it had been forecast by humble people and wily peasants observing such age-old signs as the feel of the air and the conduct of animals, and claiming that these techniques were better than foreign seismology. At Tangshan, however, there was no warning, and many of the victims were miners caught underground by the shock. There was much less talk of popular wisdom after that.[3]

On 9 September 1976 Mao Zedong died. He had suffered a stroke some two years earlier, and was variously lucid and comatose to the end. It was asserted by Hua Guofeng, and generally accepted, that Mao had assured him, 'With you in charge, I am at ease', and the new leader advertised this endorsement on billboards throughout the nation. Still more usefully, he controlled the regular army under Ye Jianying, Minister for Defence, he had support from all the military regions, and he held the loyalty of Wang Dongxing, commander of the bodyguard, twenty thousand strong, stationed in Beijing.

As Hua Guofeng took power, the rivalry of Deng Xiaoping was in abeyance, but there was a clear threat from the left. Jiang Qing and her colleagues were now deprived of the Chairman's backing, and although his death had been well expected the extreme positions of the radicals had given them small occasion to acquire allies elsewhere in the government. Faced with a crisis for which they were ill-prepared, they sought to arrange a coup against the new leadership, to be supported by popular militia in the cities and triggered by assassination of their opponents.

Hua and his followers, however, had made their own arrangements. On 6 October, Zhang Chunqiao, Wang Hongwen, and Yao Wenyuan were arrested, and troops of Wang Dongxing's 8341 unit were sent to seize Jiang Qing at her residence. On 7 October charges of conspiracy against the defeated faction, to be known as the Gang of Four, were reported to the Politburo. Its members endorsed the action, Hua Guofeng was nominated as Chairman of the Central Committee and head of the Military Commission, and on 24 October there was a rally in Tiananmen Square to celebrate his success.

So far, the transfer of power had gone as well as could be expected, but Hua Guofeng and his new regime faced major problems. Firstly, Hua claimed his position as anointed successor to Mao, but he had used his new power to arrest and depose Mao's wife and her associates, and though every care was taken to avoid linking the politics of Jiang Qing with the prestige of her late husband, the connection was well known and there were many people prepared to whisper about a Gang of Five rather than the Gang of Four.

Secondly, as the demonstrations of Tiananmen had shown, the weakness of the radicals reflected more than their isolation and unpopularity among the leadership, for great masses of the population were quite disillusioned with the policies and rhetoric of the Cultural Revolution. As they learned more about the West, and counted the costs of damage and disruption, more and more people, particularly intellectuals, managers, and workers in urban industries affected by foreign trade, realized that the turmoil and miseries of the last ten years had produced almost nothing of value. As the new regime sought to establish itself, the Communists were faced with the realization that the very source of their legitimacy, that revolutionary achievement embodied by Mao Zedong, had lost authority and respect among the people.

So the death of Mao brought critical change to the government of China. Hitherto the leadership had been able to inspire the nation with exhortations and promises for the future. Now, however, tolerance was ended, faith was gone, and there was a mood of cynical distrust. What was wanted was a competent administration to deal with the real problems of the country; and above all, where Maoists had sought to change human conduct to fit their ideals, government in future should deal with the situation as it was, not as it ought to be.

In this respect, Hua Guofeng stood between two strands of legitimacy. He claimed the inheritance of Mao Zedong, but he also sought approval as the leader of a new pragmatism. Both sources of authority, moreover, were flawed. Hua was tainted by his association with the recent past, and he could never match Mao's status as a prophet of revolution. And if the nation needed pragmatic leadership, the record of Deng Xiaoping was far more impressive.

From the moment Hua achieved power, therefore, he was under continual pressure to restore Deng to favour and grant him authority. In July 1977 the Central Committee confirmed the earlier decisions of the Politburo to make Hua Chairman of the Party and Chairman of the Military Commission, and at the same meeting, on Hua Guofeng's recommendation in obvious exchange, Deng Xiaoping was restored as a member of the Standing Committee of the Politburo, Vice-Chairman of the Central Committee, first Deputy Premier of the State Council, Vice-Chairman of the Military Commission and Chief of the General Staff in the People's Liberation Army.[4] These were the positions he had held before under the patronage of Zhou Enlai, and they gave him high authority in the government, in the party, and over the armed services.

Born in 1904, Deng Xiaoping was now more than seventy years old, and the vicissitudes of the last several years, humiliation under the Cultural Revolution followed by swift uncertainties of restoration and disgrace, made it impossible for him to challenge Hua Guofeng directly. What he could do, however, was arrange that the government of the future would be dominated by his own associates. With this end in view, and supported by such substantial leaders as Defence Minister Ye Jianying and Vice-Premier Li Xiannian, Deng pressed against the remaining representatives of the left. The Gang of Four were formally expelled from the Party, but others such as Wang Dongxing, commander of Mao's bodyguard, and Wu De, the mayor of Beijing who used the Tiananmen Incident as a means to attack Deng, were also removed from power. Hua himself was not openly attacked, but he was isolated, and most management positions were taken by Deng's people.

In similar fashion, the pragmatists gradually developed debate about the thought of Mao Zedong. Deng himself is probably best known for the saying, in 1961, that 'Whether a cat is white or black, if it catches mice it's a good cat.' More formally, however, in 1978 he borrowed a saying of Mao's, that truth can be found only from practical experience, implying that Mao's opinions and policies, like any others, must be tested on that principle, and judged by their results: so the Thought of Mao Zedong no longer provided absolute truth; and if results, as in the Cultural Revolution, were not satisfactory, then the ideology could be criticized accordingly.

Gradually the legend of Mao and the influence of his philosophy brought down from the heights of former adulation. At the Eleventh Party Congress in August 1977 Chairman Hua declared a formal end to the Cultural Revolution, rejecting the Maoist ideal of continual rectification and reform. As time passed, the Red Guards were officially disbanded, copies of the Little Red Book were no longer seen, official portraits of Mao were printed more dimly, and as the red and white placards on the walls of buildings lost their brilliant colours, his calligraphy and quotations also faded from view. Even the mausoleum in Tiananmen Square, with Mao's body embalmed in Leninist style, while providing a popular place for tourist pilgrims to pay their respects, also emphasized the fact that he was dead.

In his speech at the celebrations for the thirtieth anniversary of the foundation of the People's Republic, on 1 October 1979, Ye Jianying gave official voice to the implied criticisms of the late Chairman. It was in no way a full repudiation, for Ye and his colleagues realized the importance of Mao as a symbol of the party's achievements, but he did claim that Mao Zedong Thought was based upon the experience of the people as a whole, not upon Mao's personal genius, that it was wrong to worship any one leader, and that Mao, like any human being, had made mistakes.

On this foundation, through 1980, Deng Xiaoping and his allies removed all but the last remnants of the ideology of the Cultural Revolution. In February, the Central Committee officially abandoned the slogan 'politics in command' for Deng's 'economics in command'. Several of Deng's opponents were dismissed, and two of his protégés, Hu Yaobang, now head of the secretariat, and Zhao Ziyang were appointed to the Standing Committee of the Politburo. In May there was a commemorative service for Liu Shaoqi, posthumously rehabilitated, while preparations were made to put Jiang Qing and her associates on public trial. And at the National People's Congress meeting in August and September, Deng was able to force Hua Guofeng from the regular government: in a neat manoeuvre, he obtained agreement that party positions should no longer be linked to official power, and he and Hua and other senior colleagues resigned their appointments. Zhao Ziyang became premier, and other posts were also filled by younger supporters of Deng Xiaoping. Those who resigned continued to hold senior positions in the party, but its regular administration was controlled by Hu Yaobang, so Deng's group had essential authority in both areas.

At the end of the year, from 20 November into January 1981, came the trial of the Gang of Four. In fact, though Jiang Qing, Zhang Chunqiao, Yao Wenyuan, and Wang Hongwen were at the centre of the attack, indictments were laid also against Chen Boda and five military commanders accused of supporting a coup in favour of Lin Biao. Furthermore, at the beginning of the proceedings the prosecution announced the names of six others who would have been summonsed if they had still been alive: Lin Biao, his wife, his son, and another conspirator, and also Kang Sheng and Xie Fuzhi, former chiefs of security.

In a long indictment, both groups were accused of plotting against the party and the state and causing the persecution and death of thousands of innocent people, and each was accused of planning a *coup d'état*, in 1971 or 1976. Most of the defendants had little to say, but Chen Boda admitted conspiracy to destroy Liu Shaoqi and other leaders in 1967, and Wang Hongwen gave evidence of plans to disgrace Zhou Enlai and Deng Xiaoping in 1974. Jiang Qing, by contrast, gave an impressive performance in her own defence, and though she admitted her role in the death of Liu Shaoqi, and it was shown that she had maintained her own torture chamber, she continued to claim that she had been justified and correct in her actions.

This was, of course, a political trial, a public but cautious purging of the Cultural Revolution, so the guilty verdict was inevitable, but there was debate at the highest level on the sentences. In the end, Jiang Qing and Zhang Chunqiao were sentenced to death, but execution was suspended;[5] Wang Hongwen was given life imprisonment, and the other accused up to 20 years. Though modern Chinese leadership has seldom been reluctant to impose the death penalty on its citizens, the men in power were unwilling to set a precedent. There remained, moreover, the shadow of Mao Zedong. The prosecution remarked that he had been misled, and had failed to see through the machinations of the conspirators. No one wished to go further or press the point too hard, for too many understood that the fault lay not only with the ambitious radicals, but also with the structure of power and the leader whom all had served so readily.

The trial brought also the political demise of Hua Guofeng. Though he had certainly destroyed the Gang of Four in 1976, he had also been Minister for Public Security during the later period of the Cultural Revolution, and to that extent he was implicated in the record of wrongful arrests and executions. By the beginning of 1981 his authority was gone, and at the Central Committee meeting in June of that year Hu Yaobang replaced him as Chairman of the party, while Deng Xiaoping took over as head of the Military Commission. Hua Guofeng became a junior vice-chairman, one of the first Communists to be driven from power without disgrace or humiliation.

At that same Sixth Plenum of the Eleventh Central Committee, the party made formal decision on its recent experiences. The 'Resolution on Certain Questions in the History of Our Party Since the Founding of the People's Republic', adopted on 29 June and published in *Beijing Review* on 6 July, repudiated the Cultural Revolution, declaring that it had defied the true Thought of Mao and of Marxism–Leninism, and that it had inflicted great and unwarranted damage on the nation, the party, and innocent individuals. In particular, Mao was blamed for his leftist errors and his personal style of leadership, and Hua Guofeng was criticized as one of the followers who carried out orders too blindly, for his attempts to establish a personality cult of his own, and for his continuing defence of the Cultural Revolution.

Two days later, on 1 July, Hu Yaobang gave an address in celebration of the sixtieth anniversary of the foundation of the Chinese Communist Party. Again he pointed out Mao's errors, his arrogance and loss of contact with reality, but he praised him as a source of wisdom and strength in the long revolutionary struggle, and referred to the Thought of Mao Zedong as a major contribution to Marxist philosophy.

So the new regime rejected Mao as an individual leader, but claimed to be preserving the true tradition of his revolutionary ideals. It was a delicate balance of argument, but it worked very well for Deng Xiaoping and his allies, and they had maintained the debate with skill. What was required now, as Hu Yaobang observed, was to revive and develop the party for the future of China, to involve young people of ability in the work, and to encourage national unity for real development under socialism. Certainly the power of the Communist Party remained unrivalled, and it had reformed itself from the extremist policies of the past. It remained to be seen how the people would follow that lead along a new, strange path.

Rational Reform

The new economic regime had been proclaimed as early as 1977, when the Eleventh Congress of the Party approved the Four Modernizations — of agriculture, industry, science and technology, and national defence — and wrote them into the constitution. In the following year, the National Congress likewise adopted the Four Modernizations into the national constitution, and Hua Guofeng announced a Ten-Year Plan for development.

During the late 1950s, the government had spoken with pride of 'Ten Great Years' of progress since 1949. The records had been distorted and falsified by the effects of the Great Leap Forward at the end of the decade, but there had certainly been impressive development in the period of the First Five-Year Plan. Twenty years later, however, the new leadership referred to the Cultural Revolution as 'Ten Years of Great Catastrophe', and there is small doubt that they were correct.

In many respects, Hua Guofeng's Ten-Year Plan represented a last attempt to control the economic future by recognized Communist methods. The programme was ambitious, but since the Plan was supposed to cover the period from 1976 to 1985, it was out of date even as it was announced, and within twelve months it was effectively suspended for a three-year period of 'adjustment and consolidation'. In practice, development through the 1980s owed far less to central control than to the liberalization which the new regime permitted and encouraged.

It was some time before there was proper information on the economic situation. The damage done to the statistical-gathering apparatus at the time of the Great Leap Forward had not been fully repaired in the early 1960s, and thereafter the enthusiasms of cadres had expressed themselves with meaningless comparisons of gains 'since Liberation' in 1949 and 'since the Revolution' from the middle of the 1960s. More realistically, some judgements could be based on reports of volume and value from China's foreign trading partners, and there had been assessments by foreign China-watchers, by organizations in Taiwan and by the United States Central Intelligence Agency. Gradually, through the 1980s, the new administration began to publish reliable statistics, for current accounts and also for previous years, so that the government, the people, and outside observers could consider recent history and prospects for the future.[6]

For China, basic calculations must start with people. In 1982 the first real census for thirty years indicated that the population had passed one billion: the official figure was 1,008,175,288, but the true number was certainly higher; apart from the administrative difficulties of counting so many people over such a wide-flung territory, government policy limiting the size of families provides an incentive for parents to conceal their children, and in many rural areas this can be done quite successfully.

With some vicissitudes caused by populist argument that the more Chinese there were the better,[7] the government has generally sougt

Fig.3 Population of the People's Republic, 1949–1987

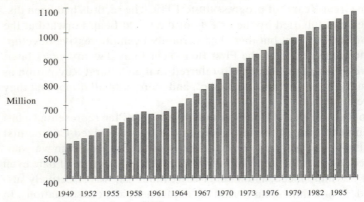

Source: *Statistical Yearbook of China*

to restrain child-bearing, and has claimed a measure of success. From the beginning of Communist rule to the mid-1960s, the rate of increase was more than 2 per cent, but it then fell below that figure and by the late 1970s it was less than 1.5. Sex outside marriage has always been discouraged, and young people have long been urged that they should marry only in their late twenties. In 1982 the policy of one child per family was formally established, with rewards in money and preferences for those who followed it and penalties, such as fines, rationing, and public criticism for those who refused to accept restrictions. In the cities, this appears to have been successful, but in the countryside, where physical labour is essential to survival, there have been numerous reports of children concealed from the authorities and, in sad traditional style, of female infants being killed so that the couple can try again and hope for a strong son to help in the fields.[8]

Eventually, it may be expected that the rate of increase will slow, but it is certain that the numbers will continue to grow for many years. The frightening thing about demography is the way that personal decisions, taken throughout the current population, have critical influence upon the future. In China, the baby boom which followed the peace of the late 1940s brought a great cohort of young women to reach child-bearing age in the second half of the 1960s, and while this group has not yet ended its period of fertility, the children born in 1970 are also ready to have children. In more general terms, despite a steady improvement in health care and an increase in average life expectancy,[9] China still has a young population, with an average age of about twenty-five, and consequent strong pressure on educational resources at every level from primary schools and literacy classes to universities and practical training.[10]

Against this background, of a population which at least doubled between the 1950s and the 1980s, other measures of the economy must be judged, and one of the most important is food grain. Recent Chinese compilations confirm the increase under early Communist rule from 164 million tonnes in 1952 to 195 million tonnes in 1957, followed by the disastrous fall to 143.5 million tonnes in 1960 in the aftermath of the Great Leap Forward. By 1965 production was almost restored to 1957 figures, and it increased to 240 million tonnes in 1970 and to 285 million in 1975. On the other hand, as practical planners looked at the value of capital and labour which had been invested in agriculture, the returns appeared woefully inadequate. Far too often, the demands of the state and the policies of the commune system had caused excessive colonization of marginal land and inefficient emphasis on general sowing of grain at the expense of specialist crops which might produce a better return. Production of food grains per head had increased only ten per cent between 1952 and 1978, and though cotton had risen from 1.3 to 2.2 million tonnes, and oil-bearing crops, peanuts, rape, and sesame, increased from 4.2 to 5.2 million tonnes, neither set of figures had kept pace with population growth. As Deng Xiaoping returned to power, he embarked on a programme of returning decisions to individual farmers, in the hope that local and personal initiative would prove more energetic and efficient than central planning.

The formal procedure for this great change was the Responsibility System, officially approved at the Third Plenum of the Eleventh Central Committee in December 1978, and spread through the provinces over following years. There had always been provision under the commune system for family units to be allocated private plots, but now it was ordered that households could make contracts with the commune to take a unit of land under cultivation, to plant specified crops, and to return a quota of the harvest to the central authority. The land remained formally in the hands of the government, but all management decisions, including choice of seeds, treatment of the soil, and the schedule of labour, were made by the contracting households. At first, these arrangements were for only a year or a single season, but in 1984, when the system was established over the whole country, some tenure was extended to fifteen years, to allow forward planning and to encourage investment either in conservation or in the establishment of longer-term crops such as fruit trees.

Besides this general programme, the government also encouraged contracts for particular crops, for animal husbandry, forestry, and fisheries, while some households had mixed arrangements, to cultivate land in the usual manner but also to engage in specialist production. Contracts could still be forfeited if terms were not fulfilled, but it was later agreed they could be made hereditary, and in 1987 the Thirteenth Party Congress allowed that the right to use a plot of land could be sold from one family to another. The spread and complexity of these private arrangements meant that the communes became irrelevant: to all intents and purposes the responsibility system in agriculture brought the return of private land-holding and the restoration of individual enterprise.

The disappearance of the commune system made little difference to the basic geography of the Chinese countryside. The hierarchy of commune, production brigade, and production team had originally been imposed on existing natural units of local economy, the market towns and villages, and rural administration now reverted to its former arrangement, based on the sub-county township. The real difference was in the degree of central planning and control. Farmers were still required to sell a quota of their produce to the state, but in the early years of the new programme they were encouraged by a 50 per cent increase in the prices for this compulsory procurement. Over the same period, direct government investment in agriculture declined by 50 per cent, but this was largely compensated for from private sources, supplemented and assisted by loans from the People's Bank.

The effects of the changes on productivity were dramatic. Between 1978 and 1987, grain production increased from 305 to 405 million tonnes, the output of cotton almost doubled, from 2.2 to 4.3 million tonnes, and the yield of oil-bearing crops went up three times, from 5.2 to 15.3 million tonnes. Furthermore, as organized communal labour on open public fields came to an end, the land was put to more appropriate use, individual farmers were given the right to choose their schedules of labour, and they became far more efficient. Where people in the country had formerly worked 250 to 300 days a year, the new production figures were now achieved on the basis of some 60 days a year, and the time thus gained was available for other sources of income, most notably handicrafts and domestic industry.

With these incentives and opportunities, individuals could do extremely well, and many of them did. Farm income per head rose from RMB 134 in 1978 to RMB 463 in 1987. Local prosperity was reflected in larger, better houses, often built by tradesmen travelling from one village to the next to satisfy the new rich, and there was growing demand for consumer items, clothing, watches and jewellery, furniture, sewing machines, refrigerators, and television sets, some produced locally but many bought from distant provinces or from overseas.

In the industrial centres, notably the cities, the principles of economic liberalization were essentially the same as in the country, but techniques were somewhat different. Firstly, whereas the commune system was almost completely dismantled to the benefit of private enterprise, the bulk of China's large-scale manufacturing remained in state hands. One major reason is that while the capital required to maintain and improve a farm is not so much as to be beyond private resources, for most industry, just as in the nineteenth century, government remains the only effective source for the amount of investment required; and the Communist regime was not ready to embark on the establishment of non-public institutions for raising capital such as merchant banks or a stock exchange.

With these considerations, the reform of China's industry presented a more complex problem. The centre of the difficulty was the same: production increase per head had averaged just under 10 per cent from the early 1950s to the early 1980s, and there was much room for improvement, particularly for encouraging the workers. Political disruption had demoralized both the managers and the people on the shop floor, and although in the countryside this problem could be cancelled by changing the structure of production, the factories in the cities could not be abandoned so easily.

Apart from the uncertainty and tension which remained from past troubles, there was now a long tradition of idleness, inefficiency, and lack of concern. Any Communist government seizing power in a less-industrialized country is faced with the fact that it is, by its own ideals, committed to the principle of a 'social wage': that every citizen is entitled to a fair level of support from the state. As in a traditional peasant family, the state is bound to provide full employment, whether people work productively or not.[11] The *laissez-faire* alternative, leaving the poor to fend for themselves, often in utter misery, is never attractive, but it does at least encourage enterprise, and it was largely upon that basis that the great industries of the West were first developed.

The People's Republic did not, by definition, have such an alternative, but in its first years the enthusiasm of building a new China, and the opportunities provided by the return of peace and assistance from the Soviet Union, inspired industrial and mining workers to patriotic energy and devotion. As time passed, however, it became increasingly clear, and the lesson was regularly reinforced, that good and honest work often went unrewarded, that enterprise and leadership could bring political and personal disaster, while there was always a guaranteed job and a regular salary for even the laziest time-server. In the aftermath of the determinedly egalitarian Cultural Revolution, security of tenure was rigid and many work-practices were grossly irrational and inefficient, so that visitors from the West told stories of how two young women were employed full-time to work an automatic lift, while a manager's chief job was to round up his workmen from their smoke-break behind the shed.[12]

The same problems appeared on a larger scale. Under central planning, all major and many minor enterprises were provided with capital and working material by the state, and they were likewise required to sell their goods at fixed prices, set primarily with regard to higher policy, and often unrelated to the actual costs of production. There was no competition between manufacturers, no concept of market supply and demand, and no proper check for quality. Moreover, since accounts were concluded at the end of each year, and all money was remitted to the state, there was no procedure to accumulate funds for investment, nor any sense of responsibility for losses. Like their employees, these enterprises continued to be supported by the government, regardless of the value of their work, with no incentive for good performance, no penalty for failure, and many opportunities for dishonesty.

In recent years, we have seen the results of a similar system in eastern Europe, and we have observed the devastation as each country seeks to change from the artificial structures of Communism to open markets influenced by the West. China was faced with the same problems, but its government still held power, and the problem was to introduce reform without destroying the economy and, in particular, without major social unrest and heavy inflation.

The first reforms, in the period 1979–84, were an industrial version of the responsibility system being applied to agriculture. Gradually, with local experiments, notably in Sichuan, a system was developed whereby individual enterprises agreed on a quota of goods to be remitted to their supervising body at fixed prices, with the balance of output sold on the open market, and they were given permission to retain a percentage of the profits. With this money, managers were able to invest for the future or plan welfare for their employees, while productivity and skill were rewarded by wage differentials and a bonus system.

By 1982, the responsibility system had been applied to all state-controlled industry, and the beginnings of improvement could be seen: production of finished goods in rolled steel, for example, grew from 2.2 million tonnes in 1978 to 2.9 million tonnes in 1982. On the other hand, the increase in funds now retained locally meant that higher levels of government suffered serious losses of revenue, while many plants had increased production but were still preserving inefficiencies and accumulating losses. In this respect, the changes went only one way: the enterprises benefited from new opportunities, but they continued to rely upon the backing of the state to protect them from the consequences of misfortune or error.

Through 1983 and 1984, therefore, the government sought to enforce real responsibility. From June 1983 enterprises were required to pay taxes amounting to 85 per cent of their profits, including an 'adjustment tax' on coastal regions which had particular advantages, and there were also levies by local authorities for land use, services, and return on capital. Despite the weight of these burdens, the new policy recognized the separation of government and business: hitherto, state enterprises had been provided with all the components they needed, whether land, capital, or materials, and all their earnings belonged to the state; now, by implication, they could use their untaxed surplus quite independently.

If the responsibility system was to mean anything, however, there had to be penalties for failure as well as rewards for success, and this brought the regime face to face with a central contradiction between Communism and the market: could a state-run enterprise be allowed to go bankrupt? And could the regime abandon the principle of full employment, even for inefficient workers? The first question was handled indirectly: between 1983 and 1985 there were many amalgamations of industrial enterprise, and from 1985, instead of lifetime appointments, the heads of state-controlled enterprises were given only four-year renewable terms. The answer to the second question was concealed behind rhetoric, but in practice, from the late 1970s, many people lost their jobs. On the other hand, though formal law for bankruptcy was introduced in 1988, it has not been enforced for fear of the social unrest which would follow large-scale losses of employment.

As in rural industry, however, new opportunities for work in private enterprises were now accepted and sanctioned. Because of the problems of capital accumulation, almost all of these were small, generally controlled by one person or a family, but many were extremely profitable. In 1978 there had been only 100,000 registered private enterprises, but there were almost 6 million in 1983 and 17 million in 1985. Many of the new businesses dealt in small-scale manufacturing or handicrafts, but their influence was strongest in the retail and service sectors of the economy, rising from 18 to 25 per cent of Gross Domestic Product, and particularly in free markets, of which there were 40,000 across the countryside by 1985 and 3,000 in the cities. Where formerly almost all trade had been carried out in government stores, now the base of retail trade was in private hands, and one of the most impressive and disconcerting sights of the early 1980s in China was the spectacle of a dismal, almost empty official store, surrounded by an active, lively, informal market with a multitude of goods and a host of eager traders and purchasers.

Such a spectacle, moreover, reflected the new pattern of development. As the state relinquished its direct control over agriculture and some industry, national production certainly increased, but government at every level found it more and more difficult to obtain its share of economic resources, and this meant that large-scale schemes were constantly short of money, while broad matters of concern such as water-control, power-generation, and transport networks were neglected in favour of short-term enterprise and small-scale local interests. In a single year, from 1978 to 1979, the government's budget deficit doubled, and from the beginning of the 1980s, as investment into agriculture, textiles, and light industries increased by an average of 75 per cent, that for heavy industry and mining was reduced by some 40 per cent.

Already it was becoming clear that the Chinese economy as a whole, and certainly that part of it now controlled by the state, was too limited to support the full wish-list of modernization. Many of the foreign investors who had seen the economic opening of China as a splendid opportunity now found themselves faced with heavy losses as the government walked away from hundreds of major new schemes. Japanese firms are said to have lost as much as US$1.5 billion, and a symbol of the débâcle was the great steel complex at Baoshan near Shanghai, which suffered cost overruns through mistaken planning and engineering, and was largely abandoned in 1981. Investment continued, but it was a great deal more cautious. Most joint-venture arrangements were restricted to individual projects such as hotels, single factories, and oil drilling projects, and there was a strong tendency for money to be marketed not through the cumbersome bureaucracy but rather to small private enterprises. Most important, foreigners were now reluctant to risk their fortunes on major projects for infrastructure such as power plants, shipping facilities, and railroads. For future development, these were what China needed most, but the country must pay for them from its own resources. Since the advent of the new policies had made it difficult for the state to extract the vast sums necessary from the internal economy, the government was thus committed to foreign loans and credits, and these relied upon the national balance of payments to service them.

Foreign Enterprises

From the middle 1950s, as the country recovered from the years of war and began to receive the benefits of the First Five-Year Plan, China had maintained a surplus in foreign trade. Comparatively speaking, the figures for imports and exports were not large, and they declined in relation to the economy as a whole. In 1956, for example, the first year of a favourable balance, imports were valued at RMB5.3 billion and exports at RMB5.6 billion. In the tragic year of 1960 there was a negative balance of RMB6.5 billion imports against RMB6.3 billion exports, but over the next ten years the situation was restored to about the same level. In 1970, the total value of trade was RMB11.3 billion, or US$4.6 billion; the exchange was almost precisely even, and in US currency the amount was little different from the $4.4 billion of 1959.[13]

In the early 1970s, however, as China opened up after the *rapprochement* with the United States, foreign trade expanded rapidly to levels never seen before, with energetic sales overseas, and purchases of foodstuffs and raw materials, including wheat from Canada and wheat and sugar from Australia. By the middle of the decade, two-way trade had more than trebled to just under US$15 billion, and in 1980 it was US$38 billion. Moreover, in 1974-75 and 1978-80 there was negative balance of imports and exports, and although this did not cause major concern, it did, as we have seen, persuade the government to abandon many ambitious projects.

By 1981 the situation had been turned around, and for the next couple of years, as total trade grew to US$44 billion, the value of exports more than covered the cost of imports. Some of the gain was from tourism, and also from arms sales to the Middle East, but there was major reduction of imported capital goods. On the other hand, this short-term solution left the development of the country much as it had been, and gave limited opportunities for the transfer of advanced technology.

Throughout this period, therefore, the central government attempted to encourage purchases and investment by Western and Japanese enterprises. Trade fairs increased in number and size; there was some attempt to adopt regular commercial laws and practices in such matters as trade marks, taxation agreements, conditions of labour, contractual liability, and the right to take profits out of the country; and although complaints continued about the one-sided attitude of official negotiators, local authorities were now permitted to arrange their own deals with foreigners. Once again, this last concession cost the central government some control of overall trade, and foreign reserves were run down as goods and materials came in.

To a considerable degree, therefore, the advance of foreign trade, as with the establishment of the responsibility system within the country, brought a weakening of central authority and rendered the national economy vulnerable to a multitude of local or even personal decisions, commonly taken on the basis of a short-term view, with small thought of future development. In attempting to deal with this problem, the government in Beijing sought the means to enhance local and regional initiative while also retaining or regaining patronage and ultimate control, and this policy was followed through a series of administrative expedients.

The first of these devices were the four Special Economic Zones (SEZs), established in 1979 at Shenzhen, just north of Hong Kong, at Zhuhai across the Pearl River estuary, at Shantou (Swatow) in northeastern Guangdong, and at Xiamen (Amoy) in Fujian opposite Taiwan. The inspiration for these zones is similar to that of the Export Processing Zone at Kaohsiung in Taiwan, but their major advertised purpose is to act as centres for the introduction of modern techniques into the nation as a whole. In each area, separated from the rest of the country, foreign firms are encouraged to establish factories and processing plants under conditions more open than those available elsewhere, with an advanced infrastructure of roads and shipping facilities, on the principle and assumption that the local workers and managers will learn on the job and be able to train others in the technical skills they have acquired.

After more than ten years, the Special Economic Zones have achieved considerable success, though not as much as had been hoped for. In general, there are three problems. Firstly, the quality of the work-force and the standard of the fixed facilities, while better than that within China as a whole, is not so high as that available in Hong Kong and Taiwan. As a result, foreign companies have tended to use the Zones as auxiliary sources of cheap labour and secondary work, not as places where the most advanced technology should be committed: it is significant that the most successful area has been Shenzhen, whose role is largely that of an overflow territory for Hong Kong. Secondly, though Chinese workers in the Zones have benefited from the training and experience, it has been difficult to arrange for the wider transfer of this knowledge and expertise. There has been a tendency to form a privileged enclave, separate even from the immediate hinterland, and there are several cases where products from the Zones are not sold abroad, but imported back into China, competing unfairly against local manufactures. Thirdly, the Chinese idea of a commercial concession is often quite out of touch with outside expectations. In the new province of Hainan, for example, it has been advertised proudly that there is free entry and exit for labour, for goods, and for money: leaving aside the backward condition of the island, this special arrangement does no more than match common practice in any advanced commercial territory.

The opening of Hainan in 1984, however, was only one expansion of the concept of the Special Economic Zone. Through the 1980s, as the government experimented further, fourteen coastal cities were identified as open areas for foreign investment. Later, this somewhat scattered approach was refined, and a Coastal Development Plan was based upon the major ports of Guangzhou, Xiamen, Shanghai, and Lüda, with their urban industry and the rural hinterlands which served them. Variations in official policy, however, were not always matched by practice on the ground, and there were frequent complaints from foreigners about inadequate facilities, shortages of energy supply, and an overstrained transport system.

At the same time, while most attention was concentrated upon official arrangements for major enterprises from Japan, the United States, and Europe, the opening of China had been energetically supported by smaller-scale investment from overseas Chinese and, first very quietly but later with the air of an open secret, from Taiwan. Much of the activity took place at local level, frequently on a village or clan basis, as *émigré* branches returned money they had earned abroad for investment among their own people. In the same fashion, merchant banks in Hong Kong acquired a new and important role placing and servicing investments, and they too used local contacts in Guangdong province to by-pass restrictions and deal direct with individual entrepreneurs. For many investors, first preference lay with informal arrangements in the countryside, and only second place was given to the official organizations in the major cities or the Special Economic Zones.

Just as the new regime opened the country to more regular trade, its diplomatic policy also acquired a steadier pattern. Even before the Cultural Revolution, China had sought to establish its position as a centre for world revolution and an independent leader of African, Asian, and Latin-American 'non-aligned' nations, and its maverick policy frequently harassed the world community. Following the visits of Kissinger and Nixon, however, and the acceptance of the Communist regime into the United Nations, the government largely abandoned this stand and attempted to display a constructive role in international affairs. Through the 1970s, more and more countries gave recognition, and in 1979 the Carter administration of the United States negotiated full representation, with an exchange of embassies and consulates, and Congressional approval for the People's Republic for trading status as a most-favoured nation.

In the broad field of diplomacy since the 1970s, China played a responsible role as a permanent member of the Security Council, often arguing on behalf of the developing countries of the Third World for a better position against exploitation by the advanced nations of the West, but generally maintaining a sensible, even boring, position. The broad revolutionary claims of the past have been largely abandoned, and most international activity is concentrated on the often intractable problems of East and South-east Asia, where China appears not so much as a world power but rather as one important player in a complex game of manoeuvre.

In this region, the limits of China's authority were embarrassingly demonstrated by the conflict with Vietnam. China had given consistent support to the northern Communists in their war against the South, and Chinese leaders often described their relationship with the government in Hanoi, by traditional analogy, 'as close as lips and teeth'. For centuries in the past, however, the Vietnamese had avoided too close contact with their great neighbour, and after the the fall of Saigon in 1975, united Vietnam showed no intention of accepting the status of a grateful client. On the contrary, its government confirmed links with the Soviet Union, which had always been the most effective supplier of materials and diplomatic support, and Hanoi's policy was quite independent of China.

Besides the unification of Vietnam, the collapse of American-backed anti-Communism on the Indo-Chinese peninsula had also brought the forces of the Khmer Rouge under Pol Pot to power in Kampuchea, formerly Cambodia, but this regime embarked upon a radical, bloody programme which may best be described as reform by massacre. As cities were laid waste, any possible opponents of the new regime were murdered, people as a whole starved amid the ruins of the economy, and the Vietnamese were eventually moved to intervene. Pol Pot mistakenly declared war, but his troops were swiftly driven from the field, and although they and other resistance forces have maintained guerilla warfare from the Thai border, the Vietnamese-supported government has held power since that time.

For most outsiders, and for the people of Kampuchea, the Vietnamese invasion came as a relief, but Beijing supported the radical Khmer Rouge, and other south-east Asian countries, seriously concerned at the expansion of Vietnamese authority, opposed the new regime. In February 1979, after several threats about 'teaching Vietnam a lesson', China attacked the northern provinces of Vietnam with a quarter of a million men. Unlike the fighting against India in 1962, however, this border campaign was far less swift and successful. Vietnamese local troops opposed the invaders with determination, and though Chinese forces captured the major provincial towns, they did so with difficulty, at considerable cost, and under increasing threat of Soviet intervention. Eventually, they withdrew, and both sides agreed on peace, but Vietnam continued operations in Kampuchea and also in Laos, and was only confirmed in its friendship with the Soviet Union and its ill will against China.

In fact, the chief lesson of the three-week affray was that China possessed only limited capacity for wielding military force at a distance, particularly in the difficult terrain of the south. This had largely been assumed by observers, but it was now confirmed by practice, and even as American power withdrew its guarantees from Southeast Asia, governments in Malaysia, the Philippines, and Indonesia could look upon the balance between Vietnam and China with a degree of equanimity.

There remains one other source of direct potential conflict. Both the People's Republic and the Republic on Taiwan maintain large national claims over the South China Sea, and maps printed by both regimes cover every island and reef, with a border stretched tight along the coasts of Vietnam, Borneo, and the Philippines. At the beginning of 1974 a Chinese flotilla drove a South Vietnamese force from the Paracel, Xisha, group, south of Hainan and east of Vietnam, and Beijing has also published its claim to the Spratleys, or Nansha, further south, while the government in Taipei, equally concerned, maintains a presence.

After the first engagement, the People's Republic kept control of the Paracels with fleet operations, but the occupation was never recognized by Vietnam. The Spratleys, much further afield, cover hundreds of square kilometres near the Philippines and Borneo, and there is likewise no agreement on their ownership and no clear case in international law. So far, the disputes have been primarily concerned with fishing rights and national dignity, but there are rumours of oil to be found beneath the sea, and the question must eventually be solved by peaceful convention or by force. Through the 1980s, however, though the Chinese navy, based on Hainan, increased its strength, it was faced by Soviet forces at Cam Ranh Bay, the former American installation granted them by Vietnam, and by United States bases in the Philippines. At sea, as on land, the military force of China offered no immediate threat to its neighbours.

Indeed, now that China had an established position in the international arena, internal problems of economic reform appeared far more important and interesting than any initiatives or quarrels abroad. The government of Deng Xiaoping and his colleagues was not concerned with the polemics of the past, and was prepared, where necessary, to negotiate with such nominally hostile countries as South Korea. There was precedent for this attitude from the days of cold-war rhetoric: even when Australia, for example, had followed the American line, denounced the People's Republic and recognized Taipei, profitable trade had yet been maintained in both directions, with notable imports of wheat and wool. During this new period of normalization, however, neither of the rival Chinese governments was concerned to press the point beyond some basic formalities, and Beijing in particular showed a welcome tolerance and patience.

To a large extent, this was the relaxed attitude of the person who was winning. Taiwan was generally recognized as a part of China, the old rivalry between Communist and Nationalist had lost much of its tension after the deaths of Mao Zedong and Chiang Kaishek, and in one form or another the exiles would surely return to their fold. In similar fashion, the Beijing government embarked on negotiations with the British for the hand-over of Hong Kong in 1997: the change was inevitable and natural, all things will come to the one who waits, so why make a fuss? After the excesses of the past, there was a sense of security abroad, of goodwill and confidence in future progress. Sadly, however, in June 1989, security was broken, confidence proved false, and goodwill was lost.

The Fifth Modernization

In 1974, as the pragmatists gathered their forces against the left, three young men in Guangzhou published a poster essay. They called themselves by the pseudonym Li Yizhe, compiled from characters of their names, and they argued with justice that the Cultural Revolution had offered the people an experience of freedom, protest, and democratic exchange which had been completely lacking under the Communist regime. The sad thing was, however, that after the Cultural Revolution had overthrown the oppressive system of Party and government, its leaders had sought to establish an equally authoritarian system of their own, while the constant denunciation of petty bourgeois capitalism had actually attacked the best hopes of individual initiative and personal freedom.

The critique held much truth, and though the writers were later abused and imprisoned, their manifesto became a symbol of the struggle for democracy, the third, intellectual, alternative to the twin enemies of bureaucracy and radicalism in the modern Chinese system. Four years later in 1978, as Deng Xiaoping confirmed his position against Hua Guofeng, a blank wall in central Beijing was occupied by a mass of posters, articles, and essays, in the style of the Cultural Revolution, supporting the new regime but urging consultation with the people, and this 'Democracy Wall' was supplemented by a host of pamphlets and small journals communicating the same message.

In December 1978, Wei Jingsheng, young editor of one of these journals, put a poster on Democracy Wall calling for the the 'Fifth Modernization': the official Four referred only to economic and technological reform, but future success required also the modernization of China's political and social structure, and this could be achieved not through autocratic authority but only with the introduction of true liberty and democracy.

The argument for political and intellectual freedom had been maintained with courage, but with most limited success, since the 1950s and the Hundred Flowers, and it had never been welcomed by the rulers of China. No matter how their theme was disguised behind expressions of loyalty and support for reform, the liberals could express themselves freely only at times of tension and confusion — and once the uncertainties were resolved the government once more silenced these unofficial commentators. In March 1979 Wei Jingsheng was arrested and sentenced to fifteen years imprisonment, and in 1980 Democracy Wall was closed and future posters were censored.

The fact that the critics were thus suppressed, however, did not mean they had ceased to think, and the history of persecution did not prevent new young dissidents from joining the debate. From the beginning of the 1970s, as China was opened to the outside world, the same *tiyong* problem returned as had faced the old empire: is it possible to import foreign tools and technology simply for use, yet preserve the essential body of native custom and philosophy?[14] And the force of the debate was strengthened by official talk of economic development: in Maoist days, it could be argued that the Chinese revolution was superior to the Western experience; but now China was seeking to learn from and eventually to match the foreigners at their own game.

In the first years after the Cultural Revolution, much of the literature was concerned with the misfortunes and failures of the past, with stories by new writers such as Chen Ruoxi about the injustices of mob rule, and the satirical play *Teahouse* by the veteran Lao She, on the follies of the recent past. As criticism and analysis continued, the conflict of ideals was expressed in a television series of 1988, *Elegy for the Yellow River*, which stirred debate in official and unofficial forums from *Beijing Review* to universities and street corners. The thrust of the argument was that old tradition, based upon peasant agriculture, landlocked to the soil and to the labour of irrigation, had exhausted its value, and the future of the nation lay south and east, beyond the horizons of the sea. It was an exciting concept, fuelled by the reality of economic development based upon the coastal cities, but it presented a frightening challenge to the nature and accepted history of China, and it answered the *tiyong* question in the cruellest way: the past must be abandoned and the future sought elsewhere, in a new and alien approach to the world.

The popularity and interest in this debate reflected a general uncertainty about the nature of China's recent progress and the conditions under which it could be maintained. In practical terms, though many observers still approved the pragmatic path of development, others recognized that the social and political structure of the nation was under strain even as the economy itself approached a crisis.

One symptom of the problem could be seen in the figures for foreign trade. In the early 1980s, as we have noted, a rising value of total exchange had been accompanied by a favourable balance of exports against imports. By the middle of the decade, however, the value of exports increased at enormous speed, from US$44 billion in 1983, to $54 billion in 1984, to $70 billion in 1985, and to $74 billion in 1986; and in the last three of those years the balance of trade became unfavourable, with a deficit of US$1.3 billion in 1984 rising to $15 billion in 1985 and then $12 billion in 1986. In 1987, though trade volume increased again to US$83 billion, the deficit was reduced to $3.7 billion, but even this was more than the total deficit for the whole of the decade from 1971 to 1980. By 1988 China's foreign debt was estimated at US$38 billion.

The composition of foreign trade, moreover, indicated that the country had gone only a limited distance in development. A third of export value was in primary goods,[15] and half the remainder, described as manufactured items, consisted of part-processed yarn and fabrics, together with cheap clothing and footwear. By contrast, 85 per cent of imports were manufactures, many of them consumer items such as washing machines and refrigerators, but also quantities of chemicals, and special industrial machinery such as office, electrical, and telecommunications equipment, power tools, and transport vehicles. There had certainly been some incremental change towards local secondary industry, but the movement was slow, and it was outstripped by consumer demand and current accounts.

One problem for foreign trade was the distortion between domestic prices and international values, requiring official subsidies for both imports and exports. In 1984, to lessen some of the effect, Beijing devalued the currency from RMB2.80 to RMB3.70 against the US dollar, but this move exacerbated the pressures of inflation, hitherto almost unknown in the People's Republic. Official figures spoke of 12.5 per cent in 1985, reduced to 8 and 7 per cent in the next two years, but unofficial estimates suggested 15 to 20 per cent per annum and, as the discrepancy implies, much of the real, heated economy of the nation was either outside the government's knowledge or simply beyond its control.

Indeed, the very nature of the new, freer policies meant that the government had given up much of its direct power, while the funds at its disposal to intervene in the economy were seriously restricted. At the most general level, the energies of free, local markets and private entrepreneurs paid little attention to broad questions of public works, whether for power, transport, and communications, conservation, pollution control, or social welfare, and the authorities which should be concerned in such matters were consistently short of money.

Most seriously for any future development, education expenditure
was no more than 3.75 per cent of Gross National Product. Despite
the youthful population, enrolments in primary and secondary schools
showed an absolute decline through the 1980s, and though numbers
at tertiary level did increase, they represented only 2 per cent of young
people. There were, moreover, repeated complaints of poor quality
teaching and a lack of resources and planning, with equally repeated
statements of its fundamental importance, of improvements to come,
and of the difficulties of funding.[16]

In still more immediate terms, the limits of central finance meant
that little could be done to restrain or ease the effect of growing
inequalities, both between social groups and across different regions.
Despite the progress of the past ten years, China remained one of the
poorest countries of the world, with income per head in 1988 estimated at US$320. Inside the country, moreover, there were enormous
disparities: in 1986 the highest income per capita was that of Shanghai,
RMB3,471, followed by Beijing and Tianjin, with a little more than
RMB2,000. Behind these provincial metropolises came Liaoning in
the north-east with RMB1300, then the lower Yangzi provinces of
Jiangsu and Zhejiang, with just over RMB1,000, Heilongjiang with
its northern oil-fields, and Guangdong at RMB897.[17] By contrast,
Guizhou in the south-west had an income per head of RMB406, with
Yunnan and Guangxi at RMB450, Sichuan RMB515, and Gansu,
Shaanxi, and Henan were all below RMB600. Overall, the prosperous provinces were those of the eastern seaboard and the poorest were
landlocked to the west, but it was the contrast that was most remarkable: average annual income in Guizhou was less than an eighth that
of Shanghai, and Guangxi was half that of Guangdong.[18]

There were, of course, disruptive effects from such imbalances,
one of the most obvious being considerable internal migration as
people from the poorer regions sought work in developing areas.
Much of the movement was unofficial and even illegal, but the new
small enterprises springing up were anxious for workers and asked
no questions. And at government level there was jealousy between
the territories, so that several provinces set up customs barriers to
protect themselves against competition, one way or another, from
their neighbours.[19]

Within local communities, there was tension and resentment as
some families and individuals gained the advantage of their neighbours, whether by skilful husbandry in the countryside or greater
enterprise in the cities. There was a steady demand for welfare, too

Map 7 The People's Republic of China: income per head by province, 1986

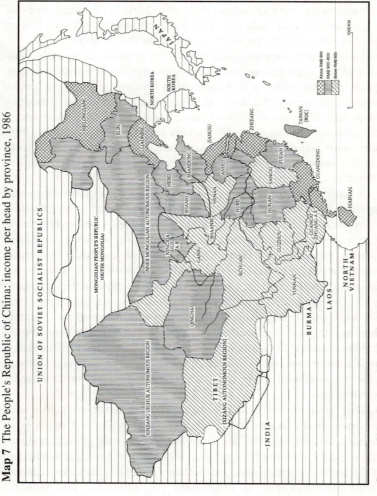

Source: *China Statistical Yearbook, 1988.*

often inadequately met, but the most obvious losers in the new competition were those on government stipends: administrators, cadres, and teachers. Just as in the old days of the Republic, though so far less dramatically, the effects of inflation took their toll on the individual's sense of worth. Some people abused their political power, and when they were found out they were harshly punished, but others, notably intellectuals, saw their livelihood and their social respect steadily reduced against the new rich of the free market.

For educated young people at university, products of a poorly funded, highly competitive structure, the contradictions between their own ideals and the realities of competition brought resentment for the apparent hypocrisy of a state which insisted upon political orthodoxy but permitted economic individualism. At the same time, their uncertain personal futures, in a society where graduates were regularly allocated work without real freedom of choice, and where salaried work was becoming increasingly unattractive, gave an immediate edge to philosophic dissatisfaction.

In December 1986 a flurry of demonstrations and strikes in universities throughout China, notably in Anhui, Shanghai, and Beijing, called for more democratic consultation and an open media. The movement was not opposed to the government, for the students claimed to support the reforms of Deng Xiaoping, but they did call for greater liberalization, and they aroused opposition both from natural conservatives and from nervous sympathizers who thought they had gone too far. In the event, the agitation was put down, a few demonstrators were briefly imprisoned, some were sent down to the countryside or otherwise re-educated, and conservatives in the central government secured the dismissal of the General Secretary of the party, Hu Yaobang, who was replaced by Zhao Ziyang. The net result was a loss to the liberal cause, and a shift by the government to a harder political line.

Deng Xiaoping still held authority without difficulty, but there were tensions in the state, and at the Thirteenth Party Congress in October–November 1987, he arranged once more a general resignation of older leaders of the regime. Deng himself became Chairman of the Military Commission, with Zhao Ziyang as first vice-chairman and Yang Shangkun as permanent vice-chairman and General Secretary.[20] Chen Yun, however, representative of the more conservative forces in the leadership, became Chairman of the Central Advisory Commission and General Secretary of the Central Discipline Inspection Commission, the censorate of the party.[21] Li Peng, a protégé of the late Zhou Enlai, became a member of the Standing Committee of the Politburo, and six months later, in April 1988, the Seventh National People's Congress confirmed him as Premier and appointed Yang Shangkun to succeed Li Xiannian as President of the People's Republic. As in the party, Deng Xiaoping was Chairman of the Central Military Commission.

The political situation remained comparatively quiet through 1988, but economic and social pressures continued, and the Gorbachev reforms in the Soviet Union, followed by growing disturbances in eastern Europe, provided an influential example for discontent. 1989, moreover, was not only the fortieth anniversary of Communist Liberation, but also awakened echoes of May Fourth 1919 and legends of the French Revolution in 1789. Early in the year the distinguished scientist Fang Lizhi sent an open letter to Deng Xiaoping urging a more liberal policy, and his words were endorsed by eighty scholars in China and abroad. In February, President Bush of the United States visited Beijing, and, despite the diplomatic compliment, the American interest in Fang Lizhi as a spokesman for reform was both embarrassing and irritating to the regime. Then, in April, Hu Yaobang died. Though aged seventy-four, he had been regarded as one of the younger leaders, a potential successor to Deng Xiaoping, and a symbol of responsive government. Somewhat like that of Zhou Enlai thirteen years before, Hu's death became the occasion for demonstrations of mourning which turned swiftly into protest rallies calling for democracy.

On this occasion, however, the conflict was not easily resolved, and the demonstrations became extraordinarily popular. By May it was estimated that more than a million people were involved in the protests in Beijing, and there were marches by university students throughout the country, with some expressions of support among industrial workers. Student leaders went on hunger strikes, their followers occupied Tiananmen Square, in the heart of the capital, with banners, placards, flags, and tents, and the obvious disorder provided the utmost embarrassment to the leadership when General Secretary Gorbachev of the Soviet Union arrived on a state visit.

The leadership was confused, but division developed between those who were prepared to talk and even negotiate with the students, and the hard-liners determined to crush them. Zhao Ziyang showed sympathy, and even visited the Tiananmen encampment, but Li Peng, evidently with approval from Deng Xiaoping and other senior leaders, proclaimed martial law on 20 May, and Zhao Ziyang was dismissed as General Secretary of the party. There was one attempt at a meeting between government officials and the youthful leaders of the protest, but the dialogue came to nothing, and the demonstrators refused to believe there was any need for a swift resolution. On 30 May they raised the standard of their rebellion: a ten-metre statue of the Goddess of Democracy, loosely based upon the symbolism of the New York Statue of Liberty.

For a few days there were desultory negotiations, but the leadership had already called in loyal troops from distant provinces to deal with the troublemakers. Organized by Yang Shangkun and commanded by his brother, his son-in-law, and his nephew, these new forces came to the capital late in May. Private citizens set up barricades and pleaded with them not to interfere, but by the beginning of June they had taken over front positions from the local garrisons and were preparing themselves for action.

During the night of 3-4 June 1989 these alien troops advanced against the demonstrators of Tiananmen. Surprise was complete, for no civilian believed so violent an assault could take place, but the soldiers cleared the area with gusto, they fired at will upon the fleeing students, and they extended the slaughter through the streets of the city. There is no full estimate of the loss of life, whether in the hundreds or thousands, but the numbers are almost irrelevant: the government had shown it would brook no opposition, while the brutality of the attack shocked the outside world and destroyed for ever the illusion of community in Communist China.

Notes

1. The square takes its name from the main southern gate [*men*] of the former imperial palace, and the phrase *tianan* is normally translated 'Heavenly Peace'.

One may note that the name of the nineteenth century Taiping rebels is rendered as 'Great Peace' (p. 10 above): *an* is best understood as 'tranquillity', and *ping* as 'pacification'.

2. It has been suggested that Hua gave valuable support to Mao at the Lushan conference in 1959, when Peng Dehuai was attacking the conduct of the Great Leap Forward. See p. 230–1 above, and *Cambridge China* 14, p. 317 [Lieberthal, 'The Great Leap Forward and the split in the Yenan leadership'].

3. To be fair, it should be noted that earthquake prediction has not advanced very far in the West, and there was considerable interest in the Shenyang experience, where peasants had observed changes in the level of water in wells and the unusual emergence of snakes from their holes in mid-winter. Such signs are more likely to be usefully noted in a basic society than in a region such as southern California, where water comes from pipes and snakes do not appear in the streets.

The difficulty seems to be that the Shenyang earthquake was a type which lent itself to such information, and the Tangshan one — probably more deeply based — did not.

4. This meeting is formally described as the Third Plenum [that is plenary ses-

sion, or full gathering] of the Tenth Central Committee [that is of the Central Committee of the Party elected at the Tenth Party Congress, which had been held in 1973].

5. Early in 1991 it was reported that Jiang Qing, seriously ill, had committed suicide.

6. Though the situation is infinitely better than it was, there are still some difficulties in dealing with Chinese statistics. Two major sources of discrepancy are, firstly, that the annual publications of the Chinese government through the 1980s often added more information on previous years, and sometimes changed figures which had been given earlier; secondly, there are many occasions when well-informed outsiders, notably US government sources, present quite different collections of figures — and the situation is not made easier by necessary calculations of exchange rates, values, and prices current at the relevant time, and formal questions as to exactly what items are included in which categories. In most of the discussion which follows, I rely upon the November 1988 edition of the *China Statistical Yearbook*, published in Beijing and Hong Kong by the State Statistical Bureau, where values are given in Chinese currency(RMB).

As an example of the disagreements which may appear, however, American government estimates for China's Gross National Product in 1952 were US$99 billion, and in 1981 US$515 billion; Chinese official figures for the same dates equate to US$52 billion and US$270 billion. Both sets of figures, however, indicate an increase of about 520 per cent, and it generally appears that even when absolute figures vary considerably, there is fair agreement on the rate of change.

7. There is an apocryphal story that during the height of the Great Leap Forward not only factories, but also maternity hospitals, were emblazoned with the slogan 'More, Bigger, Better, Faster!'

8. 80 per cent of the people live outside the great cities, but recent Chinese statistics identify 'rural' population as those living outside county towns. On such a basis of calculation, urban/rural proportions of the population are almost evenly balanced. It is in the countryside, where half the population lives, that the production, both of goods and of children, is most difficult to supervise and control.

9. During the 1950s, the national birth rate was about 35 per thousand. The death rate declined from 18 in 1950 to 11 in 1957, rose sharply under the impact of the Great Leap Forward (more than 25 in 1960), but thereafter continued to decline. By the mid-1980s, the official birth rate was 18–20 per thousand, and the death rate about 7; natural increase was between 15 and 11 per thousand.

10. One effect of the single-child-per-family policy, if it were fully successful, would be that as the younger generation came of working age, and the parents and grandparents grew old, there would be a small labouring force supporting a great overburden of aged and infirm. This, however, is a problem for the future.

11. This concept is discussed earlier, at p. 105.

12. No discussion of China's industrial problems can be complete without reference to the national airline, whose acronym, CAAC, is pronounced with feeling by Western travellers.

Features of China Airlines' operations include an inability to effect reliable bookings or fly the planes on time, ground service personnel who combine high levels of rudeness with incompetence and laziness, a general failure to use safety features such as belts, while extra passengers are accommodated on camp-stools in the aisles, and an appalling accident record, normally unpublished, which occasionally comes to light when foreigners are involved or when the disaster occurs in some such public place as Hong Kong or Guangzhou airports.

CAAC has been voted by experienced travellers for many years as the world's worst airline. Much, though perhaps not all, however, may be forgiven to the company which put a request over the public address system just before take-off, calling for passengers carrying guns or nuclear weapons to hand them over to the flight attendants, and promising they would be returned after landing.

13. There are discrepancies between the figures given in RMB and US currency, partly through the allocation of costs for freight and other charges, and also through varying rates of exchange.

14. See p. 18 above.

15. As for many years in the past, much of China's exports was in the form of food and other raw materials for consumption or processing in Hong Kong. At the same time, China is a substantial exporter of rice, but imports large quantities of wheat, the popular food grain of the north.

16. *China Facts and Figures* 12 (1990), p. 477, quoting *Beijing Review* 32.5 (30 Jan 1989), p. 23. In comparison, during the mid-1980s, education expenditure was just under 7 per cent in the United States and the Soviet Union, while 9 per cent of young people in India were being educated at tertiary level, and 30 per cent in Japan.

17. The figures for Liaoning, of course, were influenced by the metropolis of Lüda and the city of Shenyang, with the mines nearby; Jiangsu and Zhejiang reflected the expansion of Shanghai; and Guangdong relied likewise upon Guangzhou and the Special Economic Zones of Shenzhen and Zhuhai.

18. *China Statistical Yearbook,* 1988, p. 45.

19. Even in the early 1980s, manufactures from Shanghai were in high demand throughout China, at the expense of local goods. At Lanzhou in 1981, for example, the local brand of bicycle was considered far inferior to the imported model from Shanghai.

On the other side of such transactions, a major concern of local governments was to retain sufficient quantities of their produce to fulfil central government demands of quota. If they did not check exports, these tax goods would be exported for open sale on more profitable free markets.

20. Yang Shangkun was one of the Twenty-Eight Bolsheviks during the Jiangxi Soviet period (p. 125–6 above). He became chief commissar in Lin Biao's command during the civil war, and then a senior assistant to Deng Xiaoping in party administration during the early 1950s.

21. Chen Yun had been the master of economic planning during the 1950s, and opposed Mao's policies from the Great Leap Forward through the Cultural Revolution. Though he was a natural ally of Deng's attempt to eliminate the influence of Maoist economics, he was anxious about the new freedoms, and rightly concerned with the effects of unemployment, the influence of Western attitudes, and the developing weakness of ideal socialist morality.

Further Reading

Official statistics and information are given in various publications of the
    People's Republic, notably:

Beijing Review, renamed from *Peking Review* in 1979 [contains the texts of
    major reports and conferences, and also an annual summary].

China Statistical Yearbook, published annually since the early 1980s in Beijing
    and Hong Kong by the State Statistical Bureau.

Additional information and continuing analysis may be found in such outside publications as:

China Facts and Figures, edited by John L Scherer 1978-87, thereafter by
    Charles E, Greer, Gulf Breeze, Fla., Academic International Press.

The China Quarterly, London.

Far Eastern Economic Review, a journal published in Hong Kong with an annual *Asia Yearbook.*

On political figures and events:

Bonavia, David, *Verdict in Peking,* London, Hutchinson, and New York, Putnam's, 1984.

Bonavia, David, *Deng,* Hong Kong, Longman, 1989.

Chi Hsin [pseudonym], *Teng Hsiao-p'ing: a political biography,* Hong Kong, Cosmos, 1978.

Cohen, Jerome Alan, 'China's Changing Constitution' in *China Quarterly* 76 (1978).

Domes, Jürgen, 'The Gang of Four and Hua Kuo-feng: an analysis of political events in 1975-76' in *China Quarterly* 71 (1977).

Domes, Jürgen, *The Government and Politics of the PRC: a time of transition,* Boulder Colo., Westview, 1985.

Goodman, David S.G., *China's Provincial Leaders, 1949-1985,* Atlantic Highlands NJ, Humanities Press, 1986.

Manion, Melanie, 'The Cadre Management System, Post-Mao: the appointment, promotion, transfer and removal of party and state leaders',in *China Quarterly* 102 (1985).

Schram, Stuart, *The Thought of Mao Zedong,* Cambridge, Cambridge University Press, 1989.

Tiewes, Frederick C., *Leadership, Legitimacy and Conflict in China: from a charismatic Mao to the politics of succession,* Armonk NY, Sharpe, 1984.

Townsend, James R., and Womack, B., *Politics in China,* Boston, Little, Brown, 1986.

Wilson, Dick [ed.], *Mao Tse-tung in the Scales of History: a preliminary assessment organised by The China Quarterly,* Cambridge, Cambridge University Press, 1977.

Wilson, Dick, *Chou: the story of Zhou Enlai,* London, Hutchinson, 1984

Witke, Roxane, *Comrade Chiang Ch'ing,* London/Boston, Weidenfeld and Nicolson, 1977.

On reform:

Barnett, A. Doak, *China's Economy in Global Perspective,* Washington DC, Brookings, 1981.

Benewick, Robert, and Wingrove, Paul [eds.], *Reforming the Revolution: China in transition,* London, Macmillan, 1988.

Eckstein, Alexander, *China's Economic Development: the interplay of scarcity and ideology,* Ann Arbor, Michigan University Press, 1975.

Gittings, John, *China Changes Face: the road from revolution,* Oxford, Oxford University Press, 1989.

Goodman, David.S.G., Lockett, M., and Segal, G. [eds.], *The China Challenge:*

adjustment and reform, London, Royal Institute of International Affairs/Routlege and Kegan Paul, 1986.

Goodman, David S.G. [ed.], *China's Regional Development,* London, Royal Institute of International Affairs/Routlege and Kegan Paul, 1989.

Gray, Jack, and White, Gordon [eds.], *China's New Development Strategy,* London, Academic Press, 1982.

Harding, Harry, *China's Second Revolution: reform after Mao,* Washington DC, Brookings, 1987.

Lardy, Nicholas R., *Economic Growth and Distribution in China,* Cambridge, Cambridge University Press, 1978.

Lardy, Nicholas R., *Agriculture in China's Modern Economic Development,* Cambridge, Cambridge University Press, 1983.

Myers, Ramon H., *The Chinese Economy: past and present,* Belmont Ca, Wadsworth, 1980.

Prybyla, Jan S., *The Chinese Economy: problems and policies,* South Carolina University Press, [second edition] 1981.

Wang, George C. [ed. and trans.], *Economic Reform in the PRC: in which China's economists make known what went wrong, why, and what should be done about it,* Boulder Colo. Westview, 1982.

Zweig, David, 'Prosperity and Conflict in Post-Mao Rural China,' in *China Quarterly* 105 (1986).

On popular society and dissent:

China: violations of human rights, London, Amnesty International, 1984.

Death in Beijing, London, Amnesty International, 1989.

Esherick, Joseph W., and Rankin, Mary Backus, *Chinese Local Elites and Patterns of Dominance,* Berkeley, California University Press, 1990.

Goldman, Merle, *China's Intellectuals: advise and dissent,* Harvard University Press, 1981.

Goldman, Merle, Cheek, Timothy, and Hamrin, Carol Lee [eds.], *China's Intellectuals and the State: in search of a new relationship,* Harvard University Press, 1987.

Goodman, David.S.G., *Beijing's Street Voices: the poetry and politics of China's democracy movement,* London/Boston, Boyars, 1981.

Grieder, Jerome B., *Intellectuals and the State in Modern China,* New York, Free Press, 1981.

Hayhoe, Ruth, and Bastid, Marianne, *China's Education and the Industrial World: studies in cultural transfer,* Armonk NY, Sharpe, 1987.

Huang, Philip C.C., *The Peasant Family and Rural Development in the Yangzi Delta,* Stanford, Stanford University Press, 1990.

Jenner, W.J.F., and Davin, Delia (eds. and translators with others), *Chinese Lives: an oral history of contemporary China* [originally compiled by Zhang Xinxin and Sang Ye], New York, Pantheon, 1987.

Kent, Ann, *Human rights in the People's Republic of China: national and international dimensions,* Canberra, Australian National University, 1990.

Knight, Nick (ed.), 'Looking at China after Tiananmen', in *Asian Studies Review* [journal of the Asian Studies Association of Australia] 14.1 (1990).

Potter, Sulamith Heins, and Potter, Jack M., *China's Peasants: the anthropology of a revolution,* Cambridge, Cambridge University Press, 1990.

12

The Other Chinas: Taiwan
and Hong Kong

As the Chinese of the People's Republic struggled with the vagaries of their Communist masters, two smaller communities achieved economic success under alternative forms of government. In Taiwan, the old regime of the Nationalists acquired a new lease of life and, with help from American friends and firm government policies, turned the small, backward island province into a showpiece of prosperity. In Hong Kong, British administration, with Western commercial law in a very free market, developed the colony into a financial and manufacturing centre for the whole of East and South-East Asia.

Both territories are on the margin of China, but each in its own way has succeeded where the heartland has failed. Both, on the other hand, remain Chinese, and both have to come to terms with the history, the culture, and the unity of China. The Nationalists on Taiwan claim unification as their goal, and Hong Kong must accept it in 1997, but the achievements of their people raise questions for the future: can there be a China which is not Communist? Can political unity accept economic diversity? And what, after all, does it mean to be Chinese?

Taiwan: Exile and Rivalry

When the defeated, demoralized Nationalists took refuge on Taiwan in 1949, there was no reason to believe they would be able to hold out against the Communists for more than a few months. Certainly, the hundred mile straits were a significant barrier to the Communists, but the Nationalists had also sought to make a stand on Hainan in the south, and their defences there lasted only a very short time.

On Taiwan itself, the native-born islanders had been separated from China for fully fifty years, their government and schooling had been controlled by Japan, and they had small natural sympathy or respect for their new Chinese rulers. In the first years after the war, moreover, their resentment was given good cause, for the corrupt and brutal administration of the Nationalist governor Chen Yi caused rioting and massacre in Taipei and other cities. In January 1949, however, as the last Nationalist garrisons in north China were being surrounded or captured by the Communists, General Chen Cheng was appointed governor of Taiwan province, and within a few months he restored order and laid a basis for the islanders' acceptance of Nationalist rule. Most important of all, at the very beginning of his administration Chen Cheng enforced a real reduction of land rent for agriculture to a maximum of 37.5 per cent of the main crop. This programme had long been Nationalist policy, but it had never been effectively enforced. With 80 per cent of the farming population on Taiwan holding by tenancy, and former rentals as high as 50 or even 70 per cent of income, the savings which the new rate allowed showed immediate results in improved housing, new animals, and overall production. At the same time, to give the farmers security of tenure, it was made compulsory for all leases to be set out in writing and to be negotiated for a period of no less than six years.

In 1951, under the auspices of the main Nationalist government which had now established itself on the island, land reform was carried one stage further, with the distribution of land which had originally been held by the Japanese for the benefit of their own nationals who came to settle in Taiwan. At the end of the Pacific war the Japanese colonial administration, together with some 400,000 settlers, was firmly repatriated back to Japan and the land they left, some 20 per cent of the arable ground on the island, was redistributed among 140,000 farmers at a rate of three hectares of paddy or six hectares of dry farming land, sufficient in theory to support a family of six. The purchase price was based on two and a half times the value of the main crop, with repayments over 20 years and an annual interest of only 4 per cent.

In 1953, as final demonstration of the government's good intentions and determination for reform, every piece of farm-land in the country was made subject to redistribution: all private and tenanted land was purchased by the government and resold to the peasants who actually worked the soil. At the end of the programme farm tenancy had been reduced from 40 to 15 per cent, and by the late 1960s it was claimed that 90 per cent of agricultural land was tilled by the peasant proprietor. Of some 9 million people in Taiwan in 1956, over 3 million had been affected by redistribution, and during the first half of the 1950s food production increased more than 25 per cent.

The unquestioned success of land reform in Taiwan, all but unique among non-Communist countries in Asia, confirmed the authority and acceptance of Nationalist rule among the mass of Taiwanese people. Admittedly, the administration had many factors in its favour: the land seized from the Japanese meant that a great deal could be given away without immediate effect on private or political interests; and the very fact that Taiwan had long been independent and separate from China meant there were few close links between the landholders and the incoming government. Where the Nationalists in Shanghai and Nanjing had been under constant pressure from landlords, business houses, and financiers, the exiles on Taiwan had no such contacts to distort their policies, and indeed it was much in the interest of the new government to reduce the economic power of one major group which might otherwise have formed the basis for indigenous opposition. It was none the less remarkable that the formerly corrupt and inefficient regime of the mainland now appeared capable of competent government and essentially fair dealing.

The Nationalist exodus to Taiwan also gave the small province an intensity of well-trained and competent administrators. In 1947 a census count had indicated that the population of Taiwan was about 6 million, but the evacuation of the mainland two years later is estimated to have brought in 2 million civilians and 500,000 Nationalist soldiers. The soldiers, alien and discredited, long presented a problem of resettlement, but the civilians included among their numbers some of the most promising officials and intellectuals of non-Communist China, and a serious reform of the party between 1950 and 1952 improved the political organization and removed many of those who had done so much damage in the past.[1] Though there was justified criticism that native-born Taiwanese were denied political responsibility by a prejudice towards mainlanders, there is no question that the level of education and experience which the Nationalists had to draw upon during the early years of their administration on Taiwan could find no parallel in any other small developing country.

In similar fashion, the finances and international position of the government in exile gained immensely from the immigration. Chiang Kaishek and his supporters managed to bring with them not only the bulk of the old imperial art collection from the palaces at Beijing, but also the greater part of the republic's financial resources. When the British recognized the Communist government they automatically transferred the gold reserves lodged in Hong Kong and London to the credit of Beijing, but Chiang had already moved the equivalent of

US$300 million to Taiwan, and the new island economy was largely free from the disasters of inflation which had crippled the last years at Nanjing. In June 1949, after the fall of Shanghai, the government on Taiwan adopted a New Tael (NT), and the exchange rate of this currency, though not quoted on every international market, remained generally steady.[2]

In June 1950, when the invasion of South Korea by the Communist forces of the north forced the United States to reconsider its policy in east Asia, Taiwan formed a vital link in the line of defence against aggression from China and Russia. Up to that time, the administration of President Truman had affected unconcern over the fate of the Nationalists: Taiwan was excluded from any guarantees of defence and no formal recognition was given to the exiled government of Chiang Kaishek. From 1950 onwards, however, as United States forces fought Chinese Communist troops in Korea, and as American representatives developed their opposition to the entry of the People's Republic into the United Nations, the government became increasingly committed to military and economic support of the alternative regime on Taiwan. From June 1950 an economic aid program was established which averaged US$100 million a year until the early 1960s, while a military assistance group gave permanent status to the United States presence on the island and provided a constant supply of arms and advice. In 1953 the new administration of President Eisenhower raised diplomatic representation to ambassador level, and in the following year the two governments signed a mutual defence pact. In the Cold War environment, under the programme of 'containment' urged by John Foster Dulles, Taiwan offered a secure base from which pressure could be maintained against Communist countries and support could be provided to friendly governments facing insurgency. In 1954 and 1958, when the Communists opened bombardment against Jinmen and Mazu, the Americans made it plain that they would stand by their guarantee of all Nationalist-held territory.

For the short term, United States support gave the Nationalists valuable breathing space and opportunity for economic development, while the dependence on American goodwill meant that the aid which was given was used sensibly, under competent guidance. In particular, the Sino-American Joint Commission on Rural Reconstruction (JCRR) provided advice and planning to take full advantage of the land reform programme in the early 1950s: agencies of the JCRR encouraged farmers to improve their techniques of production; and the continuing interest of the foreigners discouraged any major government corruption in dealing with the new land-holders. Enemies of the Nationalists argued that the regime on Taiwan was a semi-colonial puppet, but from the point of view of the people the results of the partnership were generally superior to those provided by more ostentatiously independent local governments in Asia or Africa.

In 1953 the government announced a four-year plan to encourage the growth of industry. In contrast to policy on the mainland, the Taiwanese were urged to concentrate on light and medium industry, not requiring any great quantity of capital investment and relying so far as possible on local material. Given the agricultural base of the economy, it was natural that the first development should be in such fields as the canning of fruit and vegetables and the weaving of textiles, but from that beginning the island's industry expanded into chemicals, light manufacturing such as bicycles, sewing machines, and electrical components, and a wide variety of plastics and synthetic material. The First Four-Year Plan was a clear success, and it was followed by further plans which maintained and increased the speed of development on similar lines. By the mid-1960s the Gross National Product of Taiwan was more than US$3 billion, three times that of the early 1950s, and industrial exports had risen from 5 per cent to 50 per cent of the whole. In all of east Asia, only Japan had a higher per capita income.

If it was American power and money that had ensured the survival of the Nationalist government and the initial prosperity of the island, it was the economic growth of Japan that gave opportunity for the remarkable expansion of the economy in later years. The period of colonial rule up to 1945 had established a certain sense of interest and association, and as Japanese industry spread into new, advanced, and profitable fields — car manufactures, cameras, television sets, computers, and other electronic equipment — the great industrial companies found Taiwanese industry well suited to the high quality, labour-intensive manufacture of components for these sophisticated machines. With increasing involvement of Japanese capital and expertise the island shared in the prosperity and development of the world's fastest-growing industrial power. In 1966, after fifteen years of American economic aid, it was agreed that Taiwan was at 'take-off point' and the series of gifts and loans was formally ended.

As light industry progressed, it was followed by an expansion of heavy industry and mining. Between 1952 and 1968 coal production increased from 2.3 million to 5 million tonnes, pig iron from 10,000 to 75,000 tonnes, electric power from 1.4 billion to 9.8 billion kWh, and ship-building from 500,000 to 80 million tonnes. Almost equally impressive was the rise in rice production from less than 1.6 million tonnes to over 2.5 million tonnes, more than keeping pace with population increase, and allowing the government to claim the highest per-capita calory consumption of food in Asia. By the early 1970s, under the Fifth Four-Year Plan, the economic growth rate of Taiwan was more than 11 per cent per annum, and the Gross National Product in 1972 was NT$227.3 billion.

In this remarkable period of development during the late 1960s and early 1970s Taiwan actually become a greater trading power than mainland China. In 1966, at the beginning of the Cultural Revolution, total trade of the People's Republic of China was US$4,245 million and Taiwan's was $1,867 million; by 1971, when mainland China trade was $4,680 million, Taiwan's was $4,125 million and in 1972, although mainland trade had risen to $5,715 million, with regular export of food to Hong Kong and bulk purchases from Japan, Canada, Australia, and Europe, the figures for Taiwan were $5,975 million, of which almost half were industrial products.

On the other hand, while Taiwan had indeed become a 'model province' and a remarkable example of development, the island's position both in Asia and the world was grossly distorted by politics and strategy. In the early 1970s, with a population of no more than 15 million, Taiwan was clearly dwarfed by the 800 million people of Communist China, and even in economic terms, though her overseas trade and domestic prosperity were important, Taiwan could not compete as a potential market with her giant rival.[3] In the early 1970s, as the People's Republic once more opened its doors to development, the same optimism applied as it had a hundred years in the past, and the promise of future profits in China was more attractive than the apparently limited if less erratic trade with Taiwan.

The comparisons of population and markets, however, did not detract directly from Taiwan's position, for the two economies were not competitive. It was as allies of the Americans and rivals to the Communists that the Nationalists achieved both their chief international importance and revealed their greatest weakness among their neighbours in Asia. The government in Taipei continued to insist on its status as the Republic of China, and on the legal fiction that the mainland was merely under temporary occupation by 'rebels'.[4] With support from the Americans, Taiwan maintained an active and frequently skilful contest in diplomacy, and the Nationalists were accepted not only by associates of the United States but also by many new nations of the Third World in Asia and Africa. Most remarkable of all, as representatives from Taipei in the United Nations withstood annual votes to admit delegates from the People's Republic in their place, they also retained the permanent seat on the Security Council awarded to China as one of the five victorious allies in 1945. The situation was absurd, for the most populous nation in the world was being kept from the assembly by one of the smallest, but the influence of the United States, the memory of the war in Korea, and a general uncertainty about the ambitions of Beijing acted to prevent any change. At the same time, though Nationalist speeches in the United Nations and the general world-policy of the government were based on a rigid, often virulent anti-Communism, the people of Taiwan had earned respect for their economic achievements, and though they could not make large grants of foreign aid they did give valuable assistance to other developing countries through training and advice on small-scale industry and farm technology.

Taken as a single small country, Taiwan was a reasonable member of the international community, but there was room to criticize the government not only for its pretensions to represent all of China but also for its very possession of Taiwan. The Nationalist refugees from the mainland held control of official administration and the armed services, while native-born Taiwanese had little hope of direct political influence, for the same propaganda which insisted the mainland was occupied by rebels also asserted that the people of Taiwan province could take only a minor part in the national government. As ultimate sanction for Nationalist power, Taiwan was officially under a 'state of siege' from 1949, so martial law could override individual political rights, and in 1960, when the Constitution required that President Chiang should stand down and not seek re-election, it was likewise resolved by the National Assembly that this provision should be suspended for the duration of the Communist emergency: Chiang remained in office until his death in 1975. At the same time, though it was obviously not possible to hold elections for the mainland, the Council of Grand Justices maintained the legal fiction with a special interpretation so that the Assembly and similar bodies could continue to carry out their national functions indefinitely.

The decision of the Grand Justices, however, did provide that the quorum for the National Assembly consist of delegates able to attend meetings in Taipei, and one result was that the nominal representatives of mainland constituencies were not replaced after their retirement or death. Eventually, delegates from Taiwan province would hold the majority, and ultimately all the seats, in the Assembly. Moreover, though the Nationalist Party ensured its own position and kept close control of rival activities, from the early 1960s there were constituency elections for local government positions as high as the Provincial Assembly of Taiwan province.

In constitutional terms the situation was no more extraordinary than many other occasions in the history of the republic, and hardly so remarkable as the Communist structure on the mainland. There was, however, natural resentment among many Taiwanese, and vocal protests from dissident students and other groups in Japan. Despite their economic success, and the real involvement of Taiwanese in local government and administration, the same Nationalists who spoke abroad of democracy and freedom were compelled to maintain their minority rule on Taiwan, for if the territory were regarded as an independent nation, the legitimacy of the Republic of China, and the Nationalist government which sought to maintain it, would disappear. From quite a different viewpoint, the Communists in Beijing are equally determined that Taiwan is an integral part of China, and both parties agree that a Taiwan separatist movement would be quite intolerable. In an unsatisfactory, short-term system which nevertheless continued almost forty years, the Nationalists kept hold on power with competent administration, moral exhortation, martial law, and some use of the secret police.[5]

To the outside world, Taiwan was at first a minor irritation, and the government of Chiang Kaishek little more than a joke. To the Americans in the early years of containment, Taiwan was a base to protect the Philippines and Japan. In the late 1950s and 1960s, as her economy developed and grew, Taiwan gained admiration, but the *détente* between the United States and the Soviet Union, followed by the Sino-Soviet split, meant that the Communist bloc appeared less of an overwhelming threat, and emphasized the uncertainty of the Nationalist position. Finally, in 1971, when the American Secretary of State Henry Kissinger visited Beijing and arranged that President Nixon should follow him in the following year, such public acceptance for the People's Republic removed critical international support from the regime in Taipei. The United States did argue for the Nationalists in the United Nations, but the contradictions in their position were too obvious, and there was now a full majority for the People's Republic to take the Chinese place in the General Assembly and its permanent seat on the Security Council. The Nationalists withdrew with dignity, but their government had lost international credit, and the people of Taiwan were effectively disfranchized among the nations of the world. Through the 1970s, as one country after another transferred recognition from Taipei to Beijing, the Communists forced the expulsion of their Nationalist rivals from such bodies as UNESCO, the World Health Organisation, the Food and Agriculture Organisation, and aviation and telecommunications agencies, while there was a continuing battle of boycotts and bans in the Olympic Games and other sporting assemblies.

On 5 April 1975, at the age of eighty-eight, Chiang Kaishek died on Taiwan. He was succeeded as President by Yen Chia-kan, one of the leading organizers of Taiwan's economic success, but real power was held by Chiang Kaishek's eldest son, Chiang Ching-kuo. Born to his father's first wife in 1910, Chiang Ching-kuo had been placed in charge of Taiwan security as early as 1949, he held political positions in provincial government and in military affairs through the 1950s and 1960s, and became Minister for Defence in 1969. In 1972, as Chiang Kaishek became increasingly ill and frail, Chiang Ching-kuo was made Premier and identified as his successor. Despite the obvious family influence on his career, and the anti-Communist rhetoric with which he confirmed his leadership among the elders of the Nationalist Party, Chiang Ching-kuo was a very different man to his father, and the accession of this new ruler brought more than the symbolic end of an old political order.

Taiwan: Tiger Territory

The development of the Taiwanese economy during the years of Nationalist rule has been one of the most remarkable success stories of modern times. Other countries, notably West Germany and Japan, were able to rebuild their industry and commerce after the devastation of the Second World War, but both had the advantage of a skilled labour force and an experience of modern industry. The achievement of the small 'Asian Tigers'—Taiwan, Hong Kong, Singapore, and South Korea—has been the establishment of an advanced technology, with skilled workers, on the basis of economies which were largely undeveloped until the 1950s or later. And the record of these 'newly industrialising countries' (NICs) is still more remarkable when compared to such neighbours as the Philippines, Malaysia, Indonesia, and Thailand, where progress has been less rapid and decisive, or more distant regions of Asia and Africa, where it often appears that no advance has taken place at all. Most countries have received comparable aid and advice, and the political problems of Taiwan, South Korea, and Hong Kong have surely been at least as severe as those of any others.

Fig.4 Indicators of the Taiwan Economy, 1952–1987: logarithmic scale of percentage increase; 1952 = base 100.

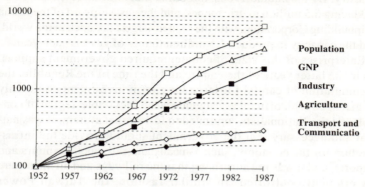

Source: *Taiwan Statistical Data Book*

Any visitor to Taiwan through the 1960s and 1970s was struck by the energy of small-scale private industry. The sidewalks of Taipei were lined with workshops, either making up components for larger manufactured goods, or repairing and reworking some salvaged items of machinery. In the early 1960s, one could see two worn-out bicycles being cannibalized to create one new one; by the latter part of that decade, the mechanics had moved on to motorcycles, and soon afterwards they were dealing with small cars. At first, motor vehicles were imported from the United States or Japan, but then they were built under licence, and by the early 1970s Taiwan had its own automobile industry. In the late 1980s, Taiwan had more than 1.5 million cars and trucks and seven million motor cycles, while Taiwanese bicycles had acquired a major share of the world market.

In one form or another, this was the model for Taiwan's industrial development: from small scale to steadily larger and more complex enterprise, and from experience in repair work and components to the construction of complete new items. In the same fashion, we have observed earlier that one of the first areas for industrial expansion was in the canning industry, processing the agricultural produce which was the original mainstay of Taiwan's exports. Based on those cans of pineapple, mushrooms, and asparagus, the metal industry developed into lightweight machinery such as refrigerators, sewing machines, and air-conditioners, while at the same time the docks of Keelung and Kaohsiung, in the north and south of the island, expanded their capacity for repair work. In due course, on the same principle as the bicycles and motor cars, Taiwan acquired steel mills and a heavy construction industry. By the middle 1980s, the China Steel Mill at Kaohsiung was producing 5.5 million metric tonnes, and the vast shipyard of the China Shipbuilding Corporation, with the second-largest dry dock in the world, had the capacity to produce 1.5 million tonnes of vessels each year.

Enterprises of this size, of course, required government support. As in the latter years of the empire and the time of the Republic, the accumulation of capital for large long-term investment could only be managed with official assistance and an inevitable degree of control, and the Nationalists continued to follow that system. It was natural and necessary that the government should provide the infrastructure for the economy, such as roads and railways, harbours and airports, and it was not surprising that the railways and airlines were government controlled. In similar fashion, the Taiwan Power Company provided for electrical generation, with coal, hydro-electricity and later nuclear plants. But there were also the China Petroleum, Taiwan Metal Mining, and Taiwan Machinery Manufacturing corporations, with BES Engineering for civil works, Chung Tai Chemical Industries and the China Petrochemical Development Corporation. In many countries, such operations would have been entrusted to private enterprise, but in Taiwan the Ministry of Economic Affairs also had direct control over sugar, alkalis, fertilizers, and aluminium. In each case, justification for government involvement was public utility, significance in defence, or the need to provide some level of partnership and supervision to aid smaller private investors.

Fig.5 Taiwan: Percentage Composition of Exports, 1952–1987

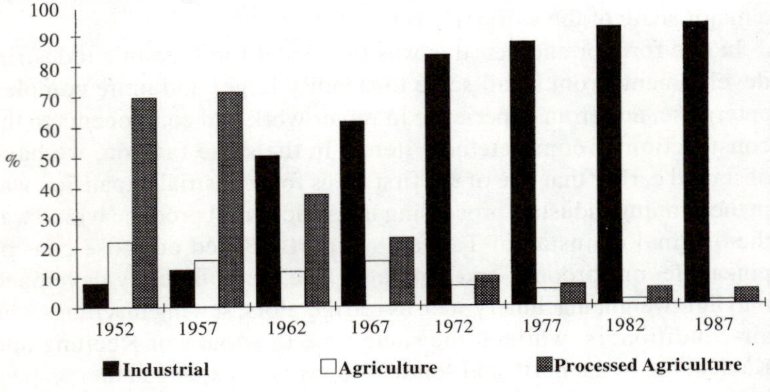

Source: *Taiwan Statistical Data Book*

Since the time of the old empire, such government sponsorship had too often served the personal interests of officials or politicians. Now, however, though appointments were influenced by politics, the administration of these enterprises was generally efficient and effective, and failure was regarded rather as a matter for embarrassment rather than concealment. Moreover, beside this heavy government involvement in the economy, there were large organizations in private hands. Many important firms owed their foundation to government favour, and particularly to the political contacts of mainland refugees, but the great Taiwan Cement Corporation and the Industrial and Mining Corporation, first established by the Japanese and seized by the Nationalists in 1945, were later sold into the hands of Taiwan-born capitalists and managers. Likewise, when land reform was enforced during the early 1950s, a proportion of the compensation was given in the form of industrial stocks. This early, American-encouraged, initiative was of limited success in appeasing the former landowners, who saw themselves stripped of economic power and political influence, but the programme did encourage the transfer of attention to industrial development, and in the years which followed, as it became clear that the mainland Nationalists were determined to retain their monopoly of political power, ambitious Taiwanese concentrated their energies on pursuing the opportunities created by an expanding economy.[6]

For individual entrepreneurs, the government was prepared to offer substantial incentives. Notably from 1960, when the Statute for Encouragement of Investment provided a five-year tax exemption for industries regarded as having significance to the economy, there has been advice and financial assistance for production, special industrial zones have been designated and developed, and in 1966 the Kaohsiung Export Processing Zone combined a free-trade area with a large industrial district, encouraging the processing of raw materials to create manufactured goods for export and providing a site for foreign investment and local employment. This, in particular, provided enormous opportunity for overseas trade, and became a model for similar projects in the People's Republic and elsewhere.

Under these policies, Taiwan developed a multitude of small plants and factories producing textiles and fibres, both natural and man-made, plastic goods of every size and quality, electrical components, including television equipment and computers, and the traditional processed foodstuffs, including canned goods and tea. A Science-Based Industrial Park was established in 1980 to encourage research and development, and by the end of the decade Taiwan was manufacturing a million microcomputers and four million television sets. Besides this, after clothing its own people, Taiwan exported some US$10 billion worth of textiles and clothing, and the production of its petrochemical industry was second only to that of Japan.

The achievement was extraordinary, but it had not been gained without social costs and political tension. Despite the ideals of Sun Yatsen, after experience of government in China and the conflict with Communism, followed by the influence of American-style free enterprise, the reconstructed government on Taiwan was concerned primarily with development of the economy, the search for wealth and power, and only in secondary fashion with any quality of livelihood. Unlike the Communists on the mainland, but in a style which reflects the custom of Japan, the Nationalists had limited concern for social security, and government publications advertised Taiwan as a 'paradise for the old' — because they could always rely upon their children to look after them. The fact that some people had no children to support them, and that the accumulation of a lifetime might not keep an ordinary citizen at any level of subsistence for more than a few months after retirement, was not of immediate concern. Somehow people managed, and in cases of real distress there was provision for support from the government or from the Nationalist Party, but the low expenditure on welfare payments left more money for investment and gave an incentive to work and savings.

To the surprise of many analysts and forecasters, moreover, the government in Taiwan maintained development without increasing the gap between rich and poor, and the income differential across society as a whole was reduced. From the early 1950s, the proportion of national income accruing to the benefit of the wealthiest in the community steadily declined, while opportunities for improvement of their circumstances in the lowest reaches of society consistently increased, and by 1980 it was judged that 50 per cent of the people could be classified as 'middle class'.[7]

None the less, it became generally true that growing economic prosperity was unfortunately combined with a good deal of public squalor. The city of Taipei, which sprawls in ungainly fashion around a small nineteenth-century central block, has little of value surviving from the past, and as population and industrial output grew, so did the noise, the dirt, and the pollution. On many days, looking down from the surrounding hills, the air above Taipei appears like well-used dishwater, and in much of the city the analogy comes still closer to home: though large-scale flooding has been brought under control, some areas still combine street gutters and storm-water drainage with the sewerage system, and a mistaken plunge into the Tamsui river is a serious health hazard.

Despite the general mess, however, the official public health programme has been well maintained, and Taiwan was one of the better members of the World Health Organisation. Malaria was officially declared eradicated in 1965, cholera and smallpox have disappeared, and there are active campaigns against tuberculosis and hepatitis-B which have reduced their incidence dramatically.[8] In 1987, the Bureau of Environmental Protection was up-graded to an Administration under the Department of Health, and some of the projects on its schedule include the control of water pollution and toxic chemicals, the construction of waste incinerators, and the establishment of a system for assessing environmental impact. The office has certainly a great deal of work to do, and for the time being, given the frequency of earthquakes and their effects upon underground piping, residents of Taiwanese cities are generally wise to boil their drinking-water.

When the Nationalists took over from the Japanese, they inherited a good elementary-education system, which formed the basis for the first stages of technical and economic progress, and they developed steadily from that beginning. In 1968 schooling was made compulsory for nine years, into the middle of secondary school, and although there has been serious overcrowding, and fierce competition for entry into higher levels, the standard of teaching and facilities has steadily improved, and the government now claims a literacy rate over 90 per cent. For a long time, graduates of universities thought only of travelling overseas, preferably to America, for further, proper training and a serious career, but by the late 1980s the number of those going overseas had dropped, and increasing numbers were returning for work and research in Taiwan. Besides National Taiwan University, Tsinghua and Chiaotung developed as major technical institutes, specializing in nuclear science and electronics, and by the late 1980s the National Science Council was distributing NT$100 million for research each year. Some universities, such as Chengchi, noted for its international studies, and Chinese Culture University, a private institution with connections to the Nationalist Party, were established with political associations, and have been the source of both good and bad analyses of the People's Republic. Both are now well regarded. There are a number of smaller academies of varying quality, but overall the standards can be very high, the political influence was never comparable to that of the People's Republic, and the success of the Taiwan education programme has had a great deal to do with the development of the economy.

Throughout the 1950s and 1960s, there had been no local threat to the authority of the Nationalist regime, though there was endemic tension between the establishment in exile and the new generation which was emerging to succeed them. By 1970, as Chiang Ching-kuo approached full political power, many of the former leaders had died or were no longer active, and though some old men stayed in their positions to an embarrassing age, by good fortune and good sense neither Chiang Kaishek nor his senior colleagues sought to emulate Chairman Mao and change the pattern of progress. Without such troubles as could be observed on the mainland, the development of the island had given opportunities to men of economic and administrative ability, and though the political prejudice against those born on Taiwan was still maintained, and the armed services were firmly under mainlander control, the leadership in less sensitive administrative areas was more open to talent and included numbers of native Taiwanese.

In the early 1970s, particularly under the pressure of rivalry from Beijing in the United Nations, the government sought to increase its domestic approval and maintain international support by a more liberal political attitude, accepting criticism in newspapers and a measure of public dissent. When Chiang Ching-kuo became Premier in May 1972, he first encouraged that policy, but in following years, partly to ensure an acceptable level of control, and partly out of concern at the threat of agitation from both Communists and separatists, the government became stricter. In 1977, after an unsuccessful attempt to force the success of government candidates in local elections, which resulted in a number of electoral defeats and considerable loss of face, Chiang Ching-kuo, as Chairman of the Party, instituted a programme of internal reform, but in 1979, when a group of young opponents called for basic changes in the structure of government and held demonstrations for liberal human rights, they were put down with bloodshed and imprisonment, and there were accusations of political murder, repeated on other occasions in later years.[9]

Politically, this could not have been an easy period. There was continuing uncertainty about United States intentions, firstly with the visits of Kissinger and Nixon to Beijing, then the transfer of recognition by President Carter on 1 January 1979, and the termination of the Mutual Security Treaty which followed in 1980. The Taiwan Relations Act of the United States Congress, approved in April 1979, gave informal undertakings for the island's security, but the effect of American withdrawal was felt, not only in diplomatic and military terms, but also in trade and finance. The direct expenses of American troops and officials had been an important source of foreign exchange, and the declining commitment could also be reflected by a switch in commercial interest to the reviving economy of the mainland. And even Taiwan's prosperity created problems, for as the People's Republic under Deng Xiaoping embarked on its own programme of development, newly wealthy and better educated people in Taiwan questioned once more the legitimacy of conservative Nationalist policies, and there were some who dreamed of the mainland.

Torn between the pressures for separatism and the temptations of unity, the government was also faced with a society which appeared in many respects to have lost its way. As part of political and military control, there had always been restrictions on travel, but many people now, with prosperity in their grasp, felt restricted and frustrated in their small, provincial, crowded homeland. During the early 1980s, there was economic evidence of a greater differential between rich and poor,[10] increasing tax evasion, conspicuous consumption of luxury goods, and sad scandals of meaningless gambling among wealthy women and men with nothing better to do. The old traditions, Confucianism, Taoism, and Buddhism, were still accepted and sponsored, but their links went back to the mainland, and even the art treasures of the Palace Museum and the scholarship of the universities were now echoed and matched by the new openness of the People's Republic.

Through the 1980s, however, the loss of formal United States protection brought less emotional strain and receded into perspective. The government was able to maintain contacts overseas through a series of informal offices, notably the China External Trade Development Council and the Far East Trading Company, with special provision for visas and other consular arrangements. Citizens and officials of the Republic of China travelling overseas had little more difficulty than before, visitors continued to arrive and do business, and after some confusion, as it became clear that the Taiwan connection remained valuable to all its major trading partners, the inconvenience was apparent rather than real. Private banks had no difficulty maintaining branches or agencies in Taiwan to process transactions, capital investment proceeded as before, and trade continued to grow. Japan, for example, second largest trading partner after the United States, recognized the People's Republic in 1972, but imports from Taiwan increased from US$245 million in 1971 to $694 million in 1975, and Japanese visiting Taiwan rose from 260,000 to 420,000 during the same period.[11] Similarly, as the oil-rich states of the Middle East sought outlets for their new-found wealth, Taiwan proved an excellent market: both Saudi Arabia, which maintained official relations, and Kuwait, which did not recognize Taipei, had major capital investments and developed substantial trade. There were predictable protests from Beijing at any appearance of formal government representation, but they could not break the network of contacts which the Taiwanese had established over the previous twenty years. Business and practical negotiations were generally more important than such niceties of protocol, and the finicky criticism rebounded often on the Communists themselves.[12]

Indeed, as time went by, the expulsion of Taiwan from the international arena began to appear less of a success for the People's Republic than an embarrassment to the world community as a whole: Taiwan's population of some twenty million is dwarfed by the People's Republic, but is greater than that of two-thirds the members of the United Nations, and Taiwan is well qualified to take part in international affairs. Many countries and a number of non-political bodies, such as the Olympic organization and the International Council of Scientific Unions, have shown themselves reluctant to be involved in the squabble over protocol, and while the Beijing regime, for reasons of *amour propre*, sought to have Taiwan expelled from such institutions as the World Bank and the Asian Development Bank, the People's Republic was in no position to take its rival's place. Moreover, as Taiwan became one of the chief trading economies of the world, it became important to keep contact with members of the General Agreement on Tariffs and Trade (GATT), while mainland China, despite its development, was not sufficiently open to fulfil the obligations of membership.

By the middle 1980s the situation looked increasingly secure. Politically, the government of Chiang Ching-kuo, who became President in 1978, had grown sufficiently self-confident to embark on a deliberate programme of liberal reform, including the appointment of several people born in Taiwan to high positions in the government and the party. There was growing demand for political rights, but in 1986, when leaders of an opposition group forced confrontation by announcing their formation of a Democratic Progressive Party, the government reversed its policy. Seizing back the initiative, President Chiang declared that constitutional changes would allow for loyal, legal opposition, and in July 1987, after almost forty years, the Emergency Decree authorizing martial law was lifted.[13] In October that year, with another gesture of confidence, official permission was given for citizens of Taiwan to visit the People's Republic.

Chiang Ching-kuo, who had long suffered from diabetes, died on 13 January 1988. Lee Teng-hui, who succeeded him as President, had been born near Taipei in 1923. Trained in the United States, he rose to prominence as a specialist in agricultural economics working on the Joint Commission on Rural Reconstruction, and under Chiang Ching-kuo he became Mayor of Taipei, Governor of Taiwan and then Vice-President. In the same way, as a result of Chiang Ching-kuo's support, Lin Yanggang, also born in Taiwan, with experience in the provincial department of construction, became Premier in 1984 and head of the judiciary in 1987. The Democratic Progressive Party gained 35 per cent of the vote in the 1989 elections, and several months of crisis followed in 1990, as Nationalist conservatives and radicals of the left each sought to overthrow the moderates. Lee's government, however, weathered the conflict, and though the political tensions remain, the constitutional structure for liberal reform is, at least for the time being, confirmed.

With his three great reforms, the acceptance of organized opposition, the 'Taiwanization' of the party and government, and the freeing of travel to the mainland, Chiang Ching-kuo changed the tradition of politics and opened the regime for the future. One notable benefit for the Nationalists was that as their subjects visited the People's Republic and saw the results of Communist rule, they became increasingly contented with the opportunities at home. Their reporters recorded the Tiananmen massacre in June 1989, and the Communists naturally sought to put blame on Nationalist agents, but the incident itself caused comparatively small concern in Taiwan, for it confirmed a common expectation of the government in Beijing.

On 1 May 1991, Lee Teng-hui decreed that the period of 'communist rebellion' on the mainland was ended. In doing so, he formally accepted the existence of the Beijing government, ended the state of civil war, and opened the way for the Republican constitution to take full effect in Taiwan. There are many opinions on the developments which may follow from this move, and much will depend on the response from the mainland. Both sides talk of one China, and the Communists have offered the slogan of 'one country, two systems', but their conduct at Tiananmen would make anyone cautious. In the mean time, and no doubt for many years to come, the businessmen and entrepreneurs of Taiwan have tacit approval and some official assistance to trade and invest across the straits, chiefly in Guangdong and Fujian, and the prosperity of the island province has the potential to become a powerful factor in the economic future of south-east China and the mainland as a whole.[14]

Colonial Enterprise: Hong Kong 1950–1990

The colony of Hong Kong was acquired by the British government in three transactions. The first was in 1841, as part of the settlement of the Opium War, when the island of Hong Kong was ceded in perpetuity by the Qing dynasty at the Treaty of Nanjing, and the second was in 1860, when the Anglo-French allies imposed the Treaty of Beijing, and the city of Kowloon was added to the original territory.[15] The third was in 1898, as the European powers competed for concessions from the weakened government of the Manchus; on this occasion, however, the 'New Territories', extending north of the Kowloon peninsula over an area ten times as large as the two earlier acquisitions, were the subject of a ninety-nine year lease. Since that time, the economy and administration of the New Territories have been thoroughly integrated to the other areas, so it has been agreed that when the lease comes to an end in 1997 the whole of the colony will return to China.

Hong Kong was always a useful entrepôt for trade into south China, but in the first half of the twentieth century chief British and foreign interest was given to Shanghai, where the International Settlement and the French Concession, combined with the rights of extraterritoriality, gave businessmen all the legal and political protection they required, and where communications along the Yangzi gave wider opportunities for trade. Like Shanghai, Hong Kong served as a place of refuge for opponents of the regimes which held power in China, but it was substantially less important, and when the Japanese declared war against the West in 1941 they seized Hong Kong without difficulty.

Though the Nationalist government had pressed for its return at the end of the war, the defeat of Japan restored Hong Kong to British control. As the Communists came to power in 1949, swift British recognition gave the colony a period of grace, and the reluctance of the new regime to embroil itself too swiftly in marginal aggression was only confirmed by the dangers and isolation of the Korean War. As much by default as by conscious decision, Hong Kong remained separate from China, acquiring special importance as the one open gateway for trade and information through the bamboo curtain.

At first, the new situation brought as much difficulty as opportunity. Though the Korean War boosted the Japanese economy, with some benefit to Taiwan, the strategic embargoes applied to China by the United States and its allies had a serious effect on trade, and the general isolation of China meant that Hong Kong had lost access to its traditional trading region. The people of Guangdong province had a long-recognized relationship with Hong Kong, speaking the same dialect, with clan and family ties across the border, but there was limited future for an entrepôt without a hinterland, and strict control of the land and sea borders by Chinese forces restricted the citizens and enterprise of Hong Kong to a very small territory.[16]

As refugees poured across from China, the population jumped from 1.7 million in 1945 to 2.5 million in the early 1950s. Public works, welfare arrangements, and the housing programme in particular were all but overwhelmed, while there were great problems in providing any satisfactory form of education and training. In 1953 the colonial government closed the border and restricted further immigration. Gradually, the situation improved, and many major enterprises such as reservoirs, the extension of Kai Tak airport, and the enlargement of the harbour foreshore, were carried out through schemes of land reclamation. By the 1960s, though there were still frequent water shortages, and the colony was consistently dependent on China for its supply of food, it was also benefiting from British policy, which sought to justify colonial rule by economic assistance and political development: the arrangements were aimed primarily at smoothing the transfer of power in Africa, but the programme gave incidental aid to Hong Kong.

If the territory was not to become a total dependency, however, relying always on increasingly reluctant support from London, industrial development was essential. Certainly, as a potential financial market, Hong Kong was unrivalled in the region, for government policy provided a free market for currency and goods, and British commercial law allowed the formation of limited companies and corporations, and the accumulation of capital, without the restrictions which applied in China, Taiwan, and the various south-east Asian nations. The Hong Kong & Shanghai Bank and the Chartered Bank provided Hong Kong with its currency and much of Asia with its funding; the great trading firms such as Jardine Matheson and Butterfield and Swire, maintained their operations there, while other groups, notably from the United States, established major branches and agencies. The great development of Hong Kong, however, was in its own domestic industry.

Hitherto, Hong Kong had concentrated largely on the repair and supply of ships and on local construction. The fall of Shanghai, however, brought an influx of textile workers and a rapid development of factories in that field, and this was followed by steady expansion into finished clothing, at first cheap but then of high quality. At the same time, similar small-scale workshops engaged in other areas of light industry, ranging from traditional handicrafts of wood, leather, enamel, and rubber to the new forms of plastic, nylon, items of electrical equipment, and then components for television and computers. The opening of the British market through imperial preference was essential to the beginning of this development, but by the 1960s piecegoods from Hong Kong were being exported world-wide, including the United States.

Through the 1960s, moreover, Hong Kong gained economic benefit from the Vietnam War, as a manufacturing source for basic items of equipment, a staging post for imports to the region, and a convenient centre for the rest and recreation of American and allied servicemen.[17] The Beijing government made no more than occasional protests about this association with the enemies of Communist Vietnam, but from 1966, as the Cultural Revolution developed, there was a good deal of anxiety about the possibility of a formal attack or, more insidious, civil disturbance in Hong Kong itself. For a time, there were strikes, demonstrations, and bomb explosions, but the authorities kept the trouble under control, the people as a whole showed no more enthusiasm for the new brand of Communism than they had for the earlier model, and the growing number of refugees, making their way illegally but desperately across the border, removed any illusions about the situation they left behind.

The opening of China in the 1970s, and the continuing policy of development through the 1980s, brought enormous prosperity. With its financial expertise and network of informal contacts, a skilled labour force and administrators trained in two languages, Hong Kong was the ideal base for any entrepreneur seeking to enter the China market. Money from Taiwan moved through Hong Kong for investment on the mainland, and manufacturers and financial institutions from Japan, Europe, and the United States found advice and assistance, while the free enterprise economy attracted investment in its own right from all over the world. Shenzhen and the other Special Economic Zones of China became overflow areas for investment and low-rate labour, and the whole region of the Pearl River, regardless of central Communist planning, was now open to Hong Kong enterprise. The stockmarket fall of 1987 damaged investment, and there were justified accusations of corruption, but it was possible to suggest that the unification of 1997 was as likely to allow Hong Kong's take-over of China as the other way round.

Politically, however, the situation was confused and unsatisfactory. Both the British and the Chinese governments agreed that the colony must be handed over, but the people themselves, now almost six million in number, were not consulted, and they were seriously concerned about their future. Faced with what they considered diplomatic necessity,[18] the British sought to negotiate human rights into a Basic Law, and the Chinese authorities undertook to maintain the territory as a Special Administrative Region, with privileges distinct from the rest of the People's Republic. As elsewhere in China, however, the Communists were reluctant to allow any real democracy, and they discouraged all British attempts to increase the limited forms of popular consultation, notably through local elections. The administration of Hong Kong had been brought steadily into the hands of local-born Chinese, and a series of scandals encouraged the tendency to reduce employment of British expatriates, but representatives of Beijing were increasingly active, and they emphasized their right to be involved or to interfere in more and more decisions of government.

In these circumstances, the Tiananmen affair of June 1989 was almost more serious for Hong Kong than for China as a whole. The brutality of the attack and the apparent disregard for human rights and the interests of their own citizens cast a new, harsh light on the leaders of the People's Republic, and removed all credibility from their promises for the future. Hitherto, the Communists had claimed they would govern with respect for the special nature of Hong Kong, and the only anxiety was that, despite good intentions, they might be too clumsy. Now, however, they appeared cruel and insensitive, and the tens of thousands of protesters who gathered in the streets of Hong Kong displayed a mixture of shock and anger, fear and shame, for the country they should soon be calling their own.[19]

So the political future was uneasy, and there was even doubt how well people would continue to accept the rule of a British government still determined to abandon them. Many Hong Kong Chinese, notably the wealthiest and the best qualified, had established residence or contacts overseas and were ready to leave as soon as it became necessary, and the great financial market, though quite uncertain of the future, prepared to maintain activity only for as long as it was profitable. There were still some years to wait, but it seemed most unlikely that the value of Hong Kong in Chinese hands would be so great as it might have been.

Notes

1. Chen Lifu, for example, head of the old CC Clique and a long associate of the Chiang regime, was specifically requested to stay in the United States: Jacobs, 'Chinese Nationalist Politics in Taiwan under the Two Chiangs', in Klintworth (ed.), *Modern Taiwan in the 1990s*, p. 4.

2. Until the 1970s, the exchange rate of the NT$ was tied at NT$40 = US$1. During the 1970s it was allowed to appreciate to NT$36 = US$1, but then returned to about NT$40 in 1980. In 1987 and 1988, however, as a consequence of huge trade surpluses, the currency increased by over 25%, to NT$28 = US$1.

3. To take example from Australia's experience, while trade with Taiwan grew rapidly from the mid-1960s, trade with the People's Republic fluctuated enormously. In 1966–7, Australian exports to mainland China amounted to A$128.5 million, of which A$116 million came from the sale of wheat. Similar figures were recorded in 1967–8 and 1969–70, but in the two years 1968–9 and 1970–1 wheat sales were only $58 million, while total exports fell to $67.2 million and $63.3 million respectively. In 1971–2, with no wheat agreement at all, exports to mainland China were down to $37.3 million, while Australia suffered an unfavourable balance of trade, for imports were $41.3 million. In 1972–3, with a Labor government in power, figures returned to their position of two years earlier, but they were still less than half of those for the late 1960s. There were obvious dangers in such volatility, but the potential appeared equally great.

4. It may be observed that the situation of the government of the Republic on Taiwan is in some respects analogous to that of Free China in Chongqing during the war against Japan. In both cases, the Nationalists could claim that they were the true government, driven from their capital and deprived of much of their territory by a bandit enemy, but maintaining defiance on Chinese soil. With the passage of time, however, the attempt at legal fiction became less and less convincing.

5. In a celebrated case of 1960, the government arrested and sentenced the newspaper publisher Lei Cheng to ten years imprisonment for attempting to establish a formal party of opposition. For more than twenty years thereafter, though rival candidates were permitted to contest some local elections, the Nationalists were the only acceptable political organization.

6. Gold, 'Colonial Origins of Taiwanese Capitalism', in Winckler and Greenhalgh, *Political Economy of Taiwan*, pp. 115–16, and Simon, 'External Incorporation and Internal Reform', *op. cit.*, pp. 144–8.

7. On Nationalist policy and achievements in income equalization and social security see, for example, Kuo, Ranis and Fei, *Taiwan Success Story*, p. 310; Gold, *State and Society*, p. 112; Greenhalgh, 'Supranational Processes of Income Distribution', in Winkler and Greenhalgh, *Political Economy of Taiwan*, p. 87; and Jacobs, 'Chinese Nationalist Politics', in Klintworth (ed.), *Modern Taiwan in the 1990s*, pp. 9-10, with references to fieldwork and other analyses. Compare also with note 10 below.

8. There is, however, a long-established problem with venereal disease, now including AIDS.

9. The ill-fated organization of 1979–80 was known as the 'Formosa movement', but in the longer term the dissidents called themselves *Tangwai*: being outside *(wai)* any party *(tang)* structure: learning from the example of Lei Cheng (note 5 above), they opposed themselves to the Nationalist Party, but were careful not to set up a formal political party.

10. See, for example, calculations and discussion of the GINI coefficient in *Republic of China Yearbook 1989*, pp. 262–3. By the late 1980s, however, it appeared that this trend had been reversed.

11. On these matters of representation, see, for example, Clough, *Island China*, pp. 161–6.

12. Australia, for example, has generally been willing to follow the People's Republic line, and was long reluctant to deal officially with Taiwan. It was compelled to do so, however, in matters of fishing rights on the continental shelf, and in recent years contact has become quite common. As a sample of the contradictions in policy, it may be recorded that early in 1991 books from the National Central Library placed on exhibition at the Australian National University had to have references to the Republic of China blotted out from their catalogue; but in the same week an agreement was negotiated for the government airline, Qantas, through a special ancillary company, to make direct flights to Taipei.

13. One result of the ending of martial law was that strikes, hitherto forbidden and in practice seldom experienced, were now made legal. 1988 saw numerous labour disturbances, including a work stoppage by railway engineers, slowdowns in state-run enterprises, and frequent disputes about year-end bonuses. The flurry of trouble, however, did not affect the island's longer-term progress, and one effect was to encourage the transfer of lower-wage, low-productivity industry to other countries in south-east Asia or to the Chinese mainland.

14. In 1988, per capita income in Taiwan was US$6,053, and the estimated figure in the People's Republic was US$320, a little over one twentieth that amount. (See also p. 321 above.)

15. Hong Kong is a local pronunciation of characters meaning 'Fragrant Harbour', pronounced *Xianggang* in Mandarin. Similarly, Kowloon is *Jiulong* in Mandarin: the name means 'Nine Dragons', and is based on the range of hills north of the city.

16. It may be observed that, unlike Taiwan, Hong Kong had no large areas of arable land whose production could be improved by land reform and better cultivation. In recent years there has been exploration for oil in the South China Sea, but there is no possibility that the People's Republic would tolerate the discovery and exploitation of substantial reserves in Hong Kong. As one senior official remarked, 'The last thing we want is to discover oil in our waters. Far better for the Chinese to find it — and we'll manage it for them!'

17. As a old port city, founded after all upon the opium trade, Hong Kong was well suited to supply drugs and prostitutes, and the dubious romance of that World of Suzy Wong has been celebrated in a number of novels. Hong Kong remains a centre of criminal activity, particularly associated with modern forms of the Triad gangs, who have connections all over east and south-east Asia, reaching to Australasia and across the Pacific.

18. It is hard to see how any government which maintained the principle of fulfilling its obligations under treaty could justify possession of the New Territories once the lease expired in 1997. From this point of view, regardless of any preferences and pressures, and irrespective of the difficulty of defending the place against determined Chinese attack, the British government has no option but to accept the date of 1997 as the end of its control of Hong Kong.

There had earlier been some hope that Prime Minister Thatcher, fresh from her triumph in the Falklands in 1982, would force the issue and extend British control of the colony: but Mrs Thatcher fought the Falklands War in defence of British rights in that territory; since Britain would lose its right to most of Hong Kong after 1997, it was never likely that she would seek to extend them by unlawful means.

In some contrast to the situation in Hong Kong, the small Portugese colony of Macau, on a peninsula on the opposite side of the Pearl River mouth, was less important to the Chinese authorities. Macau had no such development as Hong Kong, and was regarded primarily as a site for casinos and vacations. In the early 1970s, the revolution against the former government of Dr Salazar had brought Portugal to abandon its former empire, and although the flag remained it did so rather through courtesy, while China took predominant influence in the administration. Though the grant of Macau in the sixteenth century was made without any fixed term, it has been agreed that the territory will return to China in 1999.

19. In a strange juxtaposition, this crisis of confidence in Hong Kong came at a time when the territory was under pressure from thousands of 'boat people', escaping from Communist Vietnam to seek a new life amongst capitalist opportunity. Hong Kong was in no position to accept such numbers, but there was wide international and domestic concern at the manner in which the unfortunate arrivals were held in detention camps and pressed to return to the place they had escaped from. For some observers, the insoluble human tragedy presented a bleak model of the future which awaited many of the Hong Kong people themselves.

Further Reading

On Taiwan:

Republic of China Yearbook, published annually by the Kwang Hua Publishing Company, Taipei, successor to *China Yearbook and China Handbook,* formerly published by the official China Publishing Company of the Republic of China.

Taiwan Statistical Data Book, published annually by the Government of the Republic of China.

Clough, Ralph N., *Island China,* Harvard University Press, 1978.

Gold, Thomas B., *State and Society in the Taiwan Miracle,* Armonk, NY, Sharpe, 1986.

Hsiung, James C., and others (eds.), *Contemporary Republic of China: the Taiwan experience 1950–1980,* New York, Praeger, 1981.

Klintworth, Gary (ed.), *Modern Taiwan in the 1990s,* Canberra, Australian National University, 1991.

Knapp, Ronald G. (ed.), *China's Island Frontier: studies in the historical geography of Taiwan,* Honolulu, Hawaii University Press, 1980.

Kuo, Shirley W.Y., Ranis, Gustav, and Fei, John C.H. (eds.), *The Taiwan Success Story: rapid growth with improved distribution in the Republic of China, 1952–1979,* Boulder Colo., Westview, 1981.

Winckler, Edwin A., and Greenhalgh, Susan [eds.], *Contending Approaches to the Political Economy of Taiwan,* Armonk NY, Sharpe, 1988.

Wu, Yuan-li, *Becoming an Industrialized Nation: ROC's development on Taiwan,* New York, Praeger, 1985.

On Hong Kong:

Bonavia, David, *Hong Kong 1987: the final settlement,* Hong Kong, South China Morning Post, 1985.

Cheng, Joseph Y.S. (ed.), *Hong Kong in the 1980s,* Hong Kong, Summerson, 1982.

Domes, Jürgen, and Shaw, Yu-ming (eds.), *Hong Kong: a Chinese and international concern,* Boulder Colo., Westview, 1988.

Endacott, G.B., *A History of Hong Kong,* Oxford University Press, 1958.

Kelly, Ian, *Hong Kong: a political-geographical analysis,* London, Macmillan, 1987.

13

Cycles of Cathay

THE massacre in Tiananmen Square on 4 June 1989 sent a wave of revulsion through the nations of the West and among a number of China's Asian neighbours. Apart from the fact that the shooting took place in the centre of the capital, the new openness of the Communist regime meant that there were numbers of foreign observers on the streets of cities throughout China and, even more immediately, that television broadcasts could show many incidents — dead people in morgues, wounded being rushed to hospital, tanks in the streets, and soldiers firing shots — almost as soon as they happened, that radio reporters could get information on the spot, and that many of the students, even as events progressed, could telephone or send facsimile messages to contacts overseas. Much of what was said and written at the time has since been questioned, and the Chinese government has done its utmost to confuse matters with stories of innocent, peace-loving soldiers attacked by a few vicious hooligans, but most of the outside world has accepted that the leadership of the People's Republic ordered its troops to open fire and crush an unarmed, albeit embarrassing, group of youthful protesters. And there can be no real question that this is what happened.

Foreign reaction to the affair was in many respects exaggerated, at least in terms of the actual event and the past record of the Chinese government. It was recognized that political control of the country was firm and often brutal, and travellers of one kind or another had returned with accounts of incidental violence, whether of a mob trial and public execution glimpsed by chance on a river bank or, only a few months earlier, of Chinese troops shooting Tibetan protesters in their city of Lhasa. Such tales had been widely reported and were well known to most observers, but they had generally been pardoned or at least condoned on the excuse that China has many difficulties, and keeping such a large country in order requires a firmer hand than does a prosperous democracy. Foreign officials, from the highest levels of the United States and the Soviet Union to the junior ministers of smaller countries negotiating for trade or some form of cultural exchange, were prepared to ignore this unattractive side of the regime, partly for reasons of self-interest, but also because they believed the government under Deng Xiaoping was indeed concerned with progress and development.

From this point of view, however, the ruthless shooting of idealistic young people, no matter how foolishly they might behave, is neither good nor attractive politics. At the very least, it indicated a grossly distorted sense of priorities, and a propensity for brutality which cast an ugly light on all the incidents which had gone before and on all the promises which had been made for the future. Many outsiders who had dealt with China, and who had urged the benefits of trade and exchange upon their own countries, felt betrayed: the Chinese leaders since Mao had appeared to be responsible and sensible, imaginative and progressive, the sort of people one could do business with and whose word one could trust; they now appeared cruel, insensitive, and quite unreliable. Perhaps most important of all, they had carried out their policy in terms of their own interests and priorities, and showed minimal concern for Western opinion. The sense of disillusion was thus compounded by resentment born from neglect, and the emotions of many observers and advisers were given particular edge by the knowledge that they had simply deceived themselves, seeking and finding the China that they wanted to, not the one that was real.

In practice, most foreign nations and enterprises found their commercial connections with China too important to abandon, and although there was a quantity of diplomatic and political activity, with demonstrations of indignation at official, mass, and private levels throughout the West, most trading contacts were maintained or, if interrupted, quite soon restored.[1] The overall result of the incident was a general moral disapproval, frequently accompanied by homilies and criticism about human rights. In the short term, there was a swift and complete collapse of the tourist trade and in the longer term there has been increased reluctance to give China any favours or extra assistance in economic ventures.[2]

Elsewhere, among the less developed countries of the Third World, the affair at Tiananmen aroused little interest: it was a long way away and there are other governments like that. Relations with the Soviet Union, which had improved with the Gorbachev visit, remained cordial, and as a consequence China's relations with Vietnam and her position in South-east Asia were also enhanced.[3] Closer to home, however, the effect of the affair was devastating on relations with Taiwan and with Hong Kong.

For Taiwan, the situation was comparatively simple. Though private contact was still accepted, the official reluctance for closer connection with the mainland was confirmed with popular support, and any form of political reunion became utterly unlikely. For Hong Kong, faced with the inevitable transfer of power in 1997, there was confusion and despair. With a sense of national unity, people felt sympathy for the victims and a personal involvement in their fate, and at the same time there was natural fear for their own political rights. In the face of such intransigence from Beijing, no one could believe that the Basic Law, so cautiously negotiated by the British, would have any lasting value or give any worthwhile guarantees. The exodus already under way continued with greater pace and energy, and although Hong Kong remained a fevered marketplace there was a massive drop of investment into China.

It was in many respects at the individual level of exchange that the people of China suffered most severely from international disapproval of their government. In the region of the Pearl River delta, great numbers of small enterprises lost their markets and their financial support and were driven into bankruptcy. Many of their workers, now unemployed, were immigrants from other provinces, and many of these unwanted guests were driven harshly back to the poverty awaiting them at home. Some were collected from their lodgings at night, herded onto a truck, driven into the open country, then forced from the vehicle and told to keep walking. One indication of the difficulty for small firms was the reduction of 5 per cent in manufacture of yarns and cloth for 1990. Production of aluminium ware, bicycles, television sets, and cameras also declined, while that of larger household items, such as washing machines and refrigerators, fell 20 and 30 per cent respectively. At the upper end of the scale, production of power generating equipment, motor vehicles, railway engines, and shipping was lower, and the number of new machine tools, one of the most important items for development, fell by a third, while the profits of large-scale state enterprises, with increasing production costs and slow turn-over time, dropped almost 60 per cent.[4] By 1991, there were estimates that as many as two hundred million people were hopelessly unemployed, many of them living in shanty towns around the major cities.

Foreign exchange through 1990 reflected the belt-tightening programme which had been introduced by the government before the Tiananmen affair, and China achieved a trade surplus for the first time since 1984. Exports in 1988 had increased 18 per cent on 1987, but the rate had dropped to a 10.5 per cent increase in 1989. In 1990, however, the value was US$62 billion. By contrast, the rate of increase in imports was reduced from 27.4 per cent in 1988 to just 7 per cent in 1989, and in 1990 the value of imports fell almost 10 per cent, from US$59 to $53.4 billion.

The general programme of austerity, however, introduced to restrict inflation and reflected in the new balance of foreign trade, brought a long period of economic stagnation throughout the country and great financial difficulties to the government. The official figure for the budget deficit in 1989 was RMB9.25 billion. Plans for 1990 called for the same level to be maintained, but in fact the deficit rose more than 60 per cent to RMB15 billion. At the National People's Congress in March 1991 the Finance Minister Wang Bingquan proposed a budget deficit of RMB13 billion, but there was no strong reason to believe it could be adhered to. These figures, moreover, are based upon Chinese accounting practice; if calculations are based on procedures of the International Monetary Fund, the 1990 deficit would be more than six times greater.

Not only were the budget results and projections alarming, but the reasons were even more threatening, as Wang Bingquan spoke of large-scale tax evasion among the people and huge subsidies for state-controlled industries which consistently failed to hold to the limits prescribed for spending. Figures indicated that the cost of price subventions and of maintaining inefficient enterprises was greater than the allocations for defence, education, science, culture, and health combined.[5]

So China, tragically, was faced once more with signs of failure. The economic enterprise which began in the late 1970s had produced a short-term, febrile prosperity, but the basic problems remained: an inefficient state industrial structure, an economy and a government too weak and over-stretched to maintain the necessary pace of development, and a population which lacked any sense of responsibility to its rulers.

The reformers led by Deng Xiaoping had sought to develop the nation through the introduction of economic responsibility and free markets, but they had not been able to change the authoritarian structure of the Communist Party and the government, and the enormous subsidies for incompetent state enterprises were not only crippling the new order but were a symbol of the contradiction in their policies. Indeed, as Deng established his power in 1978 he had attempted to reform and purge the party, both of the 'old guard' whose membership and ideas dated back to the days of Liberation but who had learnt little since, and also of the newcomers, with few useful skills or training, who had been granted admission during the heady populist days of the Cultural Revolution. What Deng needed was people of managerial ability, and he introduced them gradually into the administration of the economy. His own authority, however, was too closely involved with the party, and was subject to too much uncertainty from the long struggle for power, for him to force a full purge of the membership. He did manage to eliminate a number of the extremists who were clear and direct enemies of the new dispensation, but the problem of the Communist Party has been not so much a matter of ideology but rather one of arrogance, corruption, and time-serving.

As a result, foreigners trading with China meet constant demands for bribes, and regular attempts to cheat on every deal, while year after year there are criminal cases of petty abuses of power, including embezzlement and the organization of prostitutes. Though much of the administration of China was clumsy, the people formerly took a pride in their honesty, but the organization now is both clumsy and corrupt. This sad situation has developed from disillusionment and a natural self-interest in uncertain times, while the system and the people who run it have proven too strongly entrenched for anything more than isolated, piecemeal punishment and correction.

So the protesters who called for the Fifth Modernization, political liberalization to accompany the economic reforms, were indeed correct, and Deng and his allies could appreciate the point. The very basis of their government, however, prevented the leaders from carrying out such a programme, and the fear of what might ensue if economic freedom was combined with political disorder had the power to bring nightmares: one Cultural Revolution was surely enough.

Furthermore, and here is another tragedy for China, intellectual opposition to the main stream of Chinese government has not presented any substantial alternative. Agitation for political reform has concentrated rather upon questions of accountability and proper administration within the system than upon a coherent rival ideology. The debate on the death of Chinese tradition which centred about the *Elegy for the Yellow River* in the late 1980s was impressive, and the display of the statue to the Goddess of Democracy by the students at Tiananmen was disconcerting, for this was a symbol which owed nothing to Chinese culture, traditional or Communist, but represented the ideals of the alien West. It is questionable, however, if the people who rallied round such a symbol had any clear idea what it was that they were seeking to emulate or introduce. Though they raised substantial questions, the symbols of democracy were really no more than a general expression of political protest, they were not accompanied by any coherent programme for national reform, and their supporters, in any case, had no means to develop such a programme. Ultimately, one must come to the conclusion that the affair at Tiananmen was simply a demonstration which went too far, and that the vast majority of those involved, despite good intentions, were primarily concerned with personal self-expression. The same emotions can also be expressed through pop, rock, and protest music, likewise imported from the West, and such songs and their singers have gained wide popularity among the present-day youth of China — still, in that sense, rebels without a cause.

Indeed, not only the students at Tiananmen, but also their mentors and their opponents, were trapped in a sorry tradition of modern China. In imperial times, though scholars and intellectuals had frequently caused trouble and disruption by their opposition to the government, they had generally accepted and worked within the political structures of the day. In twentieth-century China, however, there are two marked differences. Firstly, since the overthrow of the imperial system based upon Confucianism, there has been no agreed political organization: republicanism or Fascism, Communism or liberalism, all are alien to the Chinese tradition, and there has been no inherent reason to choose one or the other.

Secondly, from the last years of the old empire, perhaps most notably with the abolition of the traditional examination system in 1905, but already shown in the attitudes of the men trained overseas in the latter part of the nineteenth century, there has been a general tendency for intellectuals, writers, and scholars to oppose the government of the day. And it was not so much that the governments of China did not deserve to be criticized, but that their opponents were negative rather than constructive. The tradition of Gogol, deliberately followed by Lu Xun, maintained debate on the faults of those in power, with minimal consideration for the real difficulties which they faced.

In this way, the lineage of protest against the corrupt regime at the time of the May Fourth movement was followed by the League of Left-Wing Writers which opposed the Nationalist Republic in the 1930s and then fell foul of the far more determined Communists in the Rectification of the 1940s and the Hundred Flowers of the 1950s. In itself, this is a long history of personal tragedies, but it has also been a major misfortune for China as a whole, because political debate has been a sterile confrontation between irresponsible critics and an insensitve government, and those who hold power have been held to account too seldom for the errors and exaggerations of their ideas. Whether it be the 'New Life' movement preached by Chiang Kaishek in the 1930s, or the Thought of Mao Zedong worshipped so excitedly in the 1960s, the narrowness of official Chinese ideology, the insistence that the person of political power is also intellectually correct, and the lack of true and rigorous debate among the leadership, have been weaknesses for every government.

So Deng Xiaoping and the reformers of the 1980s were part of a Communist regime which reflected the experience of the past, both ancient and modern. Notably, though there was some codification of public law during the 1980s, the formal structures of government remained fluid, liable to change according to necessity or policy, and dominated by the personality and the private allegiances of the leaders, regardless of formal rank or position. Just as Mao had put his faith in personal allies such as Lin Biao and Chen Boda, and eventually gave great trust to his wife and her associates, so Deng Xiaoping in turn relied upon individual allies and protégés, often with years of shared experience, and he remembered his enemies with equal clarity. In this traditional network of *guanxi,* 'connections', we have observed how the army groups which moved against Tiananmen were commanded by members of the family of President Yang Shangkun, to such a degree that people spoke of 'Yang Family Village' in the People's Liberation Army.

Within any such system, the question of succession was always important, and for the government of China the ageing of the leadership made the matter critical. Deng was born in 1904, so he was in his forties when he held command during the civil war, he had senior posts in government through his fifties, and he was just over sixty when he was driven from power by the Cultural Revolution. At the time of his return in the late 1970s, he was over seventy and, like Mao Zedong earlier, he was compelled to consider who might come after him. By 1980 he had set Zhao Ziyang and Hu Yaobang in place, the first as Premier and the second as General Secretary of the party, but both men were already in their sixties. Hu Yaobang was dismissed in 1987 and died in 1989, while Zhao Ziyang was removed from office at the time of the Tiananmen affair. Li Peng, Premier in 1988, was sixty at the time of his appointment, and there were signs that a new generation was emerging, but the leadership was still not young, and Deng Xiaoping, eighty-seven years old in 1991, had colleagues of comparable age and potential influence, such as the conservative economist Chen Yun, then eighty-six, and President Yang Shangkun, who was eighty-four.

Critics of the regime might well sneer at all those old men, and observe the harshness which these veterans of civil war showed to the young people without such memories, but this situation comes from the very structure of Communist Chinese politics, where strong personalities dominate weak institutions, and where each leader has his own clique of supporters and protégés. As a result, accidents of personal relationship and personal ambition can be as important as practical policies, and just as the future in 1908 depended very greatly upon the death of the empress-dowager and the Guangxu Emperor, and the future in the 1970s was waiting for the passing of Mao Zedong, so too in the 1990s the people must wonder when the old men will die, and what adjustments of alliance and conflict may follow them.

Sadly, too, there are other patterns that must emerge from the past we know already. There is the tragedy of the Cultural Revolution, with its lost generation, now middle-aged, who were first encouraged and then betrayed, and who damaged so much of their own heritage and expectations. There is the generation of Tiananmen in 1989, and the question of how their resentment and protest will work itself through in the future. And there are the masses of China themselves, already a vast population, with children yet unborn who will come to swell their numbers.

For more than a hundred and fifty years the people of China and their leaders have sought to solve the problems of their own ecology: the unhappy balance of too many people on too little land with too basic a level of economics and technology. Though the West and Japan have occasionally contributed solutions and have sometimes appeared to offer help, such outsiders can only have marginal effect, and the answers must be sought in Chinese terms.

None of the remedies, however, have so far been successful. Within this century, the Chinese have overthrown the Manchu Qing and ended the dynastic empire; they have experimented with nationalism and republicanism, with civil war, and with varieties of Communism; they have forced themselves against the marginal lands and peoples of Manchuria, Mongolia, central Asia, Tibet and the far south-west; and they have lately attempted a free economy and talk of democracy. Yet one response after another has shown itself inadequate, and the problem of population, with its pressure upon available resources, has never stopped. There is now enormous industrial pollution, vast amounts of open country have been devastated, and even the beloved panda, symbol of the friendly face of China, is threatened with extinction as humans destroy its surviving habitat of bamboo forest.

So China is waiting once more for the next move. Some regions, particularly those of the eastern seaboard with access to the outside world, are doing better than others, and Taiwan, where a non-Communist regime, in a smaller, more viable economy and society, has allowed the people to demonstrate their natural abilities, may well serve as a model for separate development. Hong Kong, still more successful in the short term, can do little now but prepare for the worst. Given the record of Chinese government, perhaps the best hope for the people is that it may leave them alone.

For the saddest contradiction in the history of modern China is that between Chinese governments, which have often used their power cruelly, but seldom wisely or effectively, and the Chinese people, who have distinguished their history with a splendid culture, who have inspired others by their example, and who retain a constant capacity for courage, ability, intelligence, and hard work. After all the confusion and turmoil of this century, one can only regret that the future of China appears so unfairly uncertain.

Notes

1. One of the most ostentatious displays of disapproval was held in Australia, where Prime Minister Hawke attended a mourning ceremony at the national Parliament House and publicly declared his horror at the reports of slaughter. The government also declared that all Chinese citizens, notably students on short-term stay, could extend their visas for several years, and it cut off general cultural relations. Trade agreements, however, continued.

2. In the United States, there has been continuing discussion about the continuation of China's 'most-favoured-nation' status, which allows the import of Chinese goods at a low rate of duty. There are humanitarian, political, and economic arguments on both sides, and the privilege, not yet granted to the reforming Soviet Union, has so far been allowed to remain. If it was withdrawn, the effect on China's foreign trade, and on the economy as a whole, would probably be disastrous.

One increasing problem is posed by evidence that China is using *political* prisoners as 'slave-labour' to produce goods for export.

3. In reversal of the economic forces turned against China, described above, it is said that the government of Thailand was warned that excessive demonstrations about the Tiananmen affair might affect China's purchases of Thai rice, which is an important item in Thailand's overall balance of trade.

4. Figures from the *Statistical Communique of the State Statistical Bureau of the People's Republic of China on National Social and Economic Development in 1990, 22 February 1991*, supplement to *Beijing Review*, 11-17 March 1991.

5. Report and analysis from Andrew Higgins, Beijing correspondent of the British *Independent* newspaper.

Index